FOUNDATIONS OF MATHEMATICAL BIOLOGY

Volume I

Subcellular Systems

CONTRIBUTORS

D. Agin
Anthony F. Bartholomay
Howard H. Pattee
Robert Rosen

FOUNDATIONS OF MATHEMATICAL BIOLOGY

Edited by Robert Rosen

Center for Theoretical Biology
State University of New York at Buffalo
Amherst, New York

Volume I

Subcellular Systems

ACADEMIC PRESS New York and London 1972

ACADEMIC PRESS, INC.
111 Fifth Avenue, New York, New York 10003

United Kingdom Edition published by
ACADEMIC PRESS, INC. (LONDON) LTD.
24/28 Oval Road, London NW1

Library of Congress Catalog Card Number: 71-159622

PRINTED IN THE UNITED STATES OF AMERICA

To the memory of Nicolas Rashevsky

1899–1972

CONTENTS

LIST OF CONTRIBUTORS

Numbers in parentheses indicate the pages on which the authors' contributions begin.

D. AGIN, Department of Physiology, University of Chicago, Chicago, Illinois (253)

ANTHONY F. BARTHOLOMAY,* Center for Research in Pharmacology and Toxicology, School of Medicine, University of North Carolina at Chapel Hill, Chapel Hill, North Carolina (23)

HOWARD H. PATTEE, Center for Theoretical Biology, State University of New York at Buffalo, Amherst, New York (1)

ROBERT ROSEN, Center for Theoretical Biology, State University of New York at Buffalo, Amherst, New York (215)

*Present address: Department of Community Medicine, Rutgers Medical School, New Brunswick, New Jersey.

INTRODUCTION

The introductory material that follows is designed to fulfill several purposes. We wish herein to acquaint the reader with the place of mathematical biology in relation to the other biological, physical, and organizational sciences; we wish to give the reader some idea of the history and scope of the subject, and finally, we wish to describe the structure of the text itself—how and why it was put together in the manner it was.

By definition, the biologist is concerned with the nature and properties of living organisms. However, except for the most casual kinds of observation and experimentation, biology seldom has been able to generate from within itself either the technologies or the conceptual principles with which to undertake a deep study of organisms and their properties. This is not, of course, to deprecate what an intelligent and dedicated observer can discover about organisms through the use of his unaided senses and the inspired use of the canons of ordinary inference—both Darwin and Mendel, to use two examples which come readily to mind, transformed biology (and a great deal more besides) with only these tools. It is, nevertheless, a fact that the conventional instruments of contemporary biological research all have had to be imported, at a relatively recent date, into biology from the other sciences, primarily physics and chemistry. The most basic tool of biology, perhaps, is the microscope, which is unthinkable without a well-developed science of optics, and a well-developed technology for the production of lenses. More recent developments in microscopy, such as the ultraviolet or x-ray or electron microscopes, depend still more heavily on physical technologies. The basic tool of biochemistry is the radioactive tracer, a class of materials which only became available with the development of the cyclotron, and dates from the mid-1930's. The tools of the molecular biologist, which involve the purification and separation of infinitesimal quantities of material (for example, the chromatograph, the ultracentrifuge, the electrophoretic apparatus), are also importations into biology from physics and chemistry, as are the crystallographic techniques through which the structure of DNA was determined, and the devices for the study of electron spin resonance and nuclear magnetic resonance which have played, for example, such an important role in the unraveling of the mysteries of photosynthesis. These examples could be multiplied at every other level of biology.

On the conceptual side, too, many of the dominant ideas of biology were imported into it from the state of physical science and technology at any particular time. Norbert Weiner and others have pointed out that at the time of the Cartesians, the organism was conceptualized as a clockwork automaton, because it was possible at this time to construct complex automata which exhibited some of the properties that were thought characteristic of life; in the late 19th century, the organism was regarded at a heat engine or factory, while at the present time, our computers and communication systems lead us to treat the organism primarily as an information-processing system. There is much reason to believe that Mendel's work was, on the conceptual side, heavily influenced by the chemistry of his day, which had succeeded in showing that the infinite diversity of chemical molecules could be explained in terms of the reshuffling of a small number of basic atomic constituents, which were themselves preserved unchanged through all their interactions, and by belief that the physical world was governed by simple laws involving small integers. Most of the truly biological ideas that were proposed as a conceptual foundation for biology, most spectacularly the idea of a specific "vital force" (*élan vital*), or more recently the entelechies of Driesch, sooner or later have been discredited in biology (though they still form the basis for much philosophical speculation).

Now physics and chemistry, the two sciences from which biology has received the most substantial inputs on both the experimental and conceptual sides, are both concerned with the properties of matter. As such, these tools can tell us much about how the properties of organisms are determined by the way in which matter is arranged within them (we shall examine this question more deeply in Chapter 2 of Volume I). The great success which the application of these tools has had, especially over the past 25 years, in clarifying the mechanisms by which molecular processes determine the characteristics of organisms has given great weight to the reductionist idea that all the questions of biology are, in principle, effectively reducible to molecular terms, and solvable by the standard techniques of experimental and theoretical physics. As we shall see in subsequent chapters, it is far from clear that this is so, and that simply the problem of deciding whether it is so or not raises many deep questions with profound implications for physics, biology, and mathematics. But even supposing for the moment that it is so, let us see what this would mean for a "mathematical biology."

Mathematics has always been the *lingua franca* of physics. At the microscopic level, the level of molecular biology, the fundamental physical principles involved are those of quantum mechanics. The problems of electron distribution in organic molecules, which govern the chemical properties of these molecules, is one of profound mathematical difficulty, and forms one of the essential pillars of quantum biochemistry. The theory of catalysis, as adapted to biological systems, is as yet in its infancy, and is of an essentially mathematical character. The theory of absolute reaction rates likewise is developable only in mathematical terms. At a slightly higher level, we find the tools of statistical mechanics, and the transition from the properties of individual molecules to the properties of sets of molecules, with all their attendant conceptual difficulties aggravated by new kinds of purely biological constraints. The structure and properties of biological membranes involves the theory of phase transitions in a novel way. Transport and diffusion through membranes, both charged and uncharged, form an essentially mathematical kind of investigation as does the propagation of excitation, the generation of action potentials, and bioelectric phenomena in general. The kinetic theory of chemical reactions, their regulatory and switching properties, forms an entire mathematical realm in itself, as do recent developments in nonequilibrium statistical mechanics and thermodynamics. At higher levels, we have problems of elasticity and properties of the solid state manifesting themselves throughout physiology. Within the past ten years, concepts from control theory have made their way prominently into problems of biological regulation, and many physiological control mechanisms, from the regulation of temperature and respiration rate to the endocrine system, have been analyzed in these terms. The fields of ecology and population biology, so recently brought to the forefront of popular attention, are likewise feeling their impact.

The frank applications and extensions of the ideas and methods of theoretical physics to biology would take volumes simply to chronicle, and form a major and integral part of the ongoing attempt to understand, and ultimately to control, the nature of biological systems. The mathematical content of these applications is profound and pervasive, complex, and full of rich content and unique characteristics. There is nothing trivial about any of them, either in content or in form. Thus even if the reductionistic hypothesis were correct, and biological systems were *mere* physicochemical systems, there would still be a profound "mathematical biology"; mathematics would be at the heart of our understanding of biology as it has been at the heart of our understanding of the physical and chemical worlds.

But this is not all; far from it. Perhaps the richest fields for the application of mathematical thought and techniques to biology, and conceptually perhaps the most important, lie outside the framework of physics. Two examples of this should suffice. Perhaps the earliest field in biology to be extensively mathematized, and growing more intensively so every day, is mathematical genetics, the study of the dynamics of populations in time. This theory does not proceed from a physical picture of the gene, or the physical basis of replication and transmission of genetic information, but rather from a set of simple formal rules, going back to Mendel, specifying the manner in which genes are assorted through a sequence of generations, and an intuitive notion of *fitness*—a notion which has a biological and not a physical content. It is in fact a nonphysical theory, pitched at a particular phenomenological level, mathematical to its core, and capable of yielding the most important insights into the behavior of populations over time. [The fundamentals of this theory are developed in Chapter 4 of Volume II by Rescigno and Beck.]

A second example of a "nonphysical" approach to biological systems is the theory of neural nets due originally to McCulloch and Pitts (see Chapter 3 by Arbib in Volume III). Again, this theory did not begin from a physical description of the neuron, or from the physical basis of excitation and transmission of action potentials, but rather from a few simple numerical rules which characterize the behavior of the "formal neuron." The theory of "neural nets," which can be constructed from such formal neurons, has come to dominate the thinking of neurophysiologists, while at the same time the mathematical form of the theory made manifest a close, pervasive, and unsuspected relationship between the biology of the central nervous system and important areas of pure and applied mathematics: the theory of effective processes, computers, and cybernetics. Furthermore, the fact that the theory was phenomenological meant that the theory would apply to any physical system which satisfied the simple numerical rules that defined the "formal neuron." These rules could be satisfied in other kinds of biological systems besides the central nervous system, and hence the entire theory of neural nets could be applied to entirely different biological realms; to the theory of genetic control networks, the problems of differentiation and morphogenesis, and to clarification of the concept of "self-reproduction" which is at the heart of biology.

Finally, mathematics is the language of conceptualization, as well as the art of drawing inferences from hypotheses. The questions we are now able to pose regarding the properties of organisms require us to develop new kinds of conceptualizations—new *paradigms* (to use an overworked word) for the interactions that are at the heart of biology. To develop these we need experience with systems which have such interactive properties. Until recently, such experience could only be drawn from biology itself, where three billion years of evolution have resulted in systems of the greatest subtlety (systems which should methodologically be the last to be analyzed instead of the first) and with the kinds of artifacts and instrumentalities we build for ourselves (including both engineering systems and instrumentalities for social interactions, such as languages and the institutions of society). It could not be drawn from physics, which is typically defeated by strong interactions, and is most successful when dealing with questions involving either no interactions at all, or the limiting case in which the interactions are small (this is called perturbation theory). One of the ways of gathering this necessary experience and building the intuitions on which understanding is based is to construct artificial, simplified *model systems* in which such characteristics are manifested in a form simple enough to be graspable and their consequences explored. Such model systems can be physical in character, and indeed biology is replete with studies in which this kind of aim is paramount. There is, for example, an extensive literature on the study of enzyme models: non-biological catalysts whose properties, it is hoped, will throw light on the mechanisms by which real enzymes work. There is an extensive literature on model membranes, and an even larger one on "neuromimes," which has been beautifully reviewed by Harmon and Lewis [1966]. In the 1930's there was was an extensive class of studies called "biomimesis," in which simplified physical systems were constructed which exhibited simple biological properties such as motility. Much of the early (and present) literature in cybernetics was inspired by a desire to construct artificial systems which exhibited biological properties of purposiveness (Gray Walter's famous "tortoise" is a case in point, and it was no accident that Norbert Wiener's seminal book *Cybernetics* was subtitled "Control and Communication in the Animal and the Machine"). The fecundity of such approaches, besides its conceptual importance in the forming of our intuitions about biological systems, illustrates one of the most important new principles to emerge from biology: that it is possible to abstract *properties* or organizations as well as structures, and hence that these crucial biological entities can be treated as *things* as physics heretofore has treated structure alone. This kind of concept throws entirely new light on the reductionist hypothesis mentioned above; why should such model systems be useful at all if the only important thing about a system is the way in which matter is arranged in it? We believe that deep questions of this sort are opening entirely new chapters in epistemology and the philosophy of science, heretofore dominated entirely by canons drawn from physics.

But the construction of physical model systems, valuable as it is, is itself restricted by technology and technological ingenuity. The language of mathematics knows no such restrictions; it is possible to construct and explore the properties of model systems of whatever character the mind can reach, enormously facilitated nowadays by the availability of the computer. To many empirically minded biologists the construction and study of mathematical models of this kind (and even of physical model systems) has seemed a silly and sterile exercise, and to be sure it can be fraught with pitfalls for the unwary. But we believe it to be an essential activity in gaining the experience necessary to deal with organization (particularly, but not exclusively, biological organization) as a thing in itself, on its own terms. The creation and study of such simplified model worlds, and the mental processes that allow important generalizations of biological import to be drawn from them, is a uniquely mathematical form of activity. It is indeed no less empirical than the frank kinds of physical studies that form, for example, the basis of molecular biology, except that it involves experimentation with concepts and organizational properties instead of with materials. To our mind, this activity, remote as it may sometimes seem to an empirical biologist from any basis in reality, is the most important and exciting aspect of what we call "mathematical biology."

From what we have already said in this brief intruduction, it should be clear that "mathematical biology" is a many-sided kind of activity. J. H. Woodger spent many years in an attempt to clarify the formal structure of Mendelian genetics, and was led thereby to an axiomatization of the subject using the language of *Principia Mathematica.* This exercise seemed outlandish to many of his colleagues, and he was often asked why he was doing it. His reply was, "to discover new things." This reply perhaps sums up best the manner in which we can grasp what mathematical biology is about; it is an attempt to discover new things, an attempt to solve at least some of the problems raised by the biologist about the nature and properties of living organisms. These problems may be very specific, as in the attempt to derive a relation between blood pressure in an inaccessible blood vessel and other physiological data, or they may be very general, as are, say, the questions involved in the origin of life. All of these are, alike, "mathematical biology," and it is therefore cleary that the many sides of the subject arise from the many sides of biology itself. In a certain sense, mathematical biology *is* biology, as seen from a specific perspective, and shaped by the applicability of a particular kit of tools to particular biological problems.

The preceding discussion leads naturally into a discussion of the organization of the text which you now hold in your hands. The basic question was: given the diversity of such a subject, how can one best prepare a set of textual materials which would serve as an introduction to the subject as a whole? The remainder of this introduction will be devoted to this question.

The enterprise began, in fact, with an even more basic question: *could* one compile a set of texts which would give a fair view of mathematical biology? It is certainly still possible to write an *elementary* text about biology as a whole, but *mathematical* biology is not an elementary subject. Even the conceptually simplest parts of mathematical biology, which represent the application of physical principles to biological systems, are not part of elementary physics. Indeed, almost all of these applications presuppose the knowledge of a working physicist. The parts of mathematical biology that use the most elementary mathematics tend to use the deepest biology. Indeed, one of the difficulties of teaching mathematical biology in any broadly based way is the fact that one must spend so much time developing the tools before one ever gets to the specific content of applications. Given this fact, how indeed *could* one prepare a comprehensive text which would not at the same time have to also be a comprehensive text in physical principles, mathematical theories, and the foundations of biology itself?

Furthermore, mathematical biology is a subject (or set of subjects) that is itself in the process of active development. Because of its intrinsic diversity, there is no unified set of principles shared by everyone who might call himself a mathematical biologist. Therefore, a text which is comprehensive today may not be comprehensive tomorrow; one's entire view of the field may be changed at any time by unexpected developments in another area. Accordingly, we must understand the word "comprehensive" in a rather wider sense; a "comprehensive" text on mathematical biology must contain, in addition to specific results and developments, a sense of the open-endedness of the field, and a sense of the vastly different conceptual views maintained by those active in the field itself.

Given that it was valuable to make an attempt to resolve these problems, and actually prepare a body of comprehensive textual material in mathematical biology (itself perhaps a controversial assumption), the only resolution of these difficulties was to be found in a set of compromises. Such compromises are often doomed to satisfy nobody, but it is necessary for the proper understanding of the material itself to understand what the compromises were, and what they were compromises between, and why one specific form of compromise was chosen rather than another.

The first decision which had to be made concerned the very arrangement of the text itself. There were many ways in which this could have been done. For example, we could have arranged the material in terms of the mathematical methods employed, or in terms of the physical concepts used. We chose rather to arrange the material in terms of biological applications. The advantages of this choice were, to our mind, that it stressed the fact that *biology* was the subject to which we were addressed, and that it would result in a format which would be most readily usable by students of biology (to whom, in the widest sense, the text is actually addressed). The disadvantages of such a choice were likewise considerable; it would inevitably result in a great deal of duplication of material (where similar techniques were applied to problems arising at different biological levels), and, equally important, it would perhaps tend to obscure general principles that might be emerging from different classes of biological systems. We rationalized these objections as follows: duplication, while perhaps inelegant, is not really a major difficulty; redundancy is often useful in itself, and at worst is really a small price to pay for the attendant gain in flexibility. The obscuring of general principles is far more serious, but we hoped to compensate for this by including as an integral part of the text a chapter (Chapter 5, Volume III) that would be devoted explicitly to the search for unities in the diverse chapters which comprise the body of the text—unities that often manifest themselves through the mathematical form rather than through the biological content which forms the basis for the structure of the text itself.

The next decision involved the problem mentioned above: that mathematical biology, in its very nature, presumes an advanced level in the tool subjects. This problem manifests itself both in the choice of subject matter, and in the didactic level at which presentations are to be made. The compromises made here are perhaps the most serious and material—and the most open to objection—of any that were made in the preparation of the volumes. We decided to simply omit those areas (a) that presumed a massive preparation in the purely physical sciences and (b) for which there existed already good textual material pitched at the level contemplated for this text. This was a radical decision. It had the obvious advantages of not overloading the text with a great deal of preparatory material of a nonbiological nature elsewhere available, and thereby making the volumes more directly accessible to the intended audience. The corresponding disadvantages were, however, most serious; they imply (a) that the text is far from self-contained, and (b) that the portion of mathematical biology that is to many practitioners the best-developed and most significant part of the subject is badly scanted, thereby obviating any claim to comprehensiveness which can be made. Nevertheless, the choice had to be made between a variety of evils, and we hope that we have chosen the lesser evil. The condition that adequate textual material be available elsewhere was the decisive factor here. The type of material that was omitted, though bearing heavily on many of the basic questions of cell biology and physiology, can be found in books like those of Kautzmann, Eyring, Kimball and Pollissar, Morowitz, Rice, and many others (see the Bibliography at the end of this Introduction). In the hands of a knowledgeable instructor, therefore, it is hoped that the disadvantages implicit in this particular compromise can be minimized; but we are acutely aware of the attendant deficiencies, nevertheless.

A third choice had to be made in determining how the material was to be prepared. It is obviously best to have the entire text produced by a single individual. However, the very diversity of the field makes this a doubtful undertaking for any individual; in my own case it would have placed insupportable burdens upon my competence and my time. Therefore, the compromise was affected of approaching experts in the various fields of mathematical biology and requesting them to prepare chapters in the area of their expertise. The advantages are clearly that each chapter is prepared by an individual actively engaged in research in the specific area under discussion. The disadvantage is mainly stylistic; one loses the integrating character of an individual style and viewpoint. The disadvantage was rationalized as follows: we already have mentioned that there is far from unanimity among those who can be considered, or who consider themselves, to be mathematical biologists as to what is important in the field. In a developing field, one cannot be comprehensive without indicating this diversity. Therefore, it is in fact of great value to a student to have the multiplicity of viewpoints prevailing in the field manifested from the beginning, even though it may involve contradictions in emphasis and methodology from one chapter to another. It was first my intention to edit out all such contradictions, but on reflection it occurred to me that they were in fact an important didactic part of the text, and that it would present a truer picture of mathematical biology to keep such editing to a minimum. (Indeed, I was gratified, if somewhat surprised, to discover how few differences appeared from chapter to chapter, and how instructive were those that did appear.)

The above represent the three most serious compromises that determine the form and structure of the text, together with an explanation for how and why they were made. It remains to add a word on the relation between the present text and the other bibliographical materials presently available. One of the main reasons for preparing the text at all was the paucity of material in book form that can be used in the training of those who seek to solve biological problems using the tools and techniques of mathematics. With a few notable exceptions, most of what is available is scattered through an exceptionally diverse research literature, and a variety of volumes that represent the proceedings of symposia on one or another aspect of mathematical biology. We shall not attempt to analyze this literature, but rather shall append a selected and classified listing of other book material bearing on the pertinent matters either considered or omitted in the text, which should provide the student with an entry into the literature itself. The list makes no attempt at completeness, and the classification is somewhat arbitrary, but hopefully the compilation, together with the bibliographies of the individual chapters, will be found useful.

BIBLIOGRAPHY

I. General Books

1. N. Rashevsky, "Mathematical Biophysics," 3rd ed., 2 volumes. Dover, New York, 1960.

(This classic work, first appearing in 1938, has long been the touchstone of the field. It is based on the work of Rashevsky and his co-workers in all fields of biology, and set the standard both for those who followed Rashevsky's path and for those who chose others. Rashevsky produced a number of subsequent volumes on specialized topics.)

One may also consult the volumes he edited:

N. Rashevsky (ed.), Physiocomathemcatical Aspects of Biology," Course 16, Enrico Fermi International School of Physics. Academic Press, New York, 1962.

N. Rashevsky (ed.), Mathematical theories of biological phenomena, *Ann. N. Y. Acad. Sci.* **96**, No. 4, 895–1116 (1962).

2. H. L. Lucas, "Cullowhee Conference on Training in Biomathematics." North Carolina University, Chapel Hill, North Carolina, 1962.

(This little-known and rather hard-to-find book, the proceedings of a conference exploring the problem of how to run training programs in "biomathematics," contains review papers and discussions, often extremely frank, between many of the leading practitioners of mathematical biology. It gives a lively idea of the general state of mathematical biology as seen a decade ago.)

3. Mathematical problems in the biological sciences, *Proc. Symp. Appl. Math.* **14** (Amer. Math. Soc.) (1962).

(Sponsored by the American Mathematical Society, this volume is a very mixed bag, with little overlap with the Lucas volume, either in content or participants. It in general exhibits far narrower viewpoints.)

4. T. H. Waterman, and H. J. Morowitz (eds.), "Theoretical and Mathematical Biology." Ginn (Blaisdell), Boston, Massachusetts, 1965.

(This volume arose from a set of lectures presented at Yale University by a number of leading mathematical biologists and others active in the area. It is a very fine, well-balanced volume.)

5. "Some Mathematical Questions in Biology," 1968 *et seq.*

(This represents the title of a series sponsored by the American Mathematical Society and edited by M. Gerstenhaber. The contents represent invited papers presented annually at a minisymposium jointly sponsored by the American Mathematical Society and the American Academy for the Advancement of Science.)

6. C. H. Waddington (ed.), "Towards a Theoretical Biology," Vols. I, II, III, IV. Aldine, Chicago, 1968, 1969, 1970, 1971.

(This is a series of volumes representing the Proceedings of meetings held annually at the Villa Serbelloni, and gives most interesting perspectives. We believe that the student should read these volumes in reverse order, the last one first.)

II. Physical Principles in Biology

A. General

1. F. O. Schmitt, Biophysical sciences; a study program, *Rev. Mod. Phys.* **31**, No. 1, 5–10 (1959).
2. M. Marois (ed.), "Theoretical Physics and Biology." Amer. Elsevier, New York, 1969.

B. Specific

1. S. Glasstone, K. J. Laidler, and H. Eyring, "The Theory of Rate Processes." McGraw-Hill, New York, 1941.
2. F. H. Johnson, H. Eyring, and M. J. Polissar, "The Kinetic Basis of Molecular Biology." Wiley, New York, 1954.
3. W. Kauzmann, "Quantum Chemistry: An Introduction." Academic Press, New York, 1957.
4. W. Elsasser, "The Physical Foundation of Biology." Pergamon, Oxford, 1958.
5. I. Prigogine, "Introduction to the Thermodynamics of Irreversible Processes." Wiley, New York, 1961.
6. S. A. Rice, "Polyelectrolyte Solutions: A Theoretical Introduction." Academic Press, New York, 1961.
7. G. Ling, "A Physical Theory of the Living State." Ginn (Blaisdell), Boston, Massachusetts, 1962.
8. B. Pullman and M. Weissbluth (eds.), "Molecular Biophysics." Academic Press, New York, 1965.
9. W. Elsasser, "Atom and Organism: A New Approach to Theoretical Biology." Princeton Univ. Press, Princeton, New Jersey, 1966.
10. E. Schroedinger, "What Is Life?" Cambridge Univ. Press, London and New York, 1967.
11. H. J. Morowitz, "Energy Flow in Biology." Academic Press, New York, 1968.

III. Cell Biology

1. C. N. Hinshelwood, "The Chemical Kinetics of the Bacterial Cell." Oxford Univ. Press (Clarendon), London and New York, 1946.
2. B. C. Goodwin, "Temporal Organization in Cells." Academic Press, New York, 1963.
3. M. J. Apter, "Cybernetics and Development." Pergamon, Oxford, 1966.
4. A. C. Dean and C. N. Hinshelwood, "Growth, Function and Regulation in Bacterial Cells." Oxford Univ. Press, London and New York, 1966.
5. F. Heinmets, "Quantitative Cellular Biology: An Approach to the Quantitative Analysis of Life Processes." Dekker, New York, 1970.
6. R. Thom, "Stabilite structurelle et morphogenese." Benjamin, New York, 1968. (New ed., 1971.)

IV. Physiological Control Mechanisms

1. F. S. Grodins, "Control Theory and Biological Systems." Columbia Univ. Press, New York, 1963.
2. J. H. Milsum, "Biological Control Systems Analysis." McGraw-Hill, New York, 1966.
3. D. S. Riggs, "Control Theory and Physiological Feedback Mechanisms." Williams & Wilkins, Baltimore, Maryland, 1970.

V. Nervous System

1. A. S. Householder and H. D. Landahl, "Mathematical Biophysics of the Central Nervous System." Principia Press, Bloomington, Indiana, 1944.
2. C. E. Shannon and J. McCarthy, "Automata Studies" (Annals of Mathematics Studies, Vol. 34). Princeton Univ. Press, Princeton, New Jersey, 1956.
3. Mechanization of thought processes, Nat. Physical Laboratory Symp., 10th, London (H. M. Stationery Office) (1959).
4. F. Rosenblatt, "Principles of Neurodynamics." Spartan, New York, 1962.
5. S. Winograd and J. D. Cowan, "Reliable Computation in the Presence of Noise." MIT Press, Cambridge, Massachusetts, 1963.
6. M. A. Arbib, "Brains, Machines, and Mathematics." McGraw-Hill, New York, 1964.
7. E. R. Caianello (ed.), "Automata Theory" (International Spring School of Physics, 6th, Ravenna, Italy). Academic Press, New York, 1966.
8. L. D. Harmon and E. R. Lewis, Neural modeling, *Physiol. Rev.* **46**, 513 (1966).
9. L. Stark, "Neurological Control Systems—Studies in Bioengineering." Plenum, New York, 1968.

The above-listed volumes only scratch the surface of the enormous literature. There are numerous symposium volumes on mathematical aspects of the nervous system, together with many in cognate areas such as artificial intelligence, self-organization, bionics, pattern recognition, automata theory, and so forth.

VII. Taxonomy

1. R. R. Sokal and P. H. A. Sneath, "Principles of Numerical Taxonomy." Freeman, San Francisco, California, 1963.

VII. Population Biology

1. V. Volterra, "Leçons sur la théorie mathématique de la lutte pour la vie." Gauthier-Villars, Paris, 1931.
2. A. J. Lotka, "Elements of Mathematical Biology." Dover, New York, 1957.
3. E. P. Odum and H. T. Odum, "Fundamentals of Ecology," 2nd ed. Saunders, Philadelphia, Pennsylvania, 1959.
4. K. E. Watt (ed.), "Systems Analysis in Ecology," Vol. 1. Academic Press, New York, 1966.
5. E. C. Pielou, "Introduction to Mathematical Ecology." Wiley, New York, 1969.
6. N. S. Goel, S. C. Maitra, and E. W. Montroll, On the Volterra and other nonlinear models of interacting populations, *Rev. Mod. Phys.* **43**, No. 2, 231–276 (1971). Reprinted in book form by Academic Press, New York, 1971.
7. B. C. Patten (ed.), "Systems Analysis and Simulation in Ecology." Academic Press, New York, Vol. I, 1971; Vol. II, 1972.

VIII. General Mathematical Methods in Biology

1. L. von Bertalanffy, "Biophysik des Fliessgleichgewichts." Vieweg, Braunschweig, 1953.
2. R. W. Stacy and B. Waxman (eds.), "Computers in Biomedical Research," 3 volumes. Academic Press, New York, 1963–1969.

3. R. Rosen, "Optimality Principles on Biology." Butterworth, London, 1967.
4. M. Mesarovic (ed.), "Systems Theory and Biology." Springer-Verlag, Berlin and New York, 1968.
5. R. E. Kalman, P. L. Falb, and M. A. Arbib, "Topics in Mathematical System Theory." McGraw-Hill, New York, 1969.
6. R. Rosen, "Dynamical Systems Theory in Biology: Stability Theory and Its Applications," Vol. I. Wiley, New York, 1970.

IX. Journals and Periodical Volumes Emphasizing Biomathematics

1. *Biometrics:* Herbert A. David, editor.
2. *Biometrika:* D. R. Cox, editor.
3. *Biophysical Journal*, organ of the Biophysical Society: Max A. Lauffer, editor.
4. *Bulletin of Mathematical Biophysics:* N. Rashevsky, late editor-in-chief.
5. *Journal of Mathematical Psychology:* J. E. Keith Smith, editor.
6. *Journal of Theoretical Biology:* J. F. Danielli, editor-in-chief.
7. *Kybernetik:* H. B. Barlow *et al.*, editors.
8. *Mathematical Biosciences:* R. Bellman, editor-in-chief.
9. *Progress in Theoretical Biology:* R. Rosen and F. M. Snell, editors.
10. *Quarterly Reviews of Biophysics*, official organ of International Union for Pure and Applied Biophysics (IUPAB): Arne Engstrom, editor.

CONTENTS OF OTHER VOLUMES

Chapter 1

THE NATURE OF HIERARCHICAL CONTROLS IN LIVING MATTER

Howard H. Pattee

Center for Theoretical Biology
State University of New York at Buffalo
Amherst, New York

I. The Significance of Hierarchical Control

What is the primary distinction between living and nonliving matter? Is this an arbitrary and subjective distinction or can we state clear, physical and mathematical criteria for life? A few decades ago elementary biology textbooks could only approach these questions with a list of imprecise, descriptive properties characterizing life, such as reproduction, irritability, metabolism, cellular structure. Today the corresponding texts give us a detailed list of chemicals, often called the molecular basis of life. Starting with these molecular parts the texts go on to describe how properly integrated collections of these molecules perform most of the basic biological functions. For example, a special collection of amino acids covalently constrained in the proper linear sequence folds up to function as a highly specific catalyst. Similarly a special collection of such enzymes along with special ribonucleic acid molecules can function as a code which correlates the nucleotide triplets with amino acids. An even larger collection of proteins, enzymes, ribonucleic acids, and deoxyribonucleic acid molecules enclosed in a membrane can function as a self-reproducing cell, and a collection of such cells under the control of additional message molecules becomes a coherent organism which can function at many levels in a variety of exceedingly complex environments.

Since the molecules that make up the cell can now be manipulated in test tubes to perform their basic individual functions, it is often argued that the parts which make up living matter do not depend on the living state for their special properties. Or alternatively, it is said that since the parts of living systems behave as ordinary nonliving matter in all chemical details, we have finally reduced life to ordinary physics and chemistry. So then what is the answer to our question? What has become of the distinction between living and nonliving matter?

At the height of success of the revolution in molecular biology, this question simply faded away in the minds of many experimentalists who are concerned only with the detailed molecular basis of life. Now that much of the structural detail has been revealed we are beginning to see again the intricacy of the organization of these parts, and how being alive is not an inherent property of any of the structural units, but is still distinguished by the exceptional coherence of special collections of these units. This is true at many levels of organization, whether the units are monomers, copolymers, cells, organs, individuals, or societies. This coherence among parts has, of course, been given many different descriptions throughout the history of biological science, Concepts such as control, homeostasis, function, integrated behavior, goals, purposes, and even thought require coherent interactions among parts of a collection. The nature and origin of coherent, controlled collections of elements remains then a central problem for any theory of life. What is the basis for this coherence? How do coherent organizations arise from chaotic collections of matter? How does one molecule, ordinary by all chemical criteria, establish extraordinary control over other molecules in a collection? How do normal molecules become special messages, instructions, or descriptions? How does any fixed set of molecules establish an arbitrary code for reading molecular instructions or interpreting molecular descriptions? These are the types of questions which any physical or mathematical theory of life must answer. I have chosen to call this type of coherent collection a *hierarchical organization* and its behavior *hierarchical control.* What I want to discuss is the physical and logical nature of such organizations. In particular I want to emphasize the origin of hierarchical control systems at the simplest level where the necessary conditions are more easily distinguished from the incidental properties.

II. General Nature of Hierarchical Organizations

There is no mathematical or well-developed physical theory of hierarchical organization or control, so what I have to say will be largely intuitive and descriptive. At the same time I shall use wherever possible the language and laws of physics as a basis for my descriptions. In other words, I am assuming, at least as a working strategy, that there is a physical basis for hierarchical control which is derivable from, or at least reducible to, what we accept already as the basic laws of nature.†

First I would like to distinguish a control hierarchy from a structural hierarchy, since some form of structural hierarchy usually comes into existence before the control hierarchy is established. Structural hierarchies can often be distinguished by their graded size or by the way elements of a collection are grouped. For example the atom, the molecule, the crystal, and the solid can be distinguished as structural levels by criteria of grouping and number; that is, each level is made up of large collection of the units of the lower level. However, there is a more fundamental physical hierarchy of forces underlying these groupings, the strongest forces being responsible for the smallest or lowest-level structures. The strongest force holds together the nuclei of the atoms, and the weakest force holds together the largest bodies of matter. There is also a corresponding hierarchy of dynamical time scales which may be associated with the levels of forces, the shortest time being related to the strongest force and smallest structures, and the longest time related to the weakest force and largest structures. It is because of the separability of these graded levels of *numbers*, *forces*, and *time scales*, or their "partial decomposibility," as Simon [1962] calls it, that we can write

†I do not think that discussion of the exact degree of reducibility of life to physics would be helpful here, although this remains a very profound question. To help the reader interpret my later remarks, however, I should say that on the one hand, according to my idea of physics, I do not believe that molecular biologists have now reduced life to physics and chemistry, which many have claimed [see, for example, Watson, 1965; Crick, 1966], or that they will have no difficulty in doing so [see, for example, Kendrew, 1967]. I have given my reasons for this belief elsewhere [Pattee, 1968, 1969a, b]. On the other hand, I see little hope of explanatory theories or experiments arising from the other extreme attitudes, that we can only give necessary but never sufficient conditions for life [see, for example, Elsasser, 1969], and that life is a process which, while obeying all the laws of physics, can never be completely explained [see, for example, Bohr, 1958]. While there well may be some degree of truth in these attitudes, they do not appear to be a productive strategy at this time. In other words I regard a fundamental description of life as neither a simple problem close to solution, nor an irreducible problem with no solution. Rather I believe that it is a deceptively difficult problem which will take a large amount of effort before it is clarified. Sommerhoff [1959] in his book *Analytical Biology* has stated the problem with great care, but as he says, he does not claim to have a physical answer, only a clearer mathematical characterization of the difference between living and nonliving matter.

approximate dynamical equations describing one level at a time, assuming that the faster motions one level down are averaged out, and that the slower motions one level up are constant. Furthermore, since the forces and constraints between particles at one level have no special structure we may also use the approximation that one particle is typical or representative of any of the particles in the collection. It is only because of these approximations that our solution to a one- or two-body problem come close to our observations, and that many-body problems can be treated at all.

Hierarchical control systems are much more difficult, since they involve specific constraints on the motions of the individual elements. In a control hierarchy the collective upper level structures exert a specific, dynamic constraint on the details of motion on individuals at the lower level, so that the fast dynamics of the lower level cannot simply be averaged out. This amounts to a feedback path between two structural levels. Therefore, the physical behavior of a control hierarchy must take into account at least two levels at a time, and furthermore the one-particle approximation fails because the constrained subunits are atypical.

The epitome of hierarchical control in biology is the development of the multicell individual from the germ cells. Here the lower-level element is the cell itself. As a separate unit each cell has a large degree of autonomous internal activity which involves deeper hierarchical levels of control. These activities include growth and self-replication. However, as these cells form a physical aggregate, there arise new constraints which limit the freedom of the individual cells. Some of these constraints are of obvious physical origin such as the restriction of spatial freedom by neighboring cells of the collection. Such *structural* contraints may cause cells to stop growing and replicating because of simple overcrowding or lack of food. But these restrictions are not different from those found in a growing crystal. The *control* constraints, on the other hand, limit the individual cells' freedom in a very different way. We observe that as the collection of cells grows, certain groups of cells alter their growth patterns, depending on their positions in the collection, but not because of any direct physical limitation in food or space. The control constraint appears in the form of a message or instruction which turns off or on specific genes in the individual cells. This is a fairly clear example of hierarchical control, but how is this type of switching constraint to be distinguished from the constraint of simple crowding? In other words how can we distinguish physically between structural constraints and hierarchical controls?

There is no question, of course, that hierarchical constraints have a structural basis. That is, the molecules which turn off or on specific genes of the cell have definite structures which are responsible for the recognition of the target gene as well as the masking or unmasking of this gene. So why do we call this molecule a hierarchical rather than a structural constraint?

One way of expressing this difference between hierarchical and structural constraints is to say that structural constraints permanently eliminate or freeze out degrees of freedom, whereas hierarchical controls establish a time-dependent correlation between degrees of freedom. We could also say that structural constraints reduce the possible number of states available to a system of particles because of the inherent or unconditional dynamical or statistical laws of motion of the particles, whereas hierarchical constraints select from a set of possible states because of relatively fixed but conditional correlations between the particles of the collection. But obviously there are many possible conditional correlations. Which ones constitute a hierarchical control system? We feel intuitively that hierarchical control must result in some form of organized behavior of the total system, but what does it mean to say that there is organized behavior? Another way of saying this is that structural organizations often appear without having any function while hierarchical control systems always imply some form of function. But again since function is no better defined than organization, in any physical sense, this distinction is not much help. In many discussions of self-organizing systems we find the concept of *information* introduced in order to clarify this same distinction. For example, we say that hierarchical control is accomplished by information or messages which act as a constraint on a variety of possible configurations of the system. However, this language is usually ambiguous because it is also quite correct to say that a structural constraint uses information to constrain the variety of configurations. Therefore, while I believe that the concept of informational constraints is necessary in order to understand what we mean by a hierarchical control system, the problem is not clarified until we can say what type of information we are talking about. Specifically, we must distinguish between structural and hierarchical information.†

III. Hierarchical Control Implies a Language

The basic idea I want to express is that hierarchical control in living systems at all levels requires a set of coherent constraints which in some sense creates a symbolic or message content in physical structures, in other words, a set of constraints which establishes a *language structure*. Now immediately one might object to this idea on the grounds that I have said that I was trying to

†Such distinctions, of course, always need to be made and it is a common weakness of the use of information theory in biology that there is no way to evaluate objectively the significance of information. There are several precise measures of purely symbolic information [see, for example, Abrahamson, 1963]. and also purely physical entropy [see, for example, Brillouin, 1962]; but relating the symbolic information with the real physical event it stands for always requires a very complex transducer, such as a measuring device, a pattern recognizer, or an observer.

explain hierarchical organization in terms of more elementary concepts of physics and mathematics, and yet now I want to talk about symbols and language structures which appear to be more abstract and less well understood than many of the simpler integrated control systems we are trying to describe. Furthermore, it might be argued that our language structures are really the final outcome of billions of years of evolution and therefore could not have had much to do with the first question we asked, namely: What is the primary distinction between living and nonliving matter?

I prefer to turn these arguments around. I would agree that the fundamental nature of language is indeed less well understood than the nature of physical laws, but I would not agree that the apparent abstract, logical structure of language implies that language is not dependent on a physical embodiment or a molecular basis, which in every detail must follow the laws of physics and chemistry. The problem, as I see it, is that language has been studied with too much emphasis on its abstraction and too little attention to the common characteristics of actual physical constraints which are needed to support any language structure.

The second argument that language structure appears at the final outcome of billions of years of evolution is no evidence at all of the irrelevance of more primitive language constraints at the origin of biological evolution or at any hierarchical interface where a new functional level of description is necessary. On the contrary, the only generally acceptable condition for a living system capable of biological evolution that I know requires the propagation of genetic messages that could only make sense because of the integrated constraints of the genetic code and the reading and constructing mechanisms that go with it. Furthermore, as I shall explain below, the very idea of a new hierarchical level of function requires what amounts to a new description of the system, and any idea of a *description* is only meaningful within the context of a language.

I am using language here in its broadest functional sense, and I am interested in describing language structures in their most primitive form. Many classical linguists may object to this use of the word which they prefer to reserve for the unique, learned, symbolic activity of humans. Mathematical linguists also may object that without a formal definition of my idea of language this usage will not be productive. I am not thinking of language in either the anthropomorphic or formal sense, but as a *natural event* like life itself which needs to be studied in the context of natural laws from which it arose. Formalizing a language is useful for well-defined tasks, but in our case premature formalization would only be at the expense of ignoring the physical origin and basis of the natural rules and symbols of the most elementary language systems. Similarly restricting the concept of language to human communication eliminates from study the many stages of evolutionary hierarchies or symbol-manipulating systems which were responsible for creating this latest and most complex symbolic formalism of man.

The integrated records, descriptions, and instructions in cells are no less a language system because we know the molecular structure of some of their coding devices and symbol vehicles. But while many biologists more or less metaphorically think of the genetic processes as the "language of life," the full necessity of an authentic language system for the very existence of life, which I am proposing, is seldom recognized. In biological studies at all levels of organization we find the same implicit recognition of language-constrained behavior, such as references to hormones, chemotactic substances, and controllers of genetic expression as message molecules. Longuet-Higgins [1969] has emphasized this essential dependence of all levels of life on symbolic instructions by practically defining life as "programmed matter." However there is almost no discussion of why a particular chemical reaction is regarded as a message or an instruction. All the attention is on the chemical structure of the message vehicle and its interactions with its target, or on the formal, mathematical modeling of this process. What we need to know is how a molecule *becomes* a message [Pattee, 1970a].

To justify the study of hierarchical theory as a complement to this current emphasis in biology on detailed physical and chemical structure, we must show clearly why a knowledge of structure alone, however complete does not include an understanding of the basic nature of life. The justification, as we shall see, is very similar to the reason we cannot understand the basic nature of computation only by looking at the physical structure of a particular computer, or the reason we cannot understand the nature of language only by a detailed description of the symbol vehicles and rules of grammar of a particular language. In order to show how hierarchical interfaces are supported by language structures we first will look more carefully at the general properties of language and hierarchical control systems.

IV. Some Basic Properties of Language and Control Hierarchies

Both languages and hierarchies must ultimately be created and supported by material structures that are described physically as coherent collections of constraints. In human languages the rules of grammar are many levels of abstraction removed from the simplest physical constraints. Similarly, in human hierarchies the rules of tradition, custom, or legal systems apparently have nothing directly to do with physics. Nevertheless, they function as a limitation on the freedom of individual elements of a collection, and as with all symbols, at some deep level they must have a material counterpart. These rules are only at the top of a hierarchical structure of many levels, and I believe that they are far too complex to usefully discuss in any physical language [see Platt, 1969].

We want to look at the basic nature of much more primitive languages and hierarchies. In fact, I think that it is a valuable strategy to ask what is the simplest possible set of constraints on a collection of elements which would justify calling the collection a language or a control hierarchy. If we do not place such a severe limitation on our study of the nature of languages and hierarchies, we will be faced with an apparently inexhaustible complexity in which details cannot be clearly recognized as incidental "frozen accidents" or as essential conditions. That is why we choose to concentrate on the simplest cases.

However as a functional criterion for assuring ourselves that we are not oversimplifying, we shall require that the simplest languages and hierarchical organizations have an evolutionary potential. In other words, a language must be able to change continuously and persistently without at any point losing a grammatical structure defining the meaning or consequences of its descriptions. This continuous change and growth is observed in all higher natural languages, and it appears that use of the concept of language would be very difficult to justify in any system of constraints that did not have this property. Similarly any hierarchical organization that did not have the potential for establishing new levels of function and control would hardly be of biological interest. In other words, what we are saying is that life is phenomenologically distinguished from nonlife by its ability to evolve level upon level of hierarchical functions. Our problem is to understand the basic conditions that make this possible at the most primitive level.

A. Some Properties of Language

Human written languages are not associated with their particular physical representations. That is, we do not consider the type of paper, ink, or writing instruments as crucial properties of the language structure. Spoken languages, since they are more primitive, are more easily analyzed within the context of the physiological structures which make them possible [see, for example, Lenneberg, 1967]. Nevertheless, the universal properties of languages are all the more remarkable in view of their many divergent origins.

Six properties of language have been suggested by Harris [1968] in his book *Mathematical Structures of Language*. There may be some exceptions to these properties especially if we extend the meaning of language, as we propose to do, to include much more elementary symbolic control systems. Even so, these properties of higher languages serve very well as a basis for our discussion of more primitive language structures.

1. *The elements of language are discrete, preset, and arbitrary.*

The elements here can be regarded as the letters of an alphabet or the basic symbols or marks which can be arranged in patterns to form sentences, messages, or instructions. The idea of preset elements may be replaced in primitive symbol systems by the idea of stability in time of the symbol vehicles relative to the duration of the messages formed with them. The most difficult concept here is "arbitrary." Arbitrary according to the dictionary can mean chosen by the decision of an arbitrator who has such authority, or arising from caprice without reason. The mathematical connotation of an arbitrary choice is that there exists no significance to the choice except that it must be made decisively. Now in the case of primitive languages there is obviously no outside arbitrator to make the first choices of alphabets or grammars. On the other hand, we find it difficult to imagine a coherent language structure or a hierarchical organization with the potential for evolution arising solely by caprice or chance. Futhermore, from the physicist's point of view, if one chooses to consider any collection of matter in maximum detail, then the concept of arbitrariness does not apply, since the laws of motion leave no room for external arbitration or capricious choices. Arbitrariness in elementary physical systems can arise only because of ignorance of initial conditions or because of uncertainty in measurements. As we shall see when we discuss the properties of hierarchies, the very concept of arbitrariness in physics requires an *alternative description* to the description at the deepest dynamical level.

2. *Combinations of elements are linear and denumerable.*
3. *Not all combinations of elements constitute a discourse.*
4. *Operations which form or transform combinations are contiguous (that is, there is no metric as in musical notation; "distance" between symbols is just equivalent to the symbols in between them).*

These properties effectively isolate language vehicles from the ordinary limits of space and dynamical time of simple physical systems. Since the individual elements are fixed structures, they are independent of time and since they are strung together linearly, spatial restrictions on their order is not as important as it is in normal three-dimensional collections of matter. Condition 3, not all combinations of elements constitute a meaningful statement, reflects only the rules of grammar that are embodied in the special constraints of the language. These rules may also appear arbitrary to a large extent. At least they are not a direct or obvious result of any laws of nature.

Since the transformations on these combinations of elements must nevertheless be definite, it is essential that the combinations function as complete messages independent of real physical time. This leads to Condition 4, which places the operations on the elements under sequential control, and removes the dependence of their transformations on the real time of physical equations of motion.

These four conditions also create the apparent separation between formal, logical systems and physical systems, or between abstract automata and the real machines which approximate their behavior [see von Neumann, 1956]. Our problem is to explain how such properties which seem to separate symbolic operations from ordinary physical transformations can actually grow out of physical systems.

The last two properties of languages are much harder to define or understand, but are the most important properties for the type of evolution we find in living systems.

5. *The metalanguage is contained in the language (that is, the language can make statements about itself, its grammar, its symbols, or any constraint from which its grammar or symbols is formed).*

A most important type of statement which facilitates this property is the *classification*, for example, "*X* is a word," or "UGC is a codon," One could argue that classification is the most fundamental operation of logic, mathematics, and language. Classification requires a set of rules for distinguishing alternative events or structures, and in symbolic systems formation of these rules usually appears arbitrary but definite.

When speaking of real physical systems, however, the concept of classification, like arbitrariness, can only arise in the context of a measurement process or an observation. This is true because of the fundamental nature of physical laws which state that either no alternatives exist, as in classical determinism, or that every alternative must be considered as equally probable, as in quantum mechanics. Only when a measurement is performed do we have additional rules which create classifications, and these rules are not derived from the equations of motion but from the constraints of the measuring device or the observer. Therefore, the physical origin of natural or spontaneous classification rules has many of the same difficulties as the origin of language. I believe, in fact, that a good case could be made that any classification process which actually performs the classification in a physical system (that is, a measurement process) presupposes some form of language structure [Pattee, 1971].

6. *Language changes gradually and continuously without at any point failing to have a grammatical structure.*

This is very similar to the continuity principle on which we base our thinking about evolutionary processes. For example, it is difficult to believe that the genetic code arose complete, as it now exists, through an abrupt, discontinuous act of creation. Any alternative continuous process, on the other hand, must at all stages constitute a viable coding system. This implies that whatever message sequences occur, there must be a definite rule for classifying them as nonsense or not, and if not, then complete rules for translating the message into functional proteins. This does not mean that primitive messages themselves cannot be very simple, but it does set limits on the logical simplicity of the first set of constraints which form the language grammar [Pattee, 1972].

With regard to language, Harris [1968] says this evolutionary property implies that at any given time the grammatical rules must be describable correctly in at least two different ways, so that there can be functionally complete overlap between old and new descriptions. We shall see that this condition is related to the principle of descriptive and structural equivalence which is necessary for evolution in hierarchical control organizations.

B. Some Properties of Control Hierarchies

A hierarchical control system is a more concrete and mechanical concept than a language structure, and I am not suggesting that the two concepts are equivalent. What I hope to show is that they are so intimately related that one cannot exist without the other—at least the most basic parts of the other. Furthermore, I would expect what we do not fully understand about the natural origin and evolution of languages is often hidden in the constraints of a real physical hierarchical control system; and similarly, what we do not appreciate about the coherence of function in biological hierarchies is hidden in the descriptive constraints of a symbolic language structure.

1. *A control hierarchy constrains the behavior of the elements of a collection so that they perform some coherent activity.*

We are speaking here of autonomous hierarchies, so the constraints must arise within the collection itself and not from an outside authority. The concept of constraint in common language implies an enforceable limitation of freedom. The nature of constraints in physical language requires more elaboration, since constraints are not considered as a fundamental property of matter. One does not speak in physics of forces of constraint limiting the freedom of astronomical or atomic bodies, even though the forces between so-called "free" particles define the motions. In fact, as we pointed out earlier with regard to the concept of arbitrariness, the problem is that the dynamical level of description leaves no freedom at all. So what is the meaning of "additional constraints" when the dynamics leaves no alternative?

The answer is that the physical idea of a constraint is not a microsopic dynamical concept. The forces of constraint to a physicist are unavoidably associated with a new hierarchical level of description external to the system. Whenever a physicist adds an equation of constraint to the equations of motion, he is really writing in two languages at the same time, although they may appear indistinguishable in his equations. The equation-of-motion language relates the detailed trajectory or state of the system to dynamical time, whereas the constraint equation is not about the same type system at all, but another situation in which some dynamical detail has been purposely ignored, and in which the equation of motion language would be useless. In other words, forces of constraint are not the detailed forces between individual particles, but forces from collections of particles, or in some cases, from single units averaged over time. In any case the microscopic details are replaced by some form of statistical averaging process. In physics then, a constraint is a reinterpretation or reclassification of the system variables. A constraint is distinguished from what it constrains only by the fact it requires a different type of description.

Since we regard hierarchical control as a special set of constraints, it follows that a single level physical description of a hierarchical organization cannot begin to explain its behavior. Rosen [1969] has put this even more strongly, almost as a definition of hierarchy: "... the idea of a hierarchical organization simply does not arise if the same kind of system description is appropriate for all of [its activities] [p. 180]," and in other words, "... we recognize [hierarchical] structure *only* by the necessity for different kinds of system description at various levels in the hierarchy [p. 188]."

Now I do not mean to imply that the use of alternative descriptions is easy to understand and represents a physical reduction of the problem of hierarchies. On the contrary, even though physicists manage quite well to obtain answers for problems that involve the dynamics of single particles constrained by statistical averages of collections of particles, it is fair to say that these two alternative languages, dynamics and statistics, have never been combined in an entirely unified or elegant way, although many profound attempts have been made to do so. How well the dynamical and statistical descriptions have been related is, of course, a matter of opinion. The basic problem is that dynamical equations of motion are strictly reversible in time, whereas collections of matter approaching an equilibrium are irreversible. The resolutions of this problem have been central to the development of statistical mechanics, and have produced many profound arguments. For our purposes we need not judge the quality of these arguments, but only note that the resolutions always involve *alternative descriptions* of the same physical situation [see for example, Uhlenbeck and Ford, 1963]. Furthermore, the problem has proven even more obscure at the most fundamental level, namely, the interface between quantum mechanics and measurements statistics. This is known as the problem of quantum measurement, and although it has been discussed by the most competent physicists since quantum mechanics was discovered, it is still in an unsatisfactory state. Again, what is agreed is that measurement requires an *alternative description* that is not derivable directly from quantum mechanical equations of motion. The quantum measurement problem is closely related to the statistical irreversibility problem, and it too, has a long history of profound arguments central to the interpretation of quantum theory. The basic difficulty here is that a physical event, such as a collision of particles, is a reversible process, whereas the record of this event, which we call a measurement is irreversible (the record cannot precede the event). Yet if we look at the recording device in detail, it then should be reducible to reversible interactions between collections of particles. The difficulty also has to do with the fact that all mechanisms for control or recording require path-dependent, nonintegrable (nonholomonic) constraints, and thus far such extra relations between conjugate variables cannot be introduced into quantum mechanical formalism without basic difficulties [see, for example, Eden, 1951]. Again, for our discussion here it is not necessary to judge the many attempts to resolve this difficulty since as a practical matter they all involve alternative descriptions for the event and the record of the event, [for example, see von Neumann 1955] for a detailed treatment, or Wigner [1963], for a nonmathematical review of the problem. For a discussion of quantum measurement and biology see Pattee [1971].

So much for the physical basis of constraints, which in the context of biological organizations clearly needs some fundamental study. But what about coherent activity? What does this mean? Coherent usually implies a definite phase relationship between different periodic phenomena. I would like to extend the meaning of phase, which normally depends on real physical time, to include sequential order. I would also like to extend the idea to nonperiodic events. Thus, I would call any switching network or sequential machine a coherent set of constraints. This leads to the second property of control hierarchies:

2. *The coherent activity of the hierarchical control system is simpler than the detailed activities of its elements.*

This implies that some detail is selectively lost in the operation of the constraints.

The important point here is that constraints select which details of the elements are significant and which details are irrelevant for the collective behavior system. I want to stress that this selection in living systems is not dependent on the criteria invented by an outside observer as it is for artificial machines, although an outside observer may be clever enough to see the significant variables and thereby greatly simplify his description of the living system.

Hierarchical control therefore implies much more than a transition from a microscopic, deterministic description to a statistical description. Rosen

[1969] has used this transition from particle dynamics to thermodynamics as the only example known to him of an honest physical solution to the problem of how apparently independent system descriptions for different activities of the same system are actually related. This example is not, however, a hierarchical control system since the choice of the thermodynamic variables has no constraining effect on the microscopic degrees of freedom. Therefore, no matter how logical, practical, or even inescapable the choices of variables may appear, they must still be regarded as the physicists' choice and not the systems' choice.

The simplest natural example I know of a complex dynamical system which has a simple, collective activity is an enzyme molecule. The enzyme considered in maximum detail collides with molecules of all kinds with no regular, simple results. Only when a particular type collides with the enzyme will the simple, regular activity occur which we call a specific catalytic event. It is significant that just as the gas laws were discovered before the underlying dynamics, the enzymes were first discovered by their functional behavior; only, in the case of enzymes, we have not yet managed to completely explain the behavior by an underlying dynamical model. The behavior of enzymes also suggests two more very important properties of control hierarchies:

3. *Hierarchical constraints classify degrees of freedom to achieve selective behavior.*

Classification is another way to say that there has been a selective loss of detail. In dynamical description all degrees of freedom are treated equally. A constraint recognizes or selects some degrees of freedom as crucial for its collective activity and largely ignores the others. We also can say that the coherent activity of the collection is sensitive to some degrees of freedom and insensitive to others. The enzyme is a remarkably insensitive mechanism with respect to a large variety and number of nonsubstrate collisions which it must withstand. We call this its high specificity. It is also incredibly responsive to the sensitive properties of its particular substrate. The magnitude of this response we call its catalytic power.

A typical example of an artificial or externally designed hierarchical control is the traffic light whose timing responds only to sensors in the road. Such a signal system, like the enzyme, very strongly controls the rate of specific events on the basis of a few sensitive degrees of freedom, and completely ignores an enormous variety of other variables.

4. *Both the selection of sensitive degrees of freedom* (*or the choice of relevant variables*), *and the mechanism which performs the selective activity appear largely arbitrary.*

The arbitrariness of traffic signals is quite obvious. With living hierarchies there is room for differences of opinion. What we know about functional arbitrariness of enzymes is still very little. However, it does not strain our imagination to consider the possibility that an enzyme could be designed to recognize almost any substrate and catalyze almost any bond with almost any arbitrary correlation between the recognition and catalytic steps.

It is this type of *arbitrary* but *definite* constraint that correlates a structure and an operation which I would call the fundamental property distinguishing symbolic aspects of events from the physical interaction which underly these events. Clearly at least two levels of external description are necessary to describe this happening in such a system, since as we have explained, the constraint itself is not derivable from the microscopic dynamical equations of motion. The basic problem is the source of this arbitrary definiteness when there is no external observer or designer. If arbitrary, alternative correlations are physically possible, then what is it that determines which alternatives are fixed as the "rule of operation?" This is the central problem of the origin or source of hierarchical organization, for it is precisely this choice of arbitrary correlations between the elements of a collection which determines the type of coherent behavior or the integrated function of the collection.

This problem is often evaded by saying that the choice is made by some information in the form of other structures of the system such as the genetic deoxyribonucleic acid that determines which enzymes are to be constructed. But clearly the deoxyribonucleic acid is just another arbitrary but definite constraint that has informational significance only because of the arbitrary but definite enzymes and transfer ribonucleic acids of the genetic code. At present, the origin of this complex, coherent system of constraints is totally unknown. It is my guess that to understand the origin of the code we will have to understand more basic principles of the origin of language and hierarchical control systems.

The four properties of control hierarchies I have described might be called operational properties. Like the first four properties of language they serve primarily to distinguish physical processes from functional processes, or perhaps material systems from symbolic systems. More precisely these conditions separate physical behavior from the symbolic or functional behavior of material systems.

Again, as with the last two principles of language, the last two observed properties of biological hierarchies have to do with their evolution. They are, in fact, somewhat in parallel with the language properties.

5. *New hierarchical constraints can continue to appear at higher levels without destroying the existing constraints at the lower levels.*

This more or less obvious property of living organizations expresses the continuability or recursiveness of hierarchical origins [Bianchi and Hamann, 1970]. Hopefully, this recursive property suggests that if we could discover how *any* new functional organization or new classification is created spontaneously from a set of more or less disordered elements, we could generalize this discovery into a theory of hierarchical origins.

The corresponding property of languages follows from the fifth property, that natural languages contain a metalanguage. This is also a continuable or recursive property that allows us to say whatever we wish *about* what we have just said—no matter on how abstract a level we may have said it—while still retaining the same fixed and finite set of grammatical rules and arbitrary symbols. In other words, natural language always permits new classification and new interpretation of its structures, even though its substructure remains fixed. This is a most remarkable property which is not fully understood. This ability of descriptions is most carefully analyzed in the notion of *effective computability* in the theory of automata where there are very strong arguments, originating with Turing [1936], that one fixed and finite language could effectively describe all imaginable effective procedures in any language [see Minsky, 1967].

There is certainly some relation between these recursive properties of languages and hierarchical organizations; but unfortunately in both cases there are many mysterious points. In particular, the origin of this property or even the necessary conditions for the simplest cases of this property in both languages and hierarchical organizations remain unclear.

We can say something, however, about the relation between this new-interpretation property of language and new-function property of hierarchical levels. We have emphasized that from the physical viewpoint a new hierarchical level is recognized only when a new description of the system exists. Since the old description is assumed to be complete for the variables of the previous level, the new description must be based on a new classification of the variables at the previous level. But a new classification is exactly what a natural language can accomplish. Therefore, a language structure rich enough to reclassify its own symbols is, at least formally, a sufficient set of constraints to allow the creation of new hierarchical control levels.

The primeval origin problem is still with us, however, since, as we also emphasized, no language can be realized without a coherent set of material constraints to support its syntactical rules and symbol vehicles. This means that the physical embodiment of any language is itself a hierarchical set of constraints. In other words, the apparently endless variety of functions at all levels of biological organization could be generated under a fixed and finite set of coherent physical constraints which we would call a realization of a language, but obviously we cannot explain the origin of the first set of such constraints by the same generation process. To me this is the chicken–egg aspect of the matter–symbol paradox at the most physical level I can imagine it; but hopefully it is at a sufficiently well-defined level to suggest clues to its solution.

The last property of functional hierarchies I consider so essential for the origin and evolution of life that I would be inclined to elevate it to a principle of structural and descriptive equivalence. I would state it as follows:

6. *There are many physical structures that execute the same function; and there are many descriptions of the same physical structure.*

Examples of this principle are found at all levels. At the level of artificial control systems, from simple switches to entire computers, we know that there are many devices using quite different principles which perform equivalent functions. We also know that these devices can have equivalent alternative descriptions within one language, and of course also in other languages.

At the deeper and more primitive levels of molecular control hierarchies, this principle implies that the function of a genetic code can be achieved through equivalent sets of enzymatic constraints, and furthermore that the structure of one enzyme can have equivalent descriptions. Of course, the principle also implies that the same enzymatic function can be achieved through equivalent structures.

The last two implications we know are, in fact, the case. There is more than one sequence of nucleotides that will produce the identical amino acid sequence, and there is more than one amino acid sequence that will have identical enzymatic activity. There is no direct evidence that more than one genetic code could produce the identical form of life. But there is really no direct evidence against it either, since we have only one case. At least from our present understanding of the mechanism and structure of the transfer enzymes and transfer ribonucleic acids, there is no known physical, chemical, or logical reason why equivalent alternative codes could not occur in principle.

There are several ways to see why this property or principle is likely to be fundamental for the origin and evolution of hierarchies as well as languages. First, it would relieve the well-known problem of the spontaneous appearance of a particular structure which is highly unlikely as judged successful by only structural criteria. The principle replaces structural success by functional success. The corresponding reduction in the size of the search space depends on how broadly or narrowly we choose to define our function. For example, if we ask for the probability of the spontaneous occurrence of a hammer, we will find it high if almost any hard, dense object that we can lift easily will pass our functional needs. But if we also need the function of pulling out nails, the probability will drop enormously. What we must understand in the case of the origin of languages and hierarchies is nature's broadest criteria for functional success. Specifically, with origin-of-life experiments, this property suggests that too much emphasis on the similarity of molecular structures in abiogenic sythesis experiments is literally making life difficult.

The theory of evolution may also need this principle of structural and descriptive equivalence. The problem is well known: How does natural selection confer stability on all intermediate evolutionary steps leading to some integrated function? The mathematical equivalent of this problem is: How do random search and optimization procedures keep from being trapped at local maxima [see, for example, Bossert, 1967; Schutzenberger, 1967]? The formulation of this search problem usually involves an assumption that there is a purely physical configuration representing adaptedness or fitness, and for each configuration there is a value for the fitness which can be optimized by some form of search through the physical configurations. As we have seen, however, function and therefore fitness depend upon the *choice of description* of the physical configuration. Now, by the principle of equivalence, a new description need not change the local function, but in general a new description will alter the value of fitness in the neighborhood of a given function. In other words, the evolutionary search strategy may be primarily for descriptions of functions which do not lead into local traps. This is the same logic used by Harris [1968] to explain how language grammars can evolve. This must also be a continuous process; that is, at no stage of evolution can there fail to be a correct and complete description of the rules of grammar. It is observed, however, that at a time t_1 a given rule has a description D_1, and at a later time t_2 this rule has changed and has a new description D_2. Since for all times in between t_1 and t_2 there must be a complete description. it follows that D_1 and D_2 overlap. This is true for all times, from which it follows that all rules of grammar must always have at least two correct descriptions.

The point I wish to emphasize is that if life is at its foundation a set of descriptive constraints on matter, then its evolution need not be restricted to search and selection under one simple physical measure of fitness, but may have many simultaneous, partially overlapping descriptive measures on which natural selection may operate. This also suggests that instead of trying to understand complex higher learning processes by imitating an oversimplified model of evolution, we may be justified in applying some basic properties of language structures to help understand the apparently primitive evolutionary processes which may turn out to be not so simple. This does not, of course, get to the root of the problem of the origin of primitive language structure.

Let us return to the physical basis of languages. How are the properties of languages and hierarchies, outlined in this section, embodied in real, physical constraints? What are the physical conditions which satisfy these properties?

V. Physical Conditions for Language and Control Hierarchies

The fundamental general physical requirement for languages and hierarchies are constraints—in particular, fixed and finite sets of conditional constraints, Purely structural constraints, which permanently remove degrees of freedom, are necessary to support conditional or time-dependent constraints, but structural constraints alone cannot produce what we recognize as the rules or classifications necessary for languages or hierarchical controls.

The first property given for a control hierarchy was that the collection of elements performs some coherent activity. I extended the meaning of coherent to include nonperiodic variables and nondynamical (sequential) time scales. But what does this imply about the physical condition?

The loss of dynamical time in the description of a physical system means that some degrees of freedom or some detail has been ignored, usually by an averaging process (either number or time averages). However, detailed coherence in time has certainly also been lost by this process, so under what conditions do we expect sequential coherence to arise? Now sequential coherence means that events take place in a *definite order*, but this implies that there are such things as *definite events*. In a continuous statistical description we can get definite events only by threshold or trigger phenomena. Such events are also described as cooperative events, but the essential point is that they are irreversible and therefore dissipative. This means that sequential coherence is subject to noise (fluctuations). This is not the same as saying that *measurement* of sequence is uncertain, the way we say the measurement of dynamical variables, such as time, is uncertain. It means that the sequence itself is not precisely defineable. This places fundamental limits on the reliability of all hierarchical controls as well as on all realizations of formal logical systems that require sequential coherence in their symbolic transformations [Pattee, 1969a].

The second property of hierarchical control is that the collective functional activity is simpler than the underlying dynamics. This does not in itself lead to any profound physical condition. It implies however that there is some definite, regular process for averaging or ignoring the dynamical detail within the system itself. As we mentioned before, the pressure in a gas is independent of dynamical detail, but this detail is ignored only by the outside observer in the sense that there is no difference whatsoever on the dynamics because of the new description of pressure.

It is only when this property is added to the third property that simplification has physical meaning. The third property states that the constraint *classifies* the detailed degrees of freedom. This implies fixed rules of interaction that determine which degrees of freedom are effective in triggering the operation of the constraints. This is what separates *signals from noise*, and therefore this classification represents a very fundamental interface, inseparable from the more general matter–symbol interface.

What are the necessary physical conditions for a natural classification process? To classify means to distinguish between elements or events according fixed rules, but in the primitive context we are discussing, to distinguish must also imply definite physical change on the classified elements, such as marking or separating them from the collection. In other words, after the classification is completed, there must be a relatively permanent physical result which would not have occurred if the classification had not taken place. Before the classification there must be a distinguishing rule, and after the classification there must be a record to show that the rule was actually applied.

The question always arises why we cannot use this same description for a simple two-component chemical reaction, $A + B \rightleftharpoons AB$. We may assume that A collides with many other non-B molecules but does not react with them. Therefore, we could say, as above, that A has "classified" its collisions, and when it "recognizes" a B-type molecule, it forms a permanent bond with it, thereby establishing a "record" of the classification.

This alternative description may appear to be a gratuitous elaboration on what is acceptable physical or chemical language. But the basic question is whether in more complex situations, such as the enzyme catalyzed reaction, it is not equally gratuitous to say that the enzyme classifies or recognizes the substrate. In other words, is there some natural physical condition which distinguishes simple collisions from classifications in chemical reactions?

I believe that there is a condition, but just how it relates to physics remains to be explained. The condition that distinguishes collisions from classifications is precisely the same condition that separates physical interactions from symbolic constraints and events from records of events. The central condition is *arbitrariness*. As I said before, I believe it is the existence of an *arbitrary* but definite constraint correlating a structure and an operation which creates the symbolic aspect of physical events. Such constraints require an *alternative description*. This description is not to be associated with an outside observer or with his highly evolved language, but with a coherent set of constraints inside the system which fulfill the conditions of a language structure. These constraints are also arbitrary to some extent. As individual constraints they must appear as frozen accidents, but as collections they must appear integrated and functional.

VI. Conclusions

The most positive conclusion I can make is that life and language are parallel and inseparable concepts, and that the evolution of the many hierarchical levels uniquely characteristic of living organisms depend on corresponding levels of alternative descriptions within a language system. According to my picture, it is just as close to the truth to say that biological evolution is the product of natural selection within the constraints of a language as it is to say that language is the product of natural selection within the constraints of living organizations.

My most negative conclusion is that we still have too narrow and ambiguous a concept of language to come to grips with its relation to natural laws. We do not understand the physical basis of symbolic activity. Moreover, it is not at all clear at this point how difficult a problem this may turn out to be. The history of the matter–symbol paradox certainly should give us great respect for its difficulty, but I do not see how we can evade the question and still understand the physical basis of life.

References

Abramson, N. [1963]. "Information Theory and Coding." McGraw-Hill, New York.

Bianchi, L. M., and Hamann J. R. [1970]. The origin of life: Preliminary sketch of necessary and (possibly) sufficient formal conditions. *J. Theoret. Biol.* **28,** 489.

Bohr, N. [1958]. "Atomic Physics and Human Knowledge," p. 9. Wiley, New York.

Bossert, W. [1967]. Mathematical optimization: Are there abstract limits on natural selection?, *in* "Mathematical Challenges to the Neo-Darwinian Interpretation of Evolution" (P. S. Moorehead and M. M. Kaplan, eds.), p. 35. The Wistar Inst. Press, Philadelphia, Pennsylvania.

Brillouin, L. [1962]. "Science and Information Theory." Academic Press, New York.

Eden, R. J. [1951]. The quantum mechanics of non-holonomic systems, *Proc. Roy. Soc. (London) Ser. A* **205,** 564, 583.

Crick. F. H. C. [1966]. "Of Molecules and Men." Univ. of Washington Press, Seattle, Washington.

Elsasser, W. [1969]. Acausal phenomena in physics and biology: A case for reconstruction, *Amer. Sci.* **57,** 502.

Harris, Z. [1968]. "Mathematical Structures of Language." Wiley (Interscience), New York.

Kendrew, J. C. [1967]. *Sci. Amer.* **216**, no. 3, 142 [review of "Phage and the Origins of Molecular Biology" (J. Cairns, G. Stent, and J. Watson, eds.)].

Lenneberg, E. H. [1967]. "The Biological Foundations of Language." Wiley, New York.

Longuet-Higgins, C. [1969]. What biology is all about?, *in* "Towards a Theoretical Biology" (C. H. Waddington, ed.), 2 Sketches, p. 227. Edinburgh Univ. Press, Edinburgh, Scotland.

Minsky, M. [1967]. "Computation: Finite and Infinite Machines," Chapter 5. Prentice-Hall, Englewood Cliffs, New Jersey.

Pattee, H. [1968]. The physical basis of coding and reliability in biological evolution, *in* "Towards a Theoretical Biology" (C. H. Waddington, ed.), 1 Prolegomena, p. 69. Edinburgh Univ. Press, Edinburgh, Scotland.

Pattee, H. [1969a]. Physical problems of heredity and evolution, *in* "Towards a Theoretical Biology" (C. H. Waddington, ed.), 2 Sketches, p. 268. Edinburgh Univ. Press, Edinburgh, Scotland.

Pattee, H. [1969b]. Physical conditions for primitive functional hierarchies, *in* "Hierarchical Structures" (L. L. Whyte, A. G. Wilson, and D. Wilson, eds.), p. 179. American Elsevier, New York.

Pattee, H. [1970a]. How does a molecule become a message? "Communication in Development, " *Develop. Biol. Suppl.* **3,** 1.

Pattee, H. [1971]. Can life explain quantum mechanics?, *in* "Quantum Theory and Beyond" (T. Bastin, ed.), p. 307. Cambridge Univ. Press, London and New York.

Pattee, H. [1972]. *in* "Hierarchy Theory—The Challenge of Complex Systems" (H. Pattee, ed.). Braziller, New York (in press).

Platt, J. [1969]. Commentary—Part I. On the limits of reductionism, *J. History Biol.* **2,** no. 1.

Rosen, R. [1969]. Hierarchical organization in automata theoretic models of biological systems. *In* "Hierarchical Structures" (L. L. Whyte, A. G. Wilson, and D. Wilson, eds.), p. 179. American Elsevier, New York.

Schützenberger, M. P. [1967]. Algorithms and the neo-Darwinian theory of evolution, *in* "Mathematical Challenges to the Neo-Darwinian Interpretation of Evolution" (P. S. Moorehead and M. M. Kaplan, eds.), p. 73. The Wistar Inst. Press, Philadelphia, Pennsylvania.

Simon, H. A. [1962]. The architecture of complexity, *Proc. Amer. Philos. Soc.* **106**, 467.

Sommerhoff, G. [1950]. "Analytical Biology." Oxford Univ. Press, London and New York.

Turing, A. M. [1936]. On computable numbers with application to the *Entscheidungsproblem*, *Proc. London Math. Soc. Ser. 2* **42**, 230.

Unlenbeck, G. E., and Ford, G. W. [1963]. "Lectures in Statistical Mechanics," Chapter I. Amer. Math. Soc., Providence Rhode Island.

von Neumann, J. [1955]. "Mathematical Foundations of Quantum Mechanics," Chapter 5. Princeton Univ. Press, Princeton, New Jersey.

von Neumann, J. [1956]. Probabilistic logics and the synthesis of reliable organisms from unreliable components, *in* "Automata Studies" (C. E. Shannon and J. McCarthy, eds.), p. 43. Princeton Univ. Press, Princeton, New Jersey.

Watson, J. D. [1965]. "The Molecular Biology of the Gene," p. 67. Benjamin, New York.

Wigner, E. P. [1963]. The problem of measurement, *Amer. J. Phys.* **31**, 6.

Chapter 2

CHEMICAL KINETICS AND ENZYME KINETICS†

Anthony F. Bartholomay‡

Center for Research in Pharmacology and Toxicology
School of Medicine
University of North Carolina at Chapel Hill
Chapel Hill, North Carolina

I. Introduction

The notion of biological system was summarized elsewhere [Bartholomay, 1968a] by the symbolic expression $\mathcal{B}(C, S, E, F, \tau)$ in which C refers to its composition (that is, the set of its basic units of composition); S, its structural aspects; E, its environment, including the set of all parameters of state and other systems with which it communicates; F, the set of biological, chemical

†The preparation of this chapter was supported by Public Health Service Research Grant GM 13606, from the National Institute of General Medical Sciences, National Institutes of Health.

‡Present address: Department of Community Medicine, Rutgers Medical School, New Brunswick, New Jersey.

or physical activities or transformations associated with the system; and, τ its time parameters and relationships at least, implicitly involved in the specification of all other components of the system $\mathcal{B}$, which is therefore considered to be in a perpetuating dynamic state. It was also emphasized that any realization of the $\mathcal{B}$ concept is to some extent arbitrary, depending on the aims and levels of a particular study, the current state of biological knowledge, practical feasibility, and so forth. For example, in some genetic studies it might be sufficient to consider as elements of the composition set C, the set C_1 of all chromosomes in the gamete of one individual and the set C_2 of all chromosomes in the gamete of another, so that $C = C_1 \cup C_2$. In other genetic studies it might be necessary to define C in terms of the genes in the chromosomes. In yet others, it might be more relevant to resolve C down to the level of the deoxyribonucleic acid (DNA) molecules in a gene, or to the polynucleotides in a DNA molecule, or to the set of all hydrogen bonds of the DNA molecule.

It should also be remarked that this definition is meant to include all of the usually specified "systems" of biology and medicine such as an organ system, or the sympathetic nervous system, or the cardiovascular system, or a metabolic or biochemical system. And with reference to these standardized medical systems, expressions such as the "biochemistry of disease" or "molecular disease" or "electrocardiographic findings" are indications of the arbitrarily conceived biosystems which arise out of convenience or the tendency to specify an investigation of an overall biological process down to a particular meaningful level of discussion.

Thus, the problem under investigation leads to the description of a real $\mathcal{B}$ system, which amounts to the "target system" under investigation, and the designation of the system, like the designation of a probability sample space in a probability problem indicates the particular aspects and level of interest in the total system of which this is really a subsystem. We are in practice always dealing with subsystems, or with partial systems of an over-system, the ultimate total system, of course, being the universe. In the same way in chemistry, the system description might begin by designating as C the union of various sets of interacting molecules with structures specified simply as rigid, inelastic spheres. Or, it might be necessary in other cases to specify the domain of study or level of investigation by defining the system in terms of indicating primary, secondary, tertiary conformations of individual molecules.

In terms of the $\mathcal{B}$-system concept, a mathematical model is describable [Bartholomay, 1968a] as a convenient mapping; or better, "relation" μ: $\mathcal{B} \longrightarrow \mathcal{M}$ between the system $\mathcal{B}$ and a mathematical domain $\mathcal{M}$ which associates with the various components (either compositional, structural, environmental, and so on, or all of these) corresponding mathematical structures, the "mathematical deductive machinery" of which allows a mathematical interpretation or biometric framework for the interpretation of the system; or simply its translation into formal mathematical terms. A mathematical model is thus taken to be a mathematical abstraction or realization of an underlying biosystem.

In this general context we take biological kinetics to be the study of mathematical models for interpreting the dynamics and time relations of the intra- and intersystemic transformations or biological activities associated with a given system (or system "in the extended sense" as a collection of systems). And, in this sense, chemical kinetics is incorporable into biological kinetics at the level of chemical and biochemical subsystems underlying a given biological system.

Independently of its identification in biological kinetics in this chapter we conceive of "chemical kinetics" *per se* as the body of theoretical chemical, physical, and mathematical models, methods and techniques (including experimental) which are employed in predicting, analyzing, and formally characterizing the time aspects of chemical reaction systems. A few words on the treatment itself; that is, on the metatheoretic aspects, or approaches to the study of chemical kinetics as a discipline which have been considered in the plan of this chapter may be instructive here. Different levels of understanding and comprehension are delineated in any branch of knowledge; they are particularly apparent in chemistry. Some of these choices have been imposed by scientific and practical necessities; others are attributable to accidents of historical development; and still others to personal factors such as "artistic" preferences, scientific convictions, or even prejudices.

For example, in chemical kinetics there is apparent a distinction between "macroscopic" and "microscopic" kinetics. The former is also referred to as phenomenological or empirical since it has its origins in the interpretation of experimental data at the compositional level of molar concentration; for example, the estimation of rate constants from concentration–time courses of reaction. This is not to say that in such approaches no attention is paid to the chemical–physical context of such parameters. However, there are obvious limitations in using this as the level of investigation in studies aimed at the elucidation of mechanisms. The microscopic approach, on the other hand, which has this latter as its major concern therefore goes below the molar level, centering on intermolecular and individual, discrete intramolecular activities and interactions and physical forces associated with the reaction mechanism [see Bartholomay, 1964a]. The term "chemical kinetics" is identified almost completely with "microscopic kinetics," particularly by some chemical physicists who take as their fundamental axiom that reaction "rates" are in principle calculable from first quantum principles. The extent to which this is a practicable goal will be examined to some extent in this chapter. It is important in a text on mathematical biology to turn to this question particularly because the explanation or "calculation" of biological systems from first quantum principles has also been stated by some to be the proper goal for theoretical or mathematical biology [see, for example, Arley's discussion of Bartholomay's paper in Bartholomay, 1968a]. It is possible that a closer association between macroscopic and microscopic kinetics can be accomplished at the level of stochastic models (see Section III.F.2).

Another distinction which has some bearing on the presentation of chemical kinetics is the almost mutually exclusive relationship between chemical kinetics and chemical thermodynamics which is perpetuated by referring to the former as the dynamic aspect of chemical reaction and the latter simply as the static aspect. Developments in modern absolute rate theory or the transition-state theory have, however, brought these components closer together in even more than asymptotic relation to each other. Another aspect responsible for the closing gap is the advent of irreversible thermodynamics and the thermodynamics of open systems with emphasis on equilibrium-like treatments of the steady state.

Of even more recent vintage, not only in chemical kinetics but in biological kinetics as well, is the growing distinction between deterministic and stochastic models, which also can be interpreted as a significant mathematical aspect of the distinction between macroscopic and microscopic kinetics. As a direct derivative of the law of mass action and as a reflection of its time of origin, chemical kinetics quite naturally began with deterministic mathematical models and points of view, that is, with the idea and expectation that the entire future course of a reaction is entirely predictable from the initial state. In this sense, the time course would be expressible hopefully as a collection of ordinary, analytic time functions for the corresponding concentrations of each species in the reaction system, derivable as integrals of systems of differential equations featuring as coefficients, the "rate constants." The study of chemical kinetics in this sense soon became identified with the study of "rate constants." Consequently, to explain the dependency of time courses on physical and other environmental variables the rate constant became the point of entrance, the form of the mathematical equation which it helped specify being regarded in effect as an invariant feature of the kinetics of the underlying system. Thus, the important early results of Arrhenius on the temperature dependence, followed by elucidations of potential energy and activation energy of reaction, became incorporated into the fixed deterministic format like an expanding telescope corresponding to deeper and deeper factorizations of the original rate constant. This has always seemed surprising to the present author and is usually excused by appeals to the various laws of large numbers which are invoked indiscriminately either in the domain of experimental observations or in dealing with bothersome theoretical foundations of a subject to dismiss inherent random irregularities on the basis of their possibly microscopic dimensions. While this may lead to only rare errors of interpretation at the macroscopic level, in a way it has already impeded progress in the spread of microscopic kinetics up to the macroscopic level.

The basic considerations of microscopic kinetics going all the way back to the origins of the collision theory are all expressed in probabilistic terms and therefore require an appropriate and compatible mathematical treatment at all levels before reliable and objective assessment of their implications at the macroscopic level can be made. This matter has been of the highest concern to the present writer and the inclusion of some detailed elaborations of these points in this chapter therefore seemed obligatory. If, therefore, there appears to be overemphasis in this chapter on the stochastic aspects, perhaps it can be excused on the basis that unquestionably it has been underemphasized in other texts, though not in the current literature. A theory of stochastic processes is now very much with us just as it was almost totally absent at the beginning of chemical kinetics.

Finally, there is the distinction between chemical and biochemical, and in particular, enzyme kinetics. Important aspects of enzyme macromolecules which distinguish them from inorganic molecules pertain to their great complexity, size, and functionally significant structures or conformations. These are all qualities, as will be seen, that go far beyond the characteristics of the molecules to which our present knowledge of microscopic kinetics relates. The mathematical methods applied to the simplest molecules have had to be compromised by recourse to approximations, both in terms of the mathematical models or procedures and in terms of idealizations of the molecular details. On the other hand, the progress of biochemistry in the past few decades has been so phenomenal that the pace of theoretical methods for handling the deeper aspects of the kinetics of biochemical and enzyme systems seems by comparison almost at a standstill. Consequently, *ad hoc* methods have developed as the needs arose. But these have created attitudes, now hard to overcome, between biochemical kineticists and traditional kineticists who question conclusions arrived at in many cases, strictly empirically. On the other hand, the problems presented in developing an enzyme and biochemical kinetics compatible with the physical spirit of chemical kinetics are difficult enough to raise serious questions about their practical solvability, as we now understand the term. Enzyme kinetics calls also for a development of such areas as open systems theory, heterogeneous as opposed to homogeneous kinetics, detailed treatments of reactions in solution, the coupling of essentially physical processes with chemical processes.

Associated with this distinction between chemical and enzyme kinetics is the remaining, apparently insuperable obstacle separating knowledge of such systems obtained through *in vitro* studies from the reality of *in vivo* system activities. Thus, to advance to the stage where it now is, enzyme kinetics has had to indulge in radical modeling to the extent of constructing artificial chemical systems which hopefully resemble the actual situations in the cells, say. The original ultimate goal of biochemistry, which is the chemical explanation of life processes, apparently will have to be attained through long series of approximations. It is here also that the need for stochastic development becomes important—not only in consideration of the complexities and incompleteness of knowledge of the underlying systems, but also considering that in cells extremely dilute reactions are involved. Of course, it is difficult to reconcile randomness with the desideratum of an orderly, well-controlled universe. On the other hand, the theory of statistics teaches us principles for crystallizing regularity amidst irregularity. And, in man-made sciences such as the theory of automata considerations such as the construction of reliable machines from unreliable components, do raise some doubts about the tenability of a comfortable interpretation of the myriad details contained in the universe about us.

An additional adjuvant to the theoretical and experimental armamentaria in kinetics is the large-scale computer which does make it possible to add to the *in vitro* approach, an approach which has been referred to by the author as the *in numero* approach to experimentation [Bartholomay, 1964a, 1968a, b]. It is in this type of experimentation that certain details of biochemical as well as chemical kinetics of complex systems may be studied, for it is possible to isolate and block reaction steps in theoretical ways in computer simulation studies beasd on mathematical models and thus to achieve the unattainable in *in vitro* approximations to reality.

These considerations entered into the selection of material and the determination of style and length for this chapter. It was felt that the educational aims of this chapter on chemical kinetics and enzyme kinetics in a textbook on mathematical biology and the scientific obligations implied by the preceding discussion could best be met by structuring it into three main sections, the first two being devoted to topics in chemical kinetics and the last to enzyme kinetics. More specifically, Section II is devoted to an introductory-level exposition of some of the physicomathematical principles, models, and approximations underlying transition-state theory. One of the main goals is the convenient provision of background material for subsequent sections. Allusions to and discussions of topics from collision theory are also included in all of the sections of this chapter in considered relation to the main topics selected. In such cases an attempt has been made to make each subsection as complete as possible. However, if the reader is interested in additional references to chemical kinetics, a compatible and brief reference is the "earlier work" [Bartholomay, 1962a], including its bibliography. Effort was made to minimize the dependency of this chapter on the earlier work so that a certain amount of overlap between the two chapters is unavoidable. Another goal in Section II, is to emphasize the accomplishments as well as some of the difficulties of the transition-state theory. It will be seen that strictly from the methodological viewpoint mathematical biology can profit from studying the dependency on, and experience with mathematical modeling, including the approaches to the specification of underlying systems in this branch of chemical kinetics. Here also we see some aspects of the necessary interdisciplinary exchange between the traditionally kinetic and the thermodynamic aspects of rate theory. The extent to which this theory has contributed, and can contribute to the main theoretical goal of chemical kinetics; namely, the calculation of chemical reaction "rates" is considered in this section and continued into Section III.

While Section III appears from its heading to be narrowly devoted to the special case of the unimolecular reaction, it would be fairer to say that it is presented as an example of a study in depth to which many modern absolute rate theorists have contributed during the past half century. Virtually all of the modern chemical kinetics has been applied to or developed out of the study of this basic chemical mechanism beginning with the breakthrough by Lindemann of a feasible mechanistic hypothesis for such reactions. Consequently, it is thought of here more as a focusing for the various aspects of kinetics of great convenience. Thus, in this chapter we shall discuss the Lindemann–Hinshelwood theory, the subsequent RRK and RRKM (Rice–Ramsperger–Kassel–Marcus) developments, the Slater theory, the microscopic stochastic models beginning with the work of Kramers and Christiansen and extended by Montroll, Shuler, and others. We shall also discuss briefly the testing of the validity of the "equilibrium assumption" and the validation of the transition-state theory, as they relate to such studies. Finally, the work on macroscopic stochastic models by Singer, Bartholomay, Darvey, and Staff will be discussed along with the Monte Carlo computer simulation studies by Bartholomay and also by Bunker.

Section IV, confines itself mainly to developments in enzyme kinetics proceeding out of the basic Michaelis–Menten mechanism for the elementary enzyme-catalyzed reaction which has for its parallel in chemical kinetics the Lindemann–Hinshelwood mechanism. Some attention is given to the steady-state assumption in the Michaelis–Menten and Briggs–Haldane deterministic expressions, referred to by some as the "pseudo-steady state." And in this connection the recent work of Heineken, Aris, and Tsuchiya [1967a, b] is brought in as an example of mathematically more rigorous derivations of these expressions. An attempt is made here to feature the special aspects of enzyme-catalyzed reactions which present challenges to the accomplishment of the ultimate goal of chemical kinetics as well as to the construction of a more appropriate basis at the macroscopic level of kinetics. The work of Botts,

Morales, and Hill [1948, 1953; see also Morales, Botts, and Hill, 1948], which deals more directly with the active-site concept, is reviewed after a discussion of the active-site concept, choosing the lysozyme and the carboxypeptidase systems as examples of its implications in biochemical kinetics. The connection of Bartholomay's macroscopic stochastic model of the irreversible Michaelis–Menten mechanism with the microscopic aspects of chemical kinetics is emphasized and the work of Darvey and Staff [1967] extending the former model to the reversible case is also covered. The section concludes with some allusions to anticipated future requirements in biochemical kinetics, arising from present studies of more complicated biochemical systems. As an example of the usefulness of the stochastic approach in such cases, the recent work of Zimmerman *et al.* [1963] on DNA-replication kinetics is cited.

Some connections of this chapter with other chapters in the present series follow quite naturally from the important role which chemical kinetics has played in the development of biological kinetics and mathematical biology from the very beginning. This seen is most clearly in the work of A. J. Lotka, Volterra, Gauss, and Kostitzin, for example, on population kinetics, ecology and epidemiology. The deterministic mathematical models in these areas were strict analogs of corresponding mathematical models in chemical kinetics, the basic point of similarity being the analogy between the law of mass action in chemistry and the law of Malthus in population theory. There is a similar relation between the stochastic macroscopic models of biological kinetics and the stochastic macroscopic models of kinetics, both representing stochastic reformulations of the two laws just mentioned.

II. The Transition-State Theory

A. Origins of the Method

1. *The Schroedinger Wave Equation for an Electron*

The search for the most complete and fundamental explanation of chemical kinetics takes one to the microscopic quantum mechanical level as the natural physical context for discussing the intramolecular events to which macroscopic intermolecular events are coupled. This level of interpretation forms the basis for the so-called modern absolute rate theory, or, the transition-state theory which goes back approximately a half century. Extensions of the Schroedinger wave equation from the simplest case of a single electron to systems of atoms and molecules form the natural point of entrance.

Here we meet with mathematical difficulties immediately, for a closed solution of the Schroedinger equation beyond the case of the simplest electron system does not appear possible. And so, approximations must be invoked in attempts to deduce the basic properites of chemical kinetics on this basis.

At the same time, the physical formulation itself in these terms can be regarded only as a theoretical model conceived as an analog of the basic wave-theoretic model of the radiation of light. This has reference to Maxwell's explanation of the properties of light on the presumption that the radiations consist of electromagnetic disturbances obeying the general wave equation

$$\nabla^2 w = c^{-1}\, \partial^2 w/\partial t^2, \tag{1}$$

where c is the velocity of light, w is the wave amplitude represented as a function of the (x, y, z) rectangular co-ordinates of position, all being functions of time t, and ∇^2 is the Laplacian operator with respect to the (x, y, z) system. The interpretation of w^2 as the probability density function determining the probability of finding a photon in a given region of the wave related this interpretation to the particulate theory of light and reconciles it with Heisenberg's uncertainty principle.

The work of deBroglie and later experimental evidence obtained by L. H. Germer, C. Davisson, and G. P. Thomson lent weight to the hypothesis that electrons behave like photons and are accompanied by waves controlling their motions and gave rise to what might be called the electron analog of the photon model. By analogy with Eq. (1), it was thus considered meaningful to write as the "wave equation" for a single electron

$$\nabla^2 \Phi = (u^2)^{-1}\, \partial^2 \Phi/\partial t^2, \tag{2}$$

where now u is the velocity of propagation of the electronic wave ("phase velocity") and Φ is the wave amplitude. Then where λ is the length of the electron wave, ν is the frequency of the wave, h is Planck's constant, starting with the well-known relation for the energy E of the electron, $E = h\nu$ and taking $p = h/\nu$, $u = \lambda\nu$ so that $u = E/p$, Eq. (2) may be written

$$\nabla^2 \Phi = (p^2/E^2)\, \partial^2 \Phi/\partial t^2. \tag{3}$$

In case the waves are represented as standing waves, Φ can be factored into two functions, one of them Ψ, depending on (x, y, z) and not explicitly on t, and the other depending on time and frequency ν; that is,

$$\Phi = \Psi(x, y, z)(A \cos 2\pi\nu t + B \sin 2\pi\nu t), \tag{4}$$

where A and B are constants. Substituting Eq. (4), into Eq. (3), then gives

$$\nabla^2 \Psi = -(4\pi^2 p^2/h^2)\Psi \tag{5}$$

and, where V is potential energy of the electron and T is its kinetic energy, $\frac{1}{2}mV^2 = P^2/2m$, so that $E - V = P^2/2m$, Eq. (5) becomes

$$\nabla^2 \Psi + (8\pi^2 m/h^2)(E - V)\Psi = 0, \tag{6}$$

which is recognized as "the Schroedinger equation of electron wave mechanics."

2. *The Electron Wave Equation Extended to Atoms and Molecules in Reaction; Origins of the Potential Energy Surface Concept*

The relevance of this basic quantum mechanical model for the treatment of electron wave energy relative to a theory of chemical kinetics is demonstrated by the following analysis as in Glasstone *et al.* [1941, pp. 85–87], of a simple chemical reaction process, which is a classical example in transition-state theory and still undergoing further elaboration by modern chemical physicists. Consider a simple, irreversible homogeneous "bimolecular" exchange reaction: $A + B \longrightarrow$ products, in which a molecule of species A is actually an atom X and a molecule of species B is a diatomic molecule denoted for convenience by YZ, being composed of atoms Y and Z. The "products" in this case are considered to be the diatomic molecule XY and the atom Z, so that in a given case the exchange reaction is usually symbolized stoichiometrically by

$$X + YZ \longrightarrow XY + Z. \tag{7}$$

Again, for greatest simplicity let it be assumed that the atoms Y and Z in the diatomic molecule initially are joined by a single bond; specifically, by a pair of electrons with opposite signs, and that the atom X has an uncoupled electron available. As X moves close to YZ, one may hypothesize that the resulting interaction is describable as a decrease in exchange energy, resulting in a diminution of the attraction between Y and Z and leading to a decomposition of YZ into the component atoms Y and Z. But in such a case there would result an increase in potential energy of the system due to the increasing repulsion of X by YZ and the decreasing attraction between Y and Z. A point is then reached at which Z becomes repelled and the system advances to a state favoring the reaction $X + YZ \longrightarrow XY + Z$. If X moves to within the normal interatomic distance between Y and itself, Z is repelled and the potential energy of the system declines. Schematically, this is usually represented as in Fig. 1. Here arbitrarily r_{XY} is taken as the interatomic distance between X and Y. It is evident from the diagram that in some sense it is necessary for the system to acquire sufficient energy to "surmount the energy barrier" separating the initial state $(X + YZ)$ from the final state $(XY + Z)$.

The preceding diagram is an abstraction which is made definite in terms of the so-called "potential-energy surface," which in turn may be considered as a generalization of the potential-envergy curve of a simple diatomic molecule. The work of Born and Oppenheimer [1927], paved the way for the quantum mechanical energy calculations on molecules, providing an approximate method for separating electronic motion from nuclear motion in the following sense. The structure of any molecule is discussible on this basis by first considering the motions of electrons in the field of atomic nuclei to be fixed in a definite configuration: the electronic motion being much more

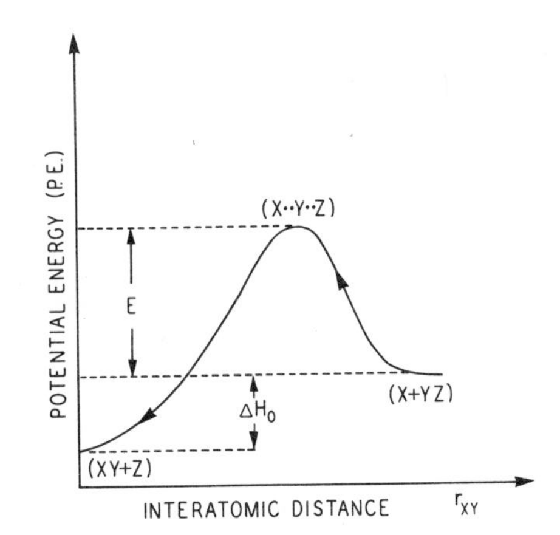

Fig. 1. Schematic representation of potential energy versus internuclear distance r_{XY} for the exchange reaction

$$X + YZ = XY + Z.$$

rapid than nuclear motion, the former can be studied by assuming the nuclei to be at rest (referring here to the classical adiabatic assumption). In this way the electronic energy of the molecule is calculable as a function of the nuclear configuration. And the configuration for the normal state of the molecule is that corresponding to the minimum value of the energy function, which insures maximum stability for the molecule. London [1928] pursued this point of view in the treatment of chemical reactions in effect, assuming that many chemical reactions are adiabatic. He proposed the calculation of such energy for chemical systems composed of three or four reactant atoms (corresponding to various interatomic distances), by taking the three or four electron quantum mechanical systems as his model, the energy calculations for which were obtained by generalization of the Schroedinger wave equation model for simple electrons (see above). This development amounts to an application of the so-called Heitler–London valence theory.†

The prototype for the Heitler–London treatment of diatomic molecules, generally referred to as the "valence–bound method" is their treatment of the hydrogen molecule, which consists of two nuclei n_1 and n_2 and two associated electrons e_1 and e_2. The calculation of the interaction energy for various values of the internulcear distance $r = d(n_1 n_2)$ reflects the following consideration. If the electronic interaction energy is calculated directly as a function of r, it is found that at large distances there is a weak attraction, which soon turns into a strong repulsion as r further diminishes, implying that the two atoms could not form a stable bond. But such a calculation ignores the resonance phenomenon. Thus, in accord with quantum mechanical principles, the calculation must be made in such a way as to allow the possibility of the exchange of places of the two electrons. Pursuing this approach an

†See Heitler and London [1927].

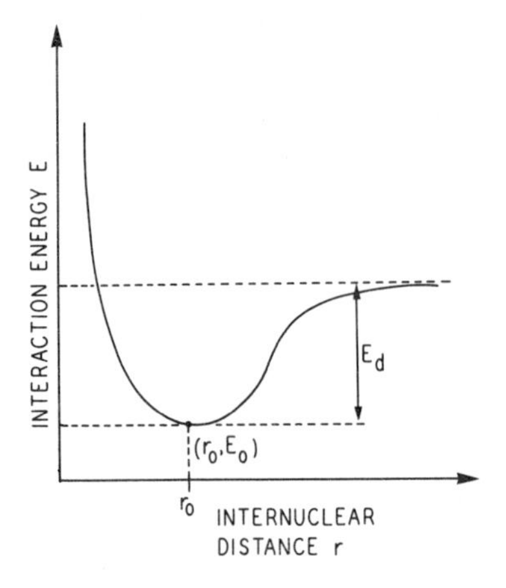

Fig. 2. Schematic representation of interaction energy E as a function of internuclear distance r for hydrogen molecule; E_d = energy of dissociation.

interaction-energy E curve is obtainable which shows a pronounced minimum at (r_0, E_0) corresponding to the formation of the stable molecule. This is pictured schematically in Fig. 2. In commenting on their calculations, Pauling [1939] observes, for example, that the calculated equilibrium distance between the nuclei is 0.05 Å larger than the observed 0.74 Å. Furthermore, the results do not agree with the expectation of the so-called virial theorem, which states that in a system of atomic nuclei and electrons, where $\bar{V}$ is the average potential energy, $\bar{T}$ the average kinetic energy, and E the total energy, a constant, the following relations hold:

$$\bar{V} = -2\bar{T}, \qquad E = -\bar{T}, \qquad \bar{V} = 2E. \tag{8}$$

Wang [1928] was able to make an improvement on the method which resulted in agreement with the virial theorem.

Again, with reference to Fig. 2, as r increases beyond r_0 the energy appears to increase, at first rapidly and in accord with the law of a simple harmonic oscillator, and then more slowly as the interatomic binding forces grow weaker. Finally, the energy asymptotically approaches a fixed limiting value corresponding to the complete dissociation of the molecule. Thus, in this figure E_d represents the energy of dissociation. However, even in this simple case it has not been possible to obtain an exact analytical expression for E as a function of r. Semiempirical equations have therefore been introduced, the constants of which can be evaluated from bond-spectroscopic data. A convenient example of such an equation is due to Morse [1929]:

$$E = E(r) = E_d \exp\{-2a(r - r_0)\} - 2E_d \exp\{-a(r - r_0)\} \tag{9}$$

(a is a constant). Substitution of this expression for the potential energy in Shroedinger's equation for an oscillator, combined with determination of the allowed values for the various energy levels of the vibrating system leads to agreement with an empirically deduced spectroscopic equation. In this manner this curve has been utilized in predicting the magnitudes of the activation energy of some simple reactions.

The Heitler–London model for the energy aspects of the hydrogen molecule system is not unique, nor can it be considered necessarily as the most acceptable basis. In this connection, Pauling [1939], makes the following points:

> Before 1927 there was no satisfactory theory of the covalent bond. The chemist had postulated the existence of the valence bond between atoms and had built up a body of empirical information about it. but his inquiries into its structure had been futile. The step taken by Lewis of associating two electrons with a bond can hardly be called the development of a theory. since it left unanswered the fundamental questions as to the nature of the interactions involved and the source of energy of the bond. Only in 1927 was the development of the covalent bond theory initiated by the work of Condon [1927] and of Heitler and London on the hydrogen molecule . . . [p. 23].†

It can well be imagined, then, that the discussions of the potential energy function of a chemical system even one step removed in complexity such as the interaction of an atom and a diatomic molecule YZ presents many difficulties and has had to rely on a number of assumptions, simplifications, and approximations to the point where work on such systems in the context of the transition-state theory of chemical physics is still current.

The General Mathematical Model of the Potential Energy Surface. The mathematical generalization of the potential energy curve corresponding to a chemically interactive molecular system containing a totality of m atoms may be described [Slater, 1959, pp. 106–109] simply as a family of $(n+1)$-dimensional hypersurfaces in $(n+1)$-space $(n > m)$, where n is the number of independent variables $(q_1, q_2, \ldots, q_n)$ needed to specify completely the relative positions of the corrsponding m nuclei. The $(n+1)$st variable is the dependent potential energy variable E and each member of the family of surfaces corresponds to a different electronic state. In practice, it is the lowest such state which is presumed to be of most importance—so that in a given application it is really the only member of the family that is constructed. Thus, one can conceive of the potential energy surface as a mathematical model described by an equation of the form

$$E = E(q_1, q_2, \ldots, q_n) \tag{10}$$

with the following mathematical constraints or properties:

P1. The surface has an absolute stable minimum $E = 0$ at $q_1 = q_2 = \cdots = q_n = 0$, that is, at the origin of the q-coordinate system.

P2. The surface has a saddle point or unstable equilibrium, call it $Q_0(q_{10}, q_{20}, \ldots, q_{n0}; E_0)$ corresponding to the dissociation configuration of least energy.

Conditions P1 and P2 imply that E cannot be simply a quadratic in $q_1, q_2, \ldots, q_n$. However, in first-order reactions near the origin E will have a quadratic component of the form

$$E = \tfrac{1}{2}\sum_{j=1}^{n}\sum_{i=1}^{n} b_{ij}q_iq_j, \tag{11}$$

and in the neighborhood of Q_0 the form

$$E = E_0 + \tfrac{1}{2}\sum_{j=1}^{n}\sum_{i=1}^{n} b_{ij}(q_i - q_{i0})(q_j - q_{j0}). \tag{12}$$

One can see, for example, from the general formulation that in the case of a three-atom suystem, where $m = 3$, there would be $n = 3$ q-variables, corresponding to the relative internuclear positions,

$$\begin{aligned} q_1 &= d(N_1, N_2) = \text{distance between X and Y nuclei,} \\ q_2 &= d(N_2, N_3) = \text{distance between Y and Z nuclei,} \\ q_3 &= d(N_1, N_3) = \text{distance between X and Z nuclei.} \end{aligned} \tag{13}$$

And so the potential energy surface would be a four-dimensional geometric expression of the potential energy as a function of the three internuclear distances. A four-atom system would require $n = 6$ q-variables corresponding to all possible internuclear distances. And, in general, in an m-atom system $\frac{1}{2}m!/(m-2)!$ such variables would be required. Consequently, simplifying assumptions must be made or fewer ultimate variables chosen in some way, so as to reduce the number of variables for purposes of practicable calculations and graphical representation. In other words, more convenient concepts or measures of internuclear "distances" must be set up.

For example, in the case of the three-atom reaction $X + YZ \rightarrow XY + Z$, it is usual to eliminate one of the three q variables by approximating the general planar configuration by a collinear configuration, that is, taking the three atoms in straight-line arrangement. Some justification for this simplification is possible from London's demonstration that a linear configuration is of lower potential energy and therefore "more probable" than a corresponding nonlinear or planar configuration, at least if the valence electrons are in so-called s-states. Thus, supposing that N_2 is located in a straight line between nuclei N_1 and N_3, referring to Eq. (13), q_1 and q_2 would be the two convenient ultimate variables to choose and the energy E would then be expressible as a function of q_1 and q_2, that is, as a three-dimensional surface. In practice, because of the lack of analytic expressions a graphical procedure is employed for correlating the characteristics of the surface with the chemical transformation process. In fact, the problem is reduced one dimension further by replacing the entire surface by a system of isoenergetic "contours" or profile plots, that is, passing to the one-parameter family $E =$ (constant) of curves parallel to the (q_1, q_2) plane in the hydrogen diatomic case. In the case of four-atom systems requiring six variables and hence a seven-dimensional spatial representation, the analytic difficulty is circumvented by passing from internuclear distance, or "configuration space" to "bond space"; the potential energy is expressed in terms of two bond energies instead of six distances [see, for example, Glasstone *et al.*, 1941, p. 121].

B. Quantum Mechanical Calculations of Energy Terms

In principle, two forces are to be considered in such atomic interactions: (1) an electrostatic part, the so-called Coulombic force; and (2) the force depending on the quantum mechanical resonance phenomenon, the "exchange force." In molecular configurations it is, of course, this latter which is of primary importance.

1. *Some Operator Theory*†

The determinations of E values are obtained out of approximations to the solution of the wave equations; specifically, the necessary components of energy are obtained as eigenvalues of E by a procedure that can be most simply demonstrated for the case of a single particle system. Consider an electron of mass m moving in a field of force corresponding to a potential function $V(x, y, z)$ where now the q's are taken simply as the three rectangular coordinates of position of the electron.

Let p_x, p_y, p_z be the conjugate momenta corresponding to x, y, z. Where h is Planck's constant, in accord with the basic postulates of quantum mechanics, let the operators for these conjugate momenta be

$$\omega_{p_x} = (h/2\pi i)\,\partial/\partial x, \qquad \omega_{p_y} = (h/2\pi i)\,\partial/\partial y, \qquad \omega_{p_z} = (2\pi i)\,\partial/\partial z. \tag{14}$$

And, where $T =$ kinetic energy $= p^2/2m = (2m)^{-1}(p_x{}^2 + p_y{}^2 + p_z{}^2)$ and V, as before the potential energy, we may introduce as an expression for the total energy of the system, the Hamiltonian function $H = T + V$ and specify it in these terms as

$$H = (2m)^{-1}(p_x{}^2 + p_y{}^2 + p_z{}^2) + V(x, y, z). \tag{15}$$

Then directly from the postulates of quantum mechanics the Hamiltonian operator $\mathcal{H}$ corresponding to H is deduced as

$$\mathcal{H} = (2m)^{-1}[(\omega_{p_x})^{(2)} + (\omega_{p_y})^{(2)} + (\omega_{p_z})^{(2)}] + V(x, y, z). \tag{16}$$

†From Glasstone *et al.* [1941, pp. 38–42].

Finally, if in a given state the total energy is known to be E, then by the postulates, the "amplitude function" ψ satisfies the equation

$$\mathcal{H}\psi = E\psi. \tag{17}$$

In other words, the energy states are interpretable mathematically as the eigenstates of the (Hermitian) Hamiltonian operator. Then, combining Eqs. (16) and (17) gives

$$[-(h^2/8\pi^2 m)\,\nabla^2 + V]\psi - E\psi = 0, \tag{18}$$

which, again, is the Schroedinger equation. So, in terms of the fundamental postulates of quantum mechanics, the Schroedinger equation is replaced by the quantum mechanical model

$$\mathcal{H}\psi - E\psi = 0, \tag{19}$$

(where $\mathcal{H} = -(h^2/8\pi^2 m)\,\nabla^2 + V$). In this way an alternative form of the mathematical model for calculating energy states has been introduced for convenience in the calculation of energy states, namely, as eigenstates of a Hamiltonian operator. It should also be noted that this procedure generalizes to the extent that we may obtain as the Hamiltonian of a system of n particles

$$\mathcal{H} = -\,(h^2/8\pi^2) \sum_{i=1}^{n} (m_i)^{-1} \nabla_i^{\,2} + V. \tag{20}$$

2. *Approximation Methods for the Hamiltonian Form of the Wave Equation*

In the absence of a complete solution of the wave equation in any form, one turns next to approximation methods, chief among which are the so-called "variation method" and the "perturbation method." These methods allow one to obtain sets of approximate eigenfunctions (ψ's) and eigenvalues (E's) of the Hamiltonian operator. We consider here the first of these, in a more general mathematical context, the so-called "Ritz variational method," or the "Rayleigh–Ritz method." It begins with a trial functional ϕ as a first approximation to the actual wave function ψ, in the form of a linear combination of, preferably, mutually orthonormal component functions $\phi_1, \phi_2, \ldots, \phi_n$ (they may or may not be chosen as eigenfunctions of some particular Schroedinger equation):

$$\phi = \sum_{i=1}^{n} c_i \phi_i \tag{21}$$

(the so-called "linear variational function"). Given any presumed solution ϕ of ψ, it has been established that the following condition must hold:

$$I = \int_a^b \bar{\phi}\, \mathcal{H}\phi \, dq \Big/ \int_a^b \bar{\phi}\phi \, dq \geqslant E; \tag{22}$$

that is, the integral I can approach, but must always remain greater than the true energy E of the system. In case ϕ is the correct wave function ψ, then $I = E$. In these terms, the goal of the variation method is to find a set of functions $\{\phi_i\}$ such that ϕ defined by Eq. (21) gives the lowest possible value

Accordingly, substituting Eq. (21) into the I integral gives

$$I = \sum_{i=1}^{n} \sum_{j=1}^{n} c_i c_j H_{ij} \Big/ \sum_{i=1}^{n} \sum_{j=1}^{n} c_i c_j \Delta_{ij}, \tag{23}$$

where

$$H_{ij} \equiv \int_a^b \bar{\phi}_i \mathcal{H} \phi_j \, dq, \tag{24}$$

$$\Delta_{ij} \equiv \int_a^b \bar{\phi}_i \phi_j \, dq, \tag{25}$$

and generally ϕ is assumed to be real-valued, so that the conjugate bars over the ϕ's may be dropped. To find the minimum value of I and obtain in this way the closest possible approximation to true energy, Eq. (23) is differentiated partially with respect to each c_i and set equal to zero. This results in a set of n homogeneous linear equations in the c's. To obtain nontrivial solutions, the coefficient determinant, the so-called "secular" determinant of the problem is set equal to zero; that is,

$$\det(H_{ij} - \Delta_{ij} I) = 0, \qquad i, j = 1, 2, \ldots, n. \tag{26}$$

As an nth degree polynomial in I, Eq. (26) has n roots, call them $I_1, I_2, \ldots, I_n$. The lowest root of these is taken as an approximation for E_0, the least eigenvalue of the Hamiltonian operator $\mathcal{H}$,

$$E_0 \approx \min_j \{I_j\}, \qquad j = 1, 2, \ldots, n. \tag{27}$$

In turn, if I is set equal to E_0 in the system of equations, the c_i coefficients ($i = 1, 2, \ldots, n$) can be determined by first dividing $(n - 1)$ of the equations by c_n, leading to explicit solutions for the $(n - 1)$ ratios $c_1/c_n, \ldots, c_{n-1}/c_n$. Then this set of ratios together with the normalizing condition for ϕ, namely,

$$\int_a^b \phi^2 \, dq \sum_{i=j=1}^{n} c_i c_j \Delta_{ij} = 1, \tag{28}$$

produces a solution for all the c's. This gives the desired approximations to that solution of the differential equations $\mathcal{H}\psi = \lambda\psi$ which corresponds to the smallest eigenvalue. Thus, the mathematical problem is formally identical with that of finding the smallest eigenvalue of an nth order matrix. This same technique, of course, allows one to associate with any energy value E_k the E_k corresponding eigenfunction ψ_k.

3. *Equipotential Contour Diagrams and Potential Energy Surfaces*†

The variation method has been applied to the system $X + YZ \rightarrow XY + Z$ of three interacting atoms to obtain eventually approximations to the potential energy function E used in determining the family $E = \lambda$ of profiles of the potential energy surface. However, it should be mentioned that in addition to the Hamiltonian equation approximation there are a considerable number of other quantum mechanical calculations involving, for example, considerations of electron spin and spin operators, acknowledging that the complete eigenfunction of an electron must include also the spin-operator eigenfunctions and the bond eigenfunctions, for example.

The resultant expression for E in the case of the triatomic X, Y, Z system is generally refererd to as the "London equation" and consists of two main components, as in the case of the single diatomic system considered earlier, namely, the electrostatic or Coulombic force Q and the exchange force, all components of which are functions of the q variables, that is, of the internuclear distances. It is usually written in the form

$$E = A + B + C - \{\tfrac{1}{2}(\alpha - \beta)^2 + (\beta - \gamma)^2 + (\gamma - \alpha)^2\}^{1/2}, \qquad (29)$$

where A, B, C are the components of Q; that is, the three coulombic energies of the three atoms taken in pairs; and α, β, γ are the exchange or resonance energies of the three possible isolated diatomic molecules, defined in terms of a system of Heitler–London exchange integrals.

The difficulties encountered in dealing with the Heitler–London integrals [see Sugiura, 1927] combined with the imprecisions and possible inaccuracies of the approximations involved have led those interested in specific problems to turn to semiempirical methods for estimating these quantities. It appears that the most extensive work along these lines has been done by Eyring and his collaborators. Initially, Eyring and Polanyi [see references to Eyring, 1931, 1932a, b, 1935a, b; Eyring and Polanyi, 1931], apparently the first to plot potential energy surfaces, considered the simple process of the conversion of para- to ortho-hydrogen ($H + H_2 \rightarrow H_2 + H$). It was in this connection that they introduced the "semiempirical method" for obtaining numerical approximations to the Heitler–London integrals on the basis of spectroscopic data for diatomic molecules XY, YZ, and XZ, combined with the additional approximation assumption that the coulombic energy is a constant fraction of the total for the interatomic distances of reactive significance for all the molecules involved. This involves use of the Morse function [Eq. (9)] in the form

$$E = D'[\exp\{-2a(r - r_0)\} - 2\exp\{-a(r - r_0)\}], \qquad (30)$$

†Following Frost and Pearson [1961, pp. 78–84].

where D' is the heat of dissociation of the molecule plus the zero-point energy; r_0 is the equilibrium interatomic distance of the normal molecule; and $a = 0.1227w_0(\mu/D')^{1/2}$, w_0 being the equilibrium vibration frequency and μ the "reduced mass" of the molecule. It is the D' and w_0 terms that are obtained spectroscopically. In this way it is possible to calculate the energy of the diatomic system for any separation (the q variables) of the two atoms.

By geometric construction or by clever contrivances using sliding rulers, the potential energies of this system have been calculated for given values of q_1 and q_2, the procedure being repeated until sufficient points are available for plotting the corresponding potential energy contour diagrams. The range of values 0.5–4.0 Å (corresponding to the activated state) is usually observed.

A typical such equipotential contour diagram consisting of curves of intersection of the reduced three dimensional surface $E = E(q_1, q_2)$ in the space $(q_1, q_2; E)$ with the family $E = \lambda_k$ $(k = 0, 1, 2, \ldots)$ of planes parallel to the q_1, q_2 plane, namely,

$$E = E(q_1, q_2), \qquad E = \lambda_k \qquad \text{(family of equipotential contours)} \qquad (31)$$

is pictured in the typical Fig. 3, where for convenience the curve corresponding to $E = \lambda_k$ has been labeled E_k. Also, for convenience in interpreting the results, in Fig. 3, it has been assumed that

$$\lambda_3 < \lambda_2 < \lambda_1 < \lambda_0 < \lambda_4 < \lambda_5 < \lambda_6 < \lambda_7. \qquad (32)$$

From diagrams such as the schematic one shown in Fig. 3, one may obtain, point by point, the data for preparing the plot of a planar cross section through the potential energy surface corresponding to the plane q_2 = constant $= q_{21}$ which would show, for fixed q_2 the variation in potential energy along the coordinate q_1, that is, E as a function of q_1. An example of this is shown in Fig. 4. As would be expected, this curve resembles the diatomic potential energy curve (Fig. 2) since with Z far removed (that is, taking q_2 quite large and fixed) the energy would depend primarily on $q_1 = d(N_1, N_2)$. E changes slowly with q_1 in the neighborhood of the plane $q_2 = q_{21}$ through $\overline{P_1P_3}$ and also in the neighborhood of the plane q_1 = constant $= q_{12}$ through $\overline{P_2P_3}$ (where these lines are assumed to correspond to large values of q_1 and q_2) because of the great distance between an atom and a molecule corresponding to these internuclear distances. A plateau would exist in the neighborhood of $P_3(q_{12}, q_{21})$, flanked with valleys comprising the neighborhoods of P_1 and P_2. The neighborhood of the saddle point at $P_0(q_{10}, q_{20})$ is spoken of as a "pass" between these valleys. As Frost and Pearson [1961, p. 81] point out, Eyring's semiempirical methods point to the possible existence of a small potential depression in the neighborhood of the saddle point which is of questionable validity. The directed, broken-line curve γ through

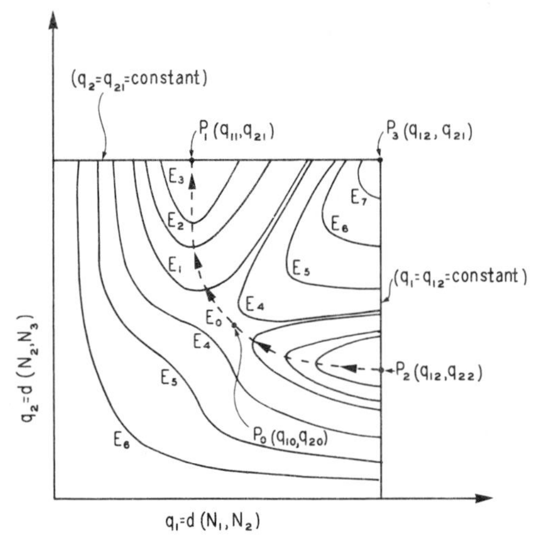

Fig. 3. Schematic equipotential potential-energy profile for linear XYZ system as a function of internuclear distances q_1 and q_2. [From Frost and Pearson, 1961.]

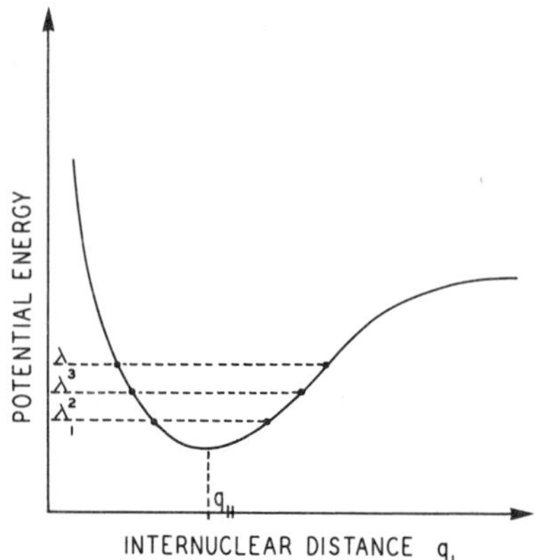

Fig. 4. Planar section $q_2 = q_{21}$ = constant through the potential-energy surface corresponding to Fig. 3. [From Frost and Pearson, 1961.]

$P_2P_0P_1$ represents a possible "directed 'path" or "trajectory" corresponding to the change in potential of the system from state (X + YZ) to (XY + Z) as it proceeds along the P_2 valley, over the saddle point P_0 and down into the P_1 valley.

In many texts a repeated point of confusion arises in connection with schematic plots such as the one pictured in Fig. 5, presented to follow the passage. In such cases the horizontal axis is labeled "reaction coordinate" with no further specification in the text for what amounts to a generic term. Obviously the "reaction coordinate" is not one of the internuclear distances, as comparison of Fig. 5 and Fig. 4 generally reveals. On the other hand, as we proceed along the path γ from P_2 through P_0 to P_1 in Fig. 3, taking note of the E levels and their relative magnitudes, we can see that such a course

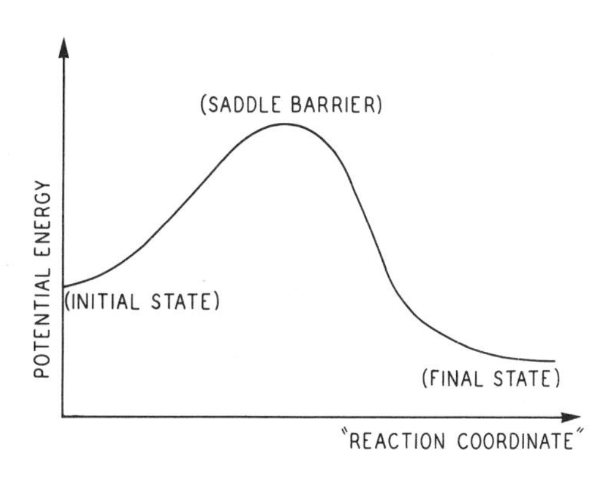

Fig. 5. Schematic plot of passage along potential-energy surface from initial to final state with "reaction coordinate" as abscissa. [From Frost and Pearson, 1961.]

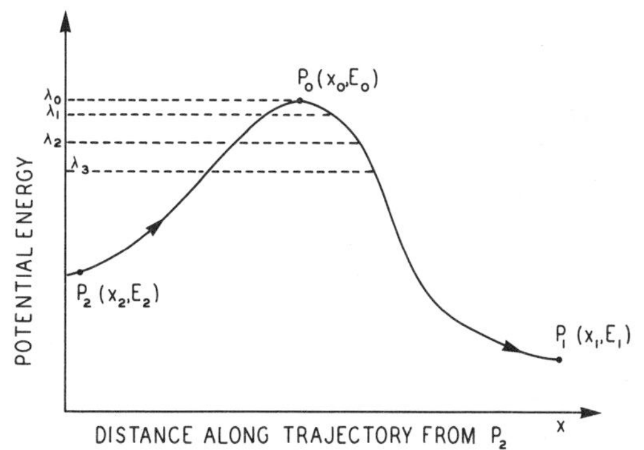

Fig. 6. Alternative plot of passage along potential-energy surface showing as abscissa x the curvilinear distance along the path from starting point.

might result. And, in fact, on this basis a diagram such as Fig. 6, might be prepared and labeled as shown so as to indicate not only the corresponding energy levels, but also a meaningful specification of the "reaction coordinate" such as $x =$ distance along trajectory from P_2.† It would, in this case be indicated in the text that the "reaction coordinate" has been selected as the curvilinear distance x of a point traversing the given path along that path from an arbitrary beginning point. For example, in Fig. 6, the coordinates (x_0, E_0) refer to the E_0, corresponding to a distance x_0 along the path chosen, beginat the point $P_2(x_2, E_2)$ in Fig. 3. The path chosen. of course, is considered to be a "most probable" potential path associated with successful completion of one of the chemical conversions involved. And, in these representations in which energy is seen to rise to a maximum level above the original and then decline to a level below the original, the height of the saddle point above the initial state $(E_0 - E_2)$ becomes an estimate of the activation energy.

In the preceding paragraph schematic passages have been discussed for the case of "successful" passage over the potential energy surface. Clearly,

†In this connection, see Eliason and Hirschfelder [1959, p. 1434].

there are infinitely many potential pathways whereby a "successful" passage, that is, one corresponding to a simple, discrete, actual transformation event in the system $X + YZ \rightarrow XY + Z$ or event corresponding to a successful pairing $T(a,b) = (c,d)$ in "molecular set-theoretic" language [see Bartholomay, 1960, 1965] of a particular atom *a* of type X and a molecule *b* of type YZ to form a molecule *c* of type XY and a molecule *d* of type Z. In the same way there are infinitely many "unsuccessful pathways" corresponding to the energy history of molecular pairings of this system. In fact, even in the case of a successful outcome the path along the surface over the barrier could be far from smooth, with many random irregularities, resulting say from vibrational and translational variations in the molecular and atomic configurations, the passage resembling a "random flight."

C. Passage of "Activated Complex" over the Potential Energy Barrier

We turn next to the problem of incorporating the notion of "activated complex" (for the moment identifying this loosely with the state of the intermolecular system in the immediate neighborhood of the saddle point or potential energy barrier) into a theory aimed at predicting the "rate" of an elementary reaction. Considering the ulnimited varieties of irregularities, oscillations, and reversals that can be described along the potential energy surface by a particular trajectory associated with the complex, it is logical to expect that the statistical–physical method must be invoked in elucidating and calculating the "rate" to be associated with the total reaction. Incidentally, an enlargement of the early views and methods of kinetics would seem to imply that the term "rate" be replaced by something like "average rate," or "reaction frequency" or ultimately even, "reaction probability," particularly when the methods of statistical physics or the theory of stochastic process are introduced. Section III, particularly, will elaborate this idea. In fact, in Section III, in retrospect (with repect to the "rate" consideration next to be made in this section), it will be seen that the rationale for stochastic models is actually implicit in the microscopic kinetics of modern absolute rate theory and transition-state theory.

Thus, looked at from the point of view of statistical mechanics, which was invented to study the motions of systems of all kinds consisting of many particles or individual objects or units, this problem, which is interpretable as the forceful propulsion of energized individual systems of molecules over a potential energy surface, may be considered at first as a statistical assembly of particles, and the methods of physical mechanics applied. It is in this context that the rate may be identified with the average velocity of the passage of "activated complexes" over the barrier. The notion of activated complex is intended to incorporate the set of energy states of the molecular *n*-tuplet (such as the XYZ triplet in the triatomic case discussed earlier) which places it at the top of the potential energy barrier (see Fig. 6), say within a small, but arbitrary horizontal distance δ along the x axis. It is understood that this set of states reflects all possible variations in other coordinates such as vibrations normal to the reaction coordinate and translations and rotations of the center of mass of the n-tuplet constituting the "activated complex." Incidentally, in derivations proceeding from this interpretation, the magnitude of δ if generally thought of as 1 Å, but it is irrelevant finally, in the sense that it does not appear in the final rate expression.

1. *The Activated Complex as a Pseudomolecule*

In characterizing the activated complex, it is necessary to consider (a) that it involves an association, of some kind, of the reactant molecules, that is, is an n-tuplet determined by the stoichiometry of the system; and (b) that the energy states of this n-tuplet must be members of a restricted subset of all possible energy states of the whole system. This latter points up the important question of whether the activated complex, that is, the energy-restricted n-tuplet is indeed an actual molecule. According to Eyring [1935 a, b], the activated complex is like an ordinary molecule, except that instead of having only the three regular translational degrees of freedom, it has a fourth, along which it first approaches the potential energy barrier, crossing it, and then disrupting. Since the activated complex, when represented as a point on a potential energy surface, is a minimum for all the internal degrees of freedom except the one degree of freedom for which it is a maximum, the theory of small vibrations used for stable compounds may be applied.

2. *Most Probable (Equilibrium) Distributions of Quantum States*

The problem of calculating the velocity of passage of the system of activated complexes through the transition state was approached originally as a problem in physical (statistical) mechanics, which derivation draws on its analogy with the mathematical model for gas molecule kinetics, the transition-state passage being thereby idealized to the point of considering the system of activated complex as a mechanical system in a "phase space." In such a case it is necessary to assume a definite form for the distribution of velocities, such as the "Maxwell–Boltzmann statistics." This represents a choice actually from three alternatives, each of which is derivable combinatorially as an "occupancy problem." The derivation of the Maxwell–Boltzmann statistics is given here as an example of the solution of this class of problems which arises in a number of other cases in chemical kinetics.

a. *The Maxwell–Boltzmann Statistics.* Suppose we have a system of n distinguishable "particles" (for example, molecules, atoms, electrons), each characterized by its own system of coordinates, calling q_ν the set of coordinates for the νth molecule. Each of these molecules will have its own wave equation of the form of Eq. (19) say. And, suppose there is at least one solution for each of these equations; that is, $g_\nu \geqslant 1$ solutions for the νth molecule. Then where $g = \sum_{\nu=1}^{n} g_\nu$, the whole set of equations generates a totality of $g \geqslant n$ eigenfunctions. Any product consisting of n eigenfunctions, one corresponding to each molecule is referred to as a "complete eigenfunction." Thus, if $\psi_{11}, \psi_{21}, \ldots, \psi_{n1}$ represents such a selection, then

$$\psi_1 = \prod_{\nu=1}^{n} \psi_{\nu 1} \tag{33}$$

is an example of a complete eigenfunction and defines an "eigenstate" of the system; any change in the assignment, to the factors of ψ by choosing a different combination of n eigenfunctions from the g eigenfunctions describes a new possible eigenstate.

Let the n (distinguishable) molecules be subdivided into a sequence of subsets or groups of sizes $n_1, n_2, \ldots, n_i, \ldots, n_N$, the elements in a particular group having "approximately the same energy." Then, where

$$n = \sum_{i=1}^{N} n_i, \tag{34}$$

$$P(N; n_1, n_2, \ldots, n_i, \ldots n_N) = n! \Big/ \prod_{i=1}^{N} n_i \tag{35}$$

gives the total number of ways of distributing n distinguishable elements into N subgroups of seizes $n_1, \ldots, n_N$. Considering that each of these groups corresponds to a different quantized energy (eigenvalue) level $\varepsilon_1, \ldots, \varepsilon_i, \ldots, \varepsilon_N$, where ε_i has g_i-fold degeneracy (so that as in the preceding paragraph in the ith group there are available g_i eigenfunctions), then there are $n_i^{g_i}$ selections open to the n_i molecules at the ε_i level; therefore,

$$P_n = n! \prod_{i=1}^{N} g_i^{n_i} \big/ (n_i!) \tag{36}$$

gives the cardinality of the set of possible eigenstates or quantum levels for the n molecules. And assuming that the most probable state is that for which P_n or $\log_e P_n$ is a maximum, this state can be determined by maximizing, say $(\log_e P_n)$ as a function of the n_i. Thus, in

$$\log_e P_n = \log_e(n!) + \sum_{i=1}^{N} [n_i \log_e g_i - \log_e(n_i!)], \tag{37}$$

utilizing the approximation

$$\log_e x! \approx x \log_e x - x \qquad \text{(for large } x\text{)}, \tag{38}$$

by assuming n and n_i large, gives

$$f(n_1, n_2, \ldots, n_i, \ldots, n_N) = \log_e P_n = n \log_e n + \sum_{i=1}^{N} [n_i \log_e g_i - n_i \log_e n_i] \tag{39}$$

with the conditions

$$n = \sum_{i=1}^{N} n_i, \qquad E = \sum_{i=1}^{N} n_i \varepsilon_i. \tag{40}$$

Then, with Lagrange's method of undetermined multipliers in mind for small variations δn_i, write

$$\delta(\log_e P_n) = \sum_{i=1}^{N} (\partial f / \partial n_i)\, \delta n_i, \tag{41}$$

so that

$$\delta f = \sum_{i=1}^{N} [\log_e g_i - 1 - \log_e n_i]\, \delta n_i, \tag{42}$$

$$\delta n = \sum_{i=1}^{N} \delta n_i, \tag{43}$$

$$\delta E = \sum_{i=1}^{N} \varepsilon_i \delta n_i. \tag{44}$$

The necessary condition for a maximum on f ($\delta f = 0$); the fact that n is a constant, giving $\delta n = 0$; and the conservation of energy requiring $\delta E = 0$, lead to

$$\sum_{i=1}^{N} [\log_e(n_i/g_i) + 1]\, \delta n_i = 0, \tag{45}$$

$$\sum_{i=1}^{N} \delta n_i = 0, \tag{46}$$

$$\sum_{i=1}^{N} \varepsilon_i \delta n_i = 0. \tag{47}$$

Multiplying Eqs. (46) and (47) by the undetermined constants α and β, respectively, gives

$$\sum_{i=1}^{N} [\log_e(n_i/g_i) + \alpha + \beta \varepsilon_i] = 0. \tag{48}$$

The arbitrariness in the δm_i then allow us to write the necessary conditions,

$$\log_e(n_i/g_i) + \alpha + \beta \varepsilon_i = 0 \tag{49}$$

so that

$$n_i = g_i / \exp\{\alpha + \beta \varepsilon_i\}, \tag{50}$$

or

$$n_i = Cg_i \exp(-\beta \varepsilon_i), \tag{51}$$

where β can be shown by laborious argument to be $(k_B T)^{-1}$. It can be shown that

$$n_i = Cg_i \exp(-\varepsilon_i / k_B T) \tag{52}$$

gives the most probable arrangement of the n molecules among N quantum states of energies $\varepsilon_1, \varepsilon_2, \ldots, \varepsilon_i, \ldots, \varepsilon_N$ in terms of the numbers n_i to be found in each state ε_i. The right-hand side of Eq. (51) is thus interpretable, with appropriate normalization, that is, setting

$$C = \left(\sum_{i=1}^{N} g_i \exp(-\varepsilon_i / k_B T \right)^{-1} \tag{53}$$

as the probability,

$$p(\varepsilon_i) = Cg_i \exp(-\varepsilon_i / k_B T) \tag{54}$$

that a given molecule will be located in eigenstate ε_i. This result is generally referred to as the Maxwell–Boltzmann statistics.

b. The Bose–Einstein Statistics. Now suppose the assumption of distinguishability is replaced by the assumption that there are n *indistinguishable* elements and assume further that only so-called "symmetric" solutions to the wave equation hold (which would be the case for atomic nuclei, electrons, photons, neutrons). In this case the association of the n_i molecules at level ε_i with the g_i eigenfunctions becomes the occupancy problem of determining how to assign $n_i \geqslant g_i$ distinguishable elements to g_i boxes. From combinational analysis it is known that

$$P_i' = \binom{n_i + g_i - 1}{n_i} = \frac{(n_i + g_i - 1)!}{n_i!(g_i - 1)!} \tag{55}$$

ways are possible; this is then used to replace Eq. (35). Correspondingly,

$$P_n' = \prod_{i=1}^{N} P_i' = \prod_{i=1}^{N} (n_i + g_i - 1)!/n_i!(g_i - 1)! \tag{56}$$

replaces Eq. (36). By similar methods this leads to the occupancy number n_i' known as the Bose–Einstein statistics (or distribution):

$$n_i' = \frac{g_i}{\exp(\epsilon_i / k_B)T - 1}. \tag{57}$$

c. The Fermi–Dirac Statistics. Finally, suppose one retains the hypothesis of *indistinguishability* but permits antisymmetric wave functions, so that in effect no two particles can have the same elementary wave function (true for systems of all fundamental particles: electrons, protons, neutrons, nuclei, and atoms containing an odd number of such particles). Combinatorially, this case amounts to the occupancy problem of determining, at a given level ϵ_i the number of ways of choosing n_i objects from g_i possibilities, where now $g_i \geqslant n_i$ so that here Eq. (35) is replaced by

$$P_i'' = \binom{g_i}{n_i} = \frac{(g_i)!}{n_i!\,(g_i - n_i)!} \tag{58}$$

so that

$$P_n'' = \prod_{i=1}^{N} P_i'' = \prod_{i=1}^{N} \frac{(g_i)!}{n_i!\,(g_i - n_i)!}, \tag{59}$$

leading to the occupancy numbers

$$n_i'' = \frac{g_i}{\exp(\varepsilon_i/k_B T) + 1}, \tag{60}$$

known as the Fermi–Dirac statistics (or distribution).

3. *Calculation of the Rate r of Reaction*

The customary choice of statistics is the Maxwell–Boltzmann case, which is sometimes referred to as the "classical statistics." Its universality results from the facts that (1) at temperatures which are not "too low" and pressures which are not "too high," the quantity (-1) in Eq. (57) may be neglected. And at "relatively high" temperatures and "relatively low" pressures the factor $(+1)$ in Eq. (60) may be neglected, so that the Bose–Einstein and Fermi–Dirac statistics appear to be well approximated in the "normal" cases by Maxwell–Boltzmann statistics. It is with these considerations in mind that Eq. (54) is taken as the probability that molecules will have energy ϵ_i. And so, in considering transition-state translations in one degree of freedom parallel to the x axis, the kinetic energy being $\epsilon = (\frac{1}{2})m\dot{x}^2$ where $\dot{x} = dx/dt$ and m is the effective mass of the activated complex for motion in the x direction, from Eq. (54),

$$p(\varepsilon) = p(\tfrac{1}{2}m\dot{x}^2) = Cg\exp(-m\dot{x}^2/2k_B T) \tag{61}$$

which has led to the expression

$$p(\dot{x})\,d\dot{x} = C\exp(-m\dot{x}^2/2k_B T)\,d\dot{x} \tag{62}$$

(for simplicity setting $g = 1$) for the probability of finding the velocity component $\dot{x}$ in the range $(\dot{x}, \dot{x} + d\dot{x})$; that is, $p(\dot{x})$ is interpretable in Eq. (62) as the probability density function of the random variable $\dot{X}$ corresponding to $\dot{x}$.

Furthermore, restricting the discussion to activated complexes with position velocity component $\dot{x} > 0$, that is, to passages to the right, we would normalize (62) by setting

$$C\int_0^\infty \exp(-m\dot{x}^2/2k_\text{B}T)\,d\dot{x} = 1, \tag{63}$$

in which case

$$C = (2m/\pi k_\text{B}T)^{1/2}, \tag{64}$$

that is,

$$p(\dot{x}) = (2m/\pi k_\text{B}T)^{1/2}\exp(-m\dot{x}^2/2k_\text{B}T). \tag{65}$$

In such a case, the mean velocity $\bar{v} = \bar{\dot{x}}$ of complexes moving toward the right is given by

$$\bar{v} = \int_0^\infty \dot{x}p(\dot{x})\,d\dot{x} = (2k_\text{B}T/\pi m)^{1/2} \tag{66}$$

from which it is concluded that the "average time τ" for successful passage through the transition state is

$$\tau = \delta/\bar{v} = \delta(\pi m/2k_\text{B}T)^{1/2}. \tag{67}$$

Conventionally, then τ^{-1} is taken as the "fraction of molecules crossing the barrier for unit volume per unit time." And where [X] is the concentration of activated complexes, that is, the number of activated complexes per unit volume lying in the length δ of the coordinate of decomposition; again conventionally $\frac{1}{2}[\text{X}]/\tau$ is taken as the number of complexes crossing over the barrier to the right per unit volume per unit time (assuming that half of the complexes move to the right). And, on the basis of these heuristic considerations the rate of reaction r is stated as

$$r = [\text{X}]/2\tau = ([\text{X}]/\delta)\,(k_\text{B}T/2\pi m)^{1/2} \tag{68}$$

Correction for the possibility that some complexes will be reflected back (to the left) across the potential energy barrier is made simply by multiplying Eq. (68) by a factor p called "the transmission coefficient." For a detailed discussion of this see Glasstone *et al.* [1941, pp. 146–150].

Critique of the Method. The conventional aspects of the derivation of the rate expression have been adversely criticized also by Slater [1959] on a number of points. In particular, he regards the separation of the positive from the negative $\dot{x}$ components as "unusual" with respect to the consequent mean-value determinations. In the earlier work [Bartholomay, 1962a, p. 53] a similar point was made. Equivocal results are therefore obtainable. For example, Frost and Pearson [1961], are in agreement with the results as presented. But Glasstone *et al.* [1941] obtain for the value of $\bar{v}$

$$\bar{v} = (k_\text{B}T/2\pi m)^{1/2} \tag{69}$$

on the basis of normalizing the probability function over the entire $\dot{x}$ — axis, taking the limits $(-\infty, \infty)$ for determining C in Eq. (62) (instead of $[0, \infty]$), at the same time restricting the range of integration in the calculation of $\bar{v}$ to the right half of the $\dot{x}$ axis; that is,

$$\bar{v} = \int_0^{\infty} \dot{x} \exp\left(-\frac{1}{2}\frac{m\dot{x}^2}{k_B T}\right) d\dot{x} \Big/ \int_{-\infty}^{\infty} \exp\left(-\frac{1}{2}\frac{m\dot{x}^2}{k_B T}\right) d\dot{x}. \tag{70}$$

In addition, Slater, in another connection criticizes the use of "average lifetime" which was introduced originally by Glasstone, Laidler, and Eyring. Certainly it is clear that this interpretation is empirical and heuristic if we think of "average lifetime" in the technical probabilistic sense, that is, as a "mean value" with respect to the probability density function. Thus, in this case the "mean lifetime" of the activated complexes would be determinable as

$$\tau = (\overline{\delta v^{-1}}) = \delta(\overline{v^{-1}}) = \int_0^{\infty} \dot{x}^{-1} p(\dot{x})\, d\dot{x}, \tag{71}$$

that is, as the mean value of the new random variable $V = \dot{X}^{-1} = f(\dot{X})$ with respect to the probability density function $p(\dot{x})$ of the random variable $\dot{X}$. Thus, he quarrels, justifiably that $(\delta/\bar{v})$ as in Eq. (67) is not really the mean δ/v. But the integral in Eq. (71), on the other hand, is nonconvergent, as he point out, in which case one has to say that the conventional derivation avoids a mathematical "trap" here.

These, as well as other criticisms of the method, are all accountable, on the basis of the necessary heterogeneous mixture of deterministic, probabilistic, and heuristic reasoning involved in early derivations of this type, which are avoidable in stochastic formulations or models. Thus, if one invokes the theory of stochastic processes then such concepts as transition probabilities and mean first passage times provide a more compatible and mathematically rigorous framework for the derivation, as has been mentioned earlier. Some stochastic validation and reformulation of this set of ideas have in fact been produced and will be discussed in Section III. The advantage in becoming familiar with the early conventional treatments is that in this way it is possible to understand the process with a minimum of involvement in somewhat more sophisticated, theoretical detail, so that in retrospect the conventional method still retains its usefulness as a simpler level of mathematical understanding.

Finally, it will be seen in Section III, that the stochastic reformulation occurred originally in terms of investigating the assumption that the Maxwell–Boltzmann statistics, that is, that the "Maxwellian distribution of velocities" continues to hold throughout the reaction process (the so-called "equilibrium hypothesis") Here the equilibrium hypothesis guarantees an invariant Maxwell–Boltzmann distribution of reactant molecules in the reactant valley of the potential-energy phase space. As Montroll and Shuler [1958] point out, it has long been recognized that the equilibrium postulate is only an approximation to reality, since any process with a nonvanishing rate must disturb an initial equilibrium; this calls for a more thorough investigation of the conditions under which this assumption may be invalid. Additional discussion of this point is postponed until Section III.

D. Equilibrium and Thermodynamic Aspects of the Transition-State Rate

Further explication of the simple rate expression given in Eq. (68) has occurred in terms of examining in greater detail the term [X]. It should be noted first of all that whereas the potential energy surface theory was discussed in close relation to the simple triatomic, bimolecular exchange reaction $X + YZ \rightarrow XY + Z$, no such restriction was implied in deriving the basic transition-state rate expression v. Thus, the activated complex molecule X, henceforth to be denoted by X* (the asterisk being appended to all quantities related to X*) was assumed to imply an expansion of the stoichiometric equation effected by interposing X* (X–Y–Z in the triatomic exchange reaction) in reversible relation with the original reactants. For example, the general irreversible ($\sum_i^r a_i$)-molecular elementary reaction, was assumed to be expanded to the form,

$$a_1A_1 + a_2A_2 + \cdots + a_rA_r \rightleftharpoons (X^*) \longrightarrow b_1B_1 + \cdots + b_sB_s. \tag{72}$$

In relation to Eq. (72), the quantity [X] appearing in Eq. (68) refers really to the "equilibrium" concentration $[X^*]_e$ if we allow, for example, that X* can be treated in the context of Eq. (72) essentially as an ordinary molecule in discussing the establishment of an equilibrium between it and the original A-reactant molecules along the lines of thermodynamic equilibrium theory. And in this case the rate expression, Eq. (68) can be carried further. But before turning to this in Section II.D.2, in the section immediately following we shall provide more background for completing the derivation by introducing some of the main features and concepts in the general thermodynamic equilibrium theory, or model, to be utilized in the derivation.

1. *Introduction to the Statistical Thermodynamic Treatment of the Equilibrium State*

Referring back to the discussion in Sections II.C.3.a,b,c of the Maxwell–Boltzmann, Bose–Einstein, and Fermi–Dirac statistics which were obtained by a maximization procedure applied to combinatorial expressions P_n, P_n', P_n'' of the related eigenstate occupancy problem, it is customary in texts on statistical mechanics to refer to the corresponding occupancy numbers $\{n_i\}$ $(i = 1, 2, \ldots, N)$ as the equilibrium set of distribution numbers, and to emphasize this property by the notation $\{n_i^*\}$. The maximization procedure applied to the corresponding P_n, P_n', P_n'' expressions led to the various sets $\{n_i^*\}$ interpreted as the numbers of molecules of each level which guarantee the largest number of distinct microscopic eigenstates for the system of n molecules. The satisfaction of this condition by the various statistics such as the Maxwell–Boltzmann statistics, taken together with the classical assumption of the equiprobability of distinct microscopic states implies that a given system is more likely to be found in a state corresponding to the set $\{n_i^*\}$ than in any other state. And it is in this sense actually that the qualifier "equilibrium" (distribution) is employed.

In this context the notion of "partition function" arises as a summary expression of how, in the equilibrium distribution, the system of particles is partitioned among the different energy levels. Thus, corresponding to the Maxwell–Boltzmann statistics defined by Eq. (54), for a given kind of energy the partition function (*Zustandsumme* or "sum over states" or "state function") Z is defined by

$$Z \equiv \sum_{i=1}^{N} g_i \exp(-\varepsilon_i/k_B T). \tag{73}$$

Because of the probabilistic interpretation given to $g_i \exp\{-\varepsilon_i/k_B T\}$ as the probability that a molecule shall have energy ε_i it follows that Z, being summed over all such states is interpretable as the total probability of occurrence of a particular molecular species. Partition functions are of central importance in thermodynamics because thermodynamic quantities such as equilibrium constants, free energies, entropies, and chemical potentials are deducible from them.

Of main concern in this section is the derivation of the partition function from the equilibrium constant associated with a reversible reaction of the form

$$a_1 A_1 + \cdots + a_r A_r \rightleftharpoons b_1 B_1 + \cdots b_s B_s . \tag{74}$$

However, this can be discussed most simply by considering mainly the simplest case

$$A \rightleftharpoons B, \tag{75}$$

that is, the reversible unimolecular reaction. And relative to Eq. (75), we begin by determining the (energy state) equilibrium distributions of the total molecular composition of the system. At a given instant let it be assumed that m molecules of A are present and n molecules of B, where $m + n =$ constant. As in the case of the derivation of the Maxwell–Boltzmann statistics, a function P_{mn} is developed, giving the number of possible eigenstates of the system that is to be maximized. Thus, the problem here may be regarded as the extension of the problem of determining the most probable states of a given molecular species, say A, to that of determining the most probable state for a system of interacting chemical species—in this case A and B.

Suppose each A molecule has the possible energies: $\varepsilon_1, \varepsilon_2, \ldots, \varepsilon_i, \ldots, \varepsilon_M$ and each B molecule has available to it a set of energies: $\varepsilon_1', \varepsilon_2', \ldots, \varepsilon_j', \ldots, \varepsilon_N'$. The A and B energy levels may be measured by independent scales (implied in the previous notation), or the lowest (zero-level) of the A molecules may be taken as the arbitrary zero of reference for both. In this latter case where ε_0 represents the difference in zero-level energies of the A and B subsystems we may write

$$\bar{\varepsilon}_j = \varepsilon_j' + \varepsilon_0, \qquad j = 1, 2, \ldots, N \tag{76}$$

In these terms the effect on the potential energy surface coordinates for the final state (the B state) may be visualized as a translation along the vertical axis of the potential energy versus x-reaction coordinate as in Fig. 7.

Now working with the B states labeled as $\bar{\varepsilon}_1, \ldots, \bar{\varepsilon}_j, \ldots, \bar{\varepsilon}_N$, let $m_1, m_2, \ldots, m_i, \ldots, m_M$ be the occupancy numbers of the A states, and $n_1, n_2, \ldots, n_j, \ldots, n_N$ those for the B states, so that

$$m = \sum_{i=1}^{M} m_i, \qquad n = \sum_{j=1}^{N} n_j, \tag{77}$$

in which case the total energy of the system is given by

$$E = \sum_{i=1}^{M} m_i \varepsilon_i + \sum_{j=1}^{N} n_j \bar{\varepsilon}_j, \tag{78}$$

and as in the case of the Maxwell–Boltzmann derivation, the total number

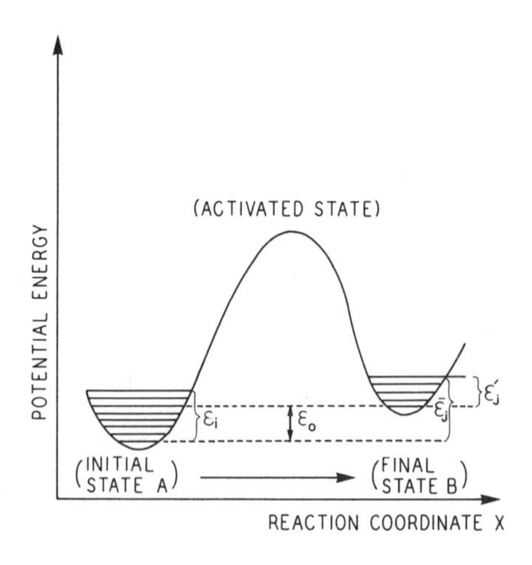

Fig. 7. Potential energy versus reaction coordinate x for reaction $A \rightleftharpoons B$, showing relations between the energy levels of the A and B molecules. [From "The Theory of Rate Processes" by S. Glasstone, K. Laidler, and H. Eyring, Fig. 52, p. 183. Copyright © 1941 McGraw-Hill Book Company. Used with permission of McGraw-Hill Book Company.]

of states is given by the function $P_{mn}(m_1, \ldots, m_M; n_1, \ldots, n_N)$,

$$P_{mn} = \left\{(m!) \prod_{i=1}^{M} \left(\frac{g_i^{m_i}}{m_i!}\right)\right\} \left\{(n!) \prod_{j=1}^{N} \left(\frac{h_j^{n_j}}{n_j}\right)\right\}. \tag{79}$$

Here now h_j gives the degeneracy number of state $\bar{\varepsilon}_j$. In tnis case the maximization problem is really a sequential pair of coupled maximizations: (1) the determination of the sets $\{m_i^*\}$ and $\{n_j^*\}$ of occupancy numbers which maximize P_{mn} for fixed m and n; (2) then the determination of the values m^* and n^* of m and n which maximize the latter expression. In other words, we seek the values $\{m_i^*\}$, $\{n_j^*\}$, m^*, n^* such that

$$P_{m^*n^*} = \max_{(m,n)}\{P_{mn}(m_1^*, \ldots, n_N^*)\}, \tag{80}$$

$$P_{m^*n^*} = \max_{(m,n)}\{ \max_{(m_1,\ldots,n_N)} P_{mn}(m_1, \ldots, m_M; n_1, \ldots, n_N\}. \tag{81}$$

As in the case of P_m, taking the natural logarithm of Eq. (79) and introducing approximations to the factorial expressions, the problem becomes one of maximizing

$$\begin{aligned} \log_e P_{mn} &= f(m_1, \ldots, n_N) \\ &= m \log_e m + n \log_e n + \sum_{i=1}^{M} m_i \log_e g_i \\ &\quad - \sum_{i=1}^{M} m_i \log_e m_i + \sum_{j=1}^{N} n_j \log_e h_j - \sum_{j=1}^{N} n_j \log_e n_j, \end{aligned} \tag{82}$$

subject to the constraints given in Eqs. (77) and (78) leading to

$$\delta E = \sum_{i=1}^{M} \varepsilon_i \, \delta m_i + \sum_{j=1}^{N} \bar{\varepsilon}_j \, \delta n_j = 0, \tag{83}$$

$$\sum_{i=1}^{M} \delta m_i = 0, \tag{84}$$

$$\sum_{j=1}^{N} \delta n_j = 0, \tag{85}$$

$$\delta f = \sum_{i=1}^{M} \log_e(g_i/m_i)\delta m_i + \sum_{j=1}^{N} \log_e(h_j/n_j)\delta n_j. \tag{86}$$

Then, (again applying the Lagrange method), multiplying Eq. (84) by α_1, Eq. (85) by α_2, Eq. (83) by β, adding to Eq. (86),

$$\delta f = \sum_{i=1}^{M} \left[\log_e\left(\frac{g_i}{m_i}\right) + \alpha_1 + \beta_i \varepsilon \right] \delta m_i + \sum_{j=1}^{N} \left[\log_e\left(\frac{h_j}{n_j}\right) + \alpha_2 + \beta \bar{\varepsilon}_j \right] \delta n_j \tag{87}$$

from which we obtain ultimately,

$$\begin{aligned} m_i^* &= g_i \exp(\alpha_1 + \beta_i \varepsilon_i) = a g_i \exp(-\varepsilon_i/k_B T), \qquad i = 1, \ldots, M, \\ n_j^* &= h_j \exp(\alpha_2 + \beta \bar{\varepsilon}_j) = b h_j \exp(-\bar{\varepsilon}_j/k_B T), \qquad j = 1, \ldots, N. \end{aligned} \tag{88}$$

Rushbrooke [1949, pp. 180–181] may be consulted for the details of justifying the use of

$$m^* = \sum_{i=1}^{M} m_i^* = a \sum_{i=1}^{M} g_i \exp(-\varepsilon_i/k_B T), \qquad n^* = \sum_{j=1}^{N} n_j^* = b \sum_{j=1}^{M} h_j \exp(-\bar{\varepsilon}_j/k_B T) \tag{89}$$

as the second set of maximizing values; that is, the values of m and n which maximize $P_{mn}(m_1^*, \ldots, n_N^*)$. The ratio

$$K \equiv \frac{n^*}{m^*} = \frac{\sum_{j=1}^{N} h_j \exp(-\bar{\varepsilon}_j/k_B T)}{\sum_{i=1}^{M} g_i \exp(-\varepsilon_i/k_B T)}, \tag{90}$$

which is the ratio of the partition function of the B subsystem to that of the A subsystem, is the "equilibrium" constant of the reversible reaction in the context of statistical thermodynamics.

2. *The Equilibrium Constant K' Arrived at via Classical Chemical Kinetics*

K is the statistical mechanical counterpart of the classical kinetics equilibrium constant, call it K', which is arrived at in the context of the classical deterministic model for the ordinary chemical kinetics of Eq. (75) as follows, where [ ] symbolizes *concentration of*

$$d[\mathrm{B}]/dt = k_1[\mathrm{A}] - k_2[\mathrm{B}] = 0, \tag{91}$$

where k_1 is the forward reaction rate and k_2 the reverse rate, assuming that the equilibrium state is characterized by the vanishing of the flow. Then with the arbitrary initial conditions,

$$[\mathrm{A}] = A_0, \qquad [\mathrm{B}] = 0 \qquad \text{at} \quad t = 0, \tag{92}$$

since

$$[\mathrm{A}] + [\mathrm{B}] = A_0, \tag{93}$$

solving Eq. (93) with Eq. (91) as a pair of simultaneous linear algebraic equations gives for the solution the values

$$A_e = k_2 A_0/(k_1 + k_2), \qquad B_e = k_1 A_0/(k_1 + k_2) \tag{94}$$

in which values A_e and B_e are referred to as the "equilibrium values" of the reactions, and their constant ratio

$$B_e/A_e = k_1/k_2, \tag{95}$$

as the "equilibrium constant"

$$K' \equiv k_1/k_2. \tag{96}$$

K and K' may be regarded simply as different forms of the equilibrium constant, if we identify n^* with B_e and m^* with A_e. This may be defended heuristically by noting first that A_e and B_e are the values of [A] and [B] toward which the reaction process tends in time,

$$A_e = \lim_{t \to \infty} [A], \tag{97}$$

$$B_e = \lim_{t \to \infty} [B] \tag{98}$$

and m^* and n^*, from the statistical thermodynamic point of view, represent in a certain sense the states of the A and B molecular subsystems which are sought out by the total reaction system. These results and considerations extend to the $(\sum_{i=1}^{r} a_i)$-molecular reversible reaction Eq. (74). For example, in this case the classical kinetic equilibrium constant K' is defined as

$$K' \equiv \prod_{j=1}^{s} B_{je}^{b_j} \Big/ \prod_{i=1}^{r} A_{ie}^{a_i}, \tag{99}$$

where B_{ie} and A_{ie} and respectively the "equilibrium concentrations" of the B and A molecules.

3. *The Classical Thermodynamic Derivation of the Equilibrium Constant*

From the viewpoint of classical (nonstatistical) thermodynamics, the condition for equilibrium in the reversible reaction, Eq. (74), is that the difference between the sum of the chemical potentials of the initial A reactant molecules and the sum of the chemical potentials of the B molecular subsystem is zero; that is, the "free energy change" vanishes:

$$\sum_{i=1}^{r} a_i \mu_i - \sum_{j=1}^{s} b_i \bar{\mu}_j = 0, \tag{100}$$

where $\mu_1, \ldots, \mu_r$ are the chemical potentials of the A_i species and $\bar{\mu}_1, \ldots, \bar{\mu}_s$ are those of the B_j species and where the term "chemical potential" [see Katchalsky and Curran, 1965, Chapter 3] in the classical thermodynamic framework of the Gibbs equation, is the "intensive" factor which relates chemical work to the number of moles transported from subsystem to subsystem. The quantity in the left-hand side of Eq. (100) is referred to as the "affinity" or driving force of the reaction; that is the affinity a is defined as

$$a \equiv a(\mu_1, \ldots, \bar{\mu}_s) = \sum_{i=1}^{r} a_i \mu_i - \sum_{j=1}^{s} b_j \bar{\mu}_j. \tag{101}$$

Thus, the equilibrium state is identifiable as the state corresponding to the values $\{\mu_{ie}\}$, $\{\bar{\mu}_{je}\}$ of the chemical potentials which are such that

$$a(\mu_{ie}, \ldots, \bar{\mu}_{se}) = 0. \tag{102}$$

The notion of chemical potential itself arises in the course of the classical characterization of a chemical reaction as a thermodynamic system. Thus, where S is the entropy parameter and V the volume parameter, and ν_i ($i = 1, 2, \ldots, r + s$) the number of moles of chemical species (combining the A and B species now), the internal energy E of the system is expressible as a function,

$$E = \phi(S, V, \{\nu_i\}), \tag{103}$$

so that in the classical thermodynamic theory of the Gibbs equation,

$$dE = (\partial E/\partial S)_{V,\nu_i}\, dS + (\partial E/\partial V)_{S,\nu_i}\, dV + \sum_{\substack{i=1\\(j\neq i)}}^{r+s} (\partial E/\partial \nu_i)_{S,V,\nu_j}\, d\nu_i. \tag{104}$$

The relation of Eq. (104) to the Gibbs equation [see *Collected Works of J. Willard Gibbs*, 1948] is seen by identifying

$$\text{Temperature} \quad T \equiv (\partial E/\partial S)_{V,\nu_i}, \tag{105}$$

$$\text{Pressure} \quad p \equiv -(\partial E/\partial V)_{S,\nu_i}, \tag{106}$$

$$\text{Chemical potential} \quad \mu_i \equiv (\partial E/\partial \nu_i)_{S,V,\nu_j}. \tag{107}$$

In other words, Eq. (104) "defines" the chemical potential as the change in internal energy due to the addition of $d\nu_i$ moels of the ith component to the system at constant S, V, ν_j ($j \neq i$).

In the same context it is possible to define entropy, volume, and chemical potential in terms of G = free energy from a similar equation,

$$dG = (\partial G/\partial T)_{p,\nu_i}\, dT + (\partial G/\partial p)_{T,\nu_i}\, dp + \sum_{\substack{i=1\\(j\neq i)}}^{r+s} (\partial G/\partial \nu_i)_{p,T,\nu_j}\, d\nu_i, \tag{108}$$

so that

$$\text{Entropy } S \equiv -(\partial G/\partial T)_{p,\nu_i}, \tag{109}$$

$$\text{Volume } V \equiv (\partial G/\partial p)_{T,\nu_i}, \tag{110}$$

$$\text{Chemical potential } \mu_i \equiv (\partial G/\partial \nu_i)_{p,T,\nu_j}. \tag{111}$$

Rewriting Eq. (108) by using Eqs. (109)–(111) gives

$$dG = -S\, dT + V\, dp + \sum_{i=1}^{r+s} \mu_i\, d\nu_i, \tag{112}$$

which at constant temperature and pressure reduces to

$$dG = \sum_{i=1}^{r+s} \mu_i\, d\nu_i. \tag{113}$$

Hence, for a single substance ($r + s = 1$) of constant composition υ, at constant temperature and pressure,

$$dG/\partial\nu = \mu = \text{constant} \tag{114}$$

so that

$$\mu = G/v \tag{115}$$

and for unit v,

$$\mu = G. \tag{116}$$

Thus, as Bray and White [1966, p. 78] point out, under these conditions chemical potential reduces to the "thermodynamic potential" or "free energy per mole," hence the term "molar free energy" preferred by some authors instead of "chemical potential" of a pure compound. And, similarly, in mixtures the chemical potential of one component is referred to as "partial molar free energy."

Taking into consideration the experimental establishment [see Katchalsky and Curran, 1965, pp. 55–57] of the fact that the osmotic pressure of dilute, ideal solutions follows the laws of ideal gases, leads to the following expression for chemical potential in the ideal case:

$$\mu_i = \mu_{i0} + RT \log_e [A_i], \tag{117}$$

where R = gas constant, T = absolute temperature, μ_{i0} = constant of integration and $[A_i]$ is the concentration of the ith species. Note that at unit concentration the chemical potential reduces to μ_{i0}. Real solutions, of course, approach the behavior given by equations such as Eq. (117), only at "infinite dilution." Deviations from the ideal situation are taken into consideration usually by introducing as a corrective factor α_i = "activity coefficient" of species A_i, considered as a solute. The activity coefficient is such that $\alpha_i[A_i]$ is the "active concentration" or "activity" f_i; that is,

$$f_i \equiv \alpha_i[A_i] \tag{118}$$

in which case the use of activity f_i in place of $[A_i]$ in Eq. (117) gives

$$\mu = \mu_{i0} + RT \log_e \alpha_i[A_i]. \tag{119}$$

Since the activity coefficient has the property that $\alpha_i \longrightarrow 1$ as $[A_i] \longrightarrow 0$. Eq. (119) reduces to Eq. (117) at infinite dilution, as required.

Using this value for chemical potential in the equilibrium condition stated in Eq. (100), gives

$$\sum_{i=1}^{r} a_i\{\mu_{i0} + RT \log_e \alpha_i[A_i]\} - \sum_{j=1}^{s} b_j\{\bar{\mu}_j + RT \log_e \beta_j[B_j]\} = 0, \tag{120}$$

that is,

$$\sum_{i=1}^{r} a_i\mu_{i0} - \sum_{j=1}^{s} b_j\bar{\mu}_{j0} + \left\{\sum_{i=1}^{r} a_iRT \log_e \alpha_i[A_i] - \sum_{j=1}^{s} b_jRT \log_e \beta_i[B_j]\right\} = 0, \tag{121}$$

or

$$\sum_{i=1}^{r} a_i \log_e \alpha_i[A_i] - \sum_{j=1}^{s} b_j \log_e \beta_j[B_j] = -(1/RT)\left\{\sum_{i=1}^{r} a_i\mu_{i0} - \sum_{j=1}^{s} b_j\bar{\mu}_{j0}\right\}. \tag{122}$$

Thus, the condition for classical thermodynamic equilibrium becomes

$$\sum_{i=1}^{r} a_i \log_e \alpha_i[A_i] - \sum_{j=1}^{s} b_j \log_e \beta_j[B_j] = C = \text{constant}, \tag{123}$$

where constant

$$C = -(1/RT)\left\{\sum_{i=1}^{r} a_i\mu_{i0} - \sum_{j=1}^{s} b_j\bar{\mu}_{j0}\right\}, \tag{124}$$

that is

$$RTC = -\left\{\sum_{i=1}^{r} a_i\mu_{i0} - \sum_{j=1}^{s} b_j\bar{\mu}_{j0}\right\}. \tag{125}$$

And K'' defined by taking $C \equiv \log_e K''$ is customarily referred to as the "classical thermodynamic equilibrium constant"; that is, K'' may be said to be defined by the expression

$$RTK'' \equiv -\left\{\sum_{i=1}^{r} a_i\mu_{i0} - \sum_{j=1}^{s} b_j\bar{\mu}_{j0}\right\}. \tag{126}$$

Substituting $RT \log_e K''$ for the right-hand side of Eq. (122), then gives the condition for equilibrium in the form

$$\log_e K'' = -\left\{\sum_{i=1}^{r} a_i \log_e \alpha_i[A_i] - \sum_{j=1}^{s} b_j \log_e \beta_j[B_j]\right\}, \tag{127}$$

that is

$$K'' = \left(\prod_{j=1}^{s} \beta_j^{b_j}[B_j]^{b_j}\right)\Big/\left(\prod_{i=1}^{r} \alpha_i^{a_i}[A_i]^{a_i}\right) \tag{128}$$

$$K'' = \left(\prod_{j=1}^{s} \beta_j^{b_j}\Big/\prod_{i=1}^{r} \alpha_i^{a_i}\right)\left(\prod_{j=1}^{s} [B_j]^{b_j}\Big/\prod_{i=1}^{r} [A_i]^{a_i}\right). \tag{129}$$

So, if the first term involving the activity coefficients is unity, K'' reduces to the "classical equilibrium" constant K' given earlier in Eq. (99). The discussion of the relation of K' to the statistical thermodynamic equilibrium constant K given in Eq. (90), taken together with the preceding derivation of K'', thus establishes the various interrelationships between the meanings of the equilibrium constants arising in the three different levels of discussion. These interrelationships will be useful in relating the final form of the transition-state rate expression to other expressions of the rate of reaction obtained in chemical kinetics.

4. *Further Considerations of the Statistical Thermodynamic Equilibrium Constant K*

Substitution of $\varepsilon_j' + \varepsilon_0$ for $\bar{\varepsilon}_j$, referring to Eq. (76), into the expression for K given by Eq. (90), leads to

$$K = \exp(-\varepsilon_0/k_BT)\left\{\sum_{j=1}^{N} h_j \exp(-\varepsilon_j'/k_BT)\bigg/\sum_{i=1}^{M} g_i \exp(-\varepsilon_i/k_BT)\right\}, \tag{130}$$

which for convenience may be written in the usual fashion as

$$K = \exp(-E_0/k_BT)(Z_B/Z_A), \tag{131}$$

where now E_0 is the symbol for the zero-level energy difference between species A and B, Z_A is the partition function for the A species and Z_B that for the B species in the reaction of Eq. (75).

Extension of K to the case of the $(\sum_{i=1}^{r} a_i)$-molecular reaction of Eq. (74) gives as the statistical counterpart of K' in Eq. (99),

$$K = \exp(-E_0/k_BT)\left\{\prod_{j=1}^{s} Z_{B_j}^{b_j}\bigg/\prod_{i=1}^{r} Z_{A_i}^{a_i}\right\}. \tag{132}$$

Here Z_{B_j} is the symbol for the partition function of species B_j; Z_{A_i}, that for species A_i; and E_0 is the difference between minimum zero-level energy of the B species and that of the A species. Actually, if the energy is expressed on a per mole basis the Boltzmann constant k_B should be replaced by R, the gas constant per mole.

The Complete Partition Function. The complete partition function for any molecular species includes terms for nuclear, electronic, vibrational rotational, and translational energy. And, as Glasstone *et al.* [1941, p. 169] point out, except for the last, the values of the various functions are calculated from spectroscopic data. The partition functions utilized in the repsent section, therefore, may be regarded as prototypes for the various distinct types. Thus, for each type of energy there would be such a function

$$\begin{aligned}
Z^{(t)} &= \text{translational partition function} = \sum_{i_1=1}^{M_1} g_{i_1}^{(t)} \exp(-\varepsilon_{i_1}/k_BT),\\
Z^{(v)} &= \text{vibrational partition function} = \sum_{i_2=1}^{M_2} g_{i_2}^{(v)} \exp(-\varepsilon_{i_2}/k_BT),\\
Z^{(r)} &= \text{rotational partition function} = \sum_{i_3=1}^{M_3} g_{i_3}^{(r)} \exp(-\varepsilon_{i_3}/k_BT),\\
Z^{(e)} &= \text{electronic partition function} = \sum_{i_4=1}^{M_4} g_{i_4}^{(e)} \exp(-\varepsilon_{i_4}/k_BT).
\end{aligned} \tag{133}$$

A discussion of the special features of each, while obviously pertinent to the present discussion, will not be undertaken here beyond noting that a different form is applied to the translational function because of the close spacing of energy levels, which calls for continuous as opposed to discrete methods of analysis. In other words, the problem of arbitrary spacing is avoided by considering the translational energy as distributed over a continuum, as opposed to quantized discrete levels. From this point of view, the summation is replaced by a so-called "phase integral," an example of which is given in the following section. Assuming that the different forms of energy distribution are independent of one another, the complete partition function is expressible as the product of the correspodning individual partition functions. Hence, the most complete form for the equilibrium constant in Eq. (131) would be

$$K = \exp(-E_0/k_B T)(Z^{(t)}_{(B)} Z^{(v)}_{(B)} Z^{(r)}_{(B)} Z^{(e)}_{(B)} / Z^{(t)}_{(A)} Z^{(v)}_{(A)} Z^{(r)}_{(A)} Z^{(e)}_{(A)}). \tag{134}$$

E. Completion of the Derivation of the Rate Expression,† Thermodynamic Aspects of the Rate Constant

We return now to the extended form of the $(\sum_{i=1}^{r} a_i)$-molecular reaction given in Eq. (72), in which the activated complex (X*) is shown interposed between the initial reactants and final products, and apply the methods of Section II.D preceding to obtain the final form of the transition-state rate expression. It has already been stressed that the activated complex X^* differs from normal molecules in certain respects. However, it is generally assumed that the same considerations which were applied to the reversible reaction may be applied by considering the mechanism

$$a_1 A_1 + \cdots + a_r A_r \rightleftharpoons X^* \tag{135}$$

as a case of Eq. (74), that is, by using the equilibrium theory of Section II.D as a model for treating Eq. (135).

From this point of view, then, the "equilibrium" between reactants and X* is expressesd in the usual fashion in terms of a classical deterministic kinetic "equilibrium constant" $(K')^*$ defined as

$$(K')^* = [X^*]_e \Big/ \prod_{i=1}^{r} [A_i]_e^{a_i}. \tag{136}$$

Then, substituting this into Eq. (68) gives

$$r = [(K')^*/\delta](k_B T/2\pi m)^{1/2} \left(\prod_{i=1}^{r} [A_i]_e^{a_i} \right). \tag{137}$$

This expression is further expanded by expressing $(K')^*$ in terms of the "partition function" of the X* molecules as follows: It has already been noted that the X* molecule differs in terms of energy from ordinary molecules

†Following Frost and Pearson [1961, pp. 88–91].

to the extent of having an additional translational degree of freedom along the x coordinate. This extra contribution to the translational partition function is calculated on the basis of an energy continuum using the "phase integral"

$$h^{-1}\int_{-\infty}^{+\infty}\int_{0}^{\delta} \exp(-m\dot{x}^2/2k_BT)\, m\, dx\, d\dot{x} = (2\pi m k_B T)^{1/2}\delta/h. \tag{138}$$

Thus, in greatly simplified form, the total partition function for the activated complex molecules will have the form

$$Z_{X^*} = Z'_{X^*}(2\pi m k_B T)^{1/2}\delta/h, \tag{139}$$

which has been factored so that Z'_{X^*} includes contributions from all degrees of freedom except that for the reaction coordinate x. Hence, the equilibrium constant K^* expressed in terms of partition functions is given by the expression

$$K^* = Z'_{X^*}(2\pi m k_B T)^{1/2}(\delta/h)\exp(-E_0^*/k_BT)\Big/\prod_{i=1}^{r} Z_{A_i}^{a_i}, \tag{140}$$

where now $E_0{}^*$ is the difference between zero-point energy per molecule of X* and that of reactants, and Z_{A_i} refers to the partition functions of the reactants A_i. Thus, the final expression for the rate r becomes

$$r = (k_BT/h)\Big(Z'_{X^*}\Big/\prod_{i=1}^{r} Z_A^{a_i}\Big)\exp(-E_0^*/k_BT)\Big(\prod_{i=1}^{r}[A_i]_e^{\alpha_i}\Big). \tag{141}$$

And so, identifying the equilibrium concentrations $[A_i]_e$ taken with respect to the equilibrium between activated complex and reactants, with the active mass concept of the law of mass action, that is, with the concentrations $[A_i]$ used in the classical deterministic model and in terms of which the rate r of reaction would be expressible for the reaction $a_1A_1 + \cdots + a_rA_r \xrightarrow{k} B$ as

$$r = (d[B]/dt) = k\prod_{i=1}^{r}[A_i]^{a_i}, \tag{142}$$

we have that the classical kinetic rate constant k is identifiable with the "rate constant" of the transition-state theory in the form

$$k \equiv (k_BT/h)\Big(Z'_{X^*}\Big/\prod_{i=1}^{r} Z_{A_i}^{a_i}\Big)\exp(-E_0{}^*/k_BT). \tag{143}$$

Note that the terms m and δ have vanished from the final expression of the rate. Finally, as noted earlier, a so-called transmission factor is introduced as a multiplier on the right-hand side of Eq. (141) as a concession to the random irregularities inherent in the passage of activated complexes across the potential-energy barrier.

The necessity for factors of this kind, added to the large number of idealizations, approximations, and arguments by analogy at times, together with the imprecision of chemical determinations at the phenomenological level as well as at the microscopic physicochemical level, make it clear that perfect agreement between theoretical and experimental kinetic observations would itself be a chance event. On the other hand, it seems to be argeed that sufficiently close agreements with experimental results and with the classical kinetic and collision-theoretic treatments have been obtained to uphold the transition-state theory. Much of the evidence, of course, has accumulated from gaseous reactions and kinetics. In the case of reactions in solution further considerations enter, though the theoretical methods which derive principally by analogy from the models of gas kinetics are considered to be applicable, at least in the spirit of reasonable idealizations and/or as first approximations to complex reaction mechanisms. However, it is also clear from the present discussion that much more remains to be done. For example, the complex and complicated nature of the mathematical and graphical methods alone, which may be anticipated from the present discussion, of dealing with reactions consisting of many reactant species of molecules of greater complexity and size make it debatable whether the goal of chemical kinetics theory, namely, the theoretical calculation of reaction rates is achievable, practically speaking. In the meantime, phenomenological, empirical, and semiempirical approaches requiring much closer cooperation between experimentalist and theoretician must be relied upon, together with the newer forms of experimentation in terms of computer methods and mathematical approaches beased on stochastic methods and the Monte Carlo method.

In Section III, an effort is made to bring together some of these newer points of view which have, in fact, been attempted on the simple unimolecular reaction. The work that has been done in testing, applying, revising, and extending transition-state theory and chemical kinetics generally, in pursuit of the deepest possible understanding of the unimolecular reaction covers actually a long period of time and is both intensive and extensive. Very general lessons may be learned from the study of the unimolecular reaction system, including the applicability and significance of the theory of stochastic processes in the study of chemical kinetics at both the microscopic and macroscopic levels and its usefulness in bridging the traditional gap between these two.

III. The Unimolecular Reaction

A. The Lindemann–Hinshelwood Mechanism

Given a reaction involving one reactant species of molecules A, except in case of "spontaneous" decompositions such as occur in radioactive disintegrations and transformations, the "kinetic order" is not always first order.

In fact, generally speaking, molecular order and kinetic order are not necessarily the same, nor is the latter always defined [see Bartholomay, 1962a]. On the other hand, a simple first-order-kinetics decomposition step appears to be the last step in a more complex mechanism which is accepted as the basis for treating unimolecular reactions which is compatible with collision-theoretic and transition-state theoretic mechanistic treatments. The unimolecular reaction mechanism referred to here may be regarded also as a primordial form or lowest limit of the transition-state theory of multimolecular and complex reaction mechanisms, including the enzyme-catalyzed reaction.

Behind the proposed mechanism around which the set of ideas to be presented in this section is built is the idea that intramolecular energy requirements for the transformation of A molecules must be preceded by intermolecular interactions of some kind. Consistent with the Arrhenius expression of the rate constant, at first it was suggested that the unimolecular rate constant k be described by the equation

$$r = \nu e^{-E/RT}, \tag{144}$$

where E is the activation (or Arrhenius) energy and R is the gas constant (= 1.98646 cal/deg mole). [Note that k_B is Boltzmann's constant = (R/N) where N is Avogadro's number, the number of molecules in one mole.] The quantity ν was postulated to be the vibration frequency of one of the bonds, k being interpreted as the rate at which sufficient energy E accumulates to break the bond, resulting in dissociation of the molecule.

But this begged the basic question of how the molecule acquires its acitvation energy. It was on this point that Lindemann [1922] invoked collision theory to suggest that the reacting A molecule acquires energy as a result of its collision with other A molecules. In order to reconcile intermolecular collisions with first-order kinetics, first of all, he stipulated that dissociation does not follow immediately on collision; there is a finite time lag before dissociation (or decomposition) occurs. Thus, in this earliest mechanistic hypothesis about the unimolecular process can be seen explicit mention of the "random gap" or "random lifetime" assumption though in later formulations, refinements, and extensions, the assumption became increasingly more obscure until recrystallized by Slater [1959], a matter which will be discussed more fully later. According to Lindemann's earliest theory, if the average time interval between activation and decomposition is long enough, the process would be kinetically of the first order.

It would appear that in these circumstances the rate constant would have the form

$$k = Z \exp(-E/k_B T), \tag{145}$$

where now Z is the collisional frequency and $\exp(-E/k_B T)$ is interpretable as the probability that a given colliding molecule acquires the requisite activation energy. But this is based on the classical interpretation that the energy is available in two "squared terms," (where by the latter terms we refer to summands for E of the form, $\frac{1}{2}m\dot{x}^2$): (1) the translational energy of each molecule along the lines of centers, and (2) the vibrational energy. On the other hand, the probability that molecules will achieve activation on collision would be increased if the energy from several squared terms (see below) could contribute to the energy of activation [Hinshelwood, 1927, 1940]. This combination of thoughts let to the "collision-theoretic model for the unimolecular reaction," later named the "Lindemann–Hinshelwood mechanism," the main features of which follow.

1. *Detailed Description of the Steps*

Consider the simple stoichiometric chemical equation

$$A = B + C, \tag{146}$$

in its more general sense, that is, not accepting the simplest mechanistic interpretations,

$$A \longrightarrow B + C, \tag{147}$$

by means of which first-order kinetics would be achievable. The following detailed mechanism can lead to first-order kinetics as well as to kinetics of higher order; or, indeed to concentration–time curves for which the term "order of kinetics" is not applicable (see earlier work).

STEP 1. ACTIVATION BY COLLISION. The reacting A molecule acquires its energy of activation in several degrees of freedom by collision with another A molecule,

$$A + A \xrightarrow{k_1} A^* + A \tag{148}$$

that is, $T_1(a_1, a_2) = (a_1{}^*, a_2)$, using molecular-set theoretic notation [Bartholomay, 1960, 1965], where $a_1, a_2 \in A$ and $a_1{}^* \in A^*$; here A^* symbolizes the "activated A species," the subset of A molecules which have acquired sufficient energy for dissociation. Thus, the activated A^* molecules correspond in this theory to the "activated complexes" in the transition-state theory of Section II. Note also, that k_1 is a bimolecular rate constant.

STEP 2. DEACTIVATION BY COLLISION.

$$A^* + A \longrightarrow A + A, \tag{149}$$

that is, $T_2(a_1{}^*, a_2) = (a_1, a_2)$. This is the reverse of the reaction in Eq. (148). And, again, k_2 is a bimolecular rate constant. Deactivation is expected to occur at the first collision of a^*, the A^* molecule, after it has formed—unless, of course, it goes on to decomposition first. It is assumed that $k_2 \gg k_1$, since the activation process is limited by the energy requirements. Thus, in molecular set-theoretic symbolism,

$$T_1 \circ T_2\,(a_1, a_2) \equiv T_2(T_1(a_1, a_2)) = T_2(a_1{}^*, a_2) = (a_1, a_2). \tag{150}$$

STEP 3. SPONTANEOUS REACTION.

$$A^* \xrightarrow{k_3} B + C. \tag{151}$$

Here, k_3 is the first-order kinetic constant, so that first-order molecularity and kinetics both are arrived at as a consequence of the acritvation step, referring to the activated species as the reactant. Hence, opposing the composition $T_1 \circ T_2$ given in Eq. (150), is the "successful route"

$$T_1 \circ T_3(a_1, a_2) \equiv T_3(T_1(a_1, a_2)) = T_3(a_1{}^*, a_2) = ((b, c); a_2), \tag{152}$$

where now $b \in$ B and $c \in$ C. ,

The deterministic mathematical model for the kinetics of the Lindemann–Hinshelwood mechanism just outlined consists of the set of nonlinear differential equations

$$d[A]/dt = -k_1[A]^2 + k_2[A^*][A], \tag{153}$$

$$d[A^*]/dt = k_1[A]^2 - k_2[A^*][A] - k_3[A^*], \tag{154}$$

$$d[B]/dt = k_3[A^*] = d[C]/dt, \tag{155}$$

with arbitrary initial condition $[A] = A_0$ at $t = 0$. If one makes the usual "pseudo-steady-state assumption,"

$$d[A^*]/dt = 0, \tag{156}$$

it is seen that at this state,

$$[A^*] = k_1[A]^2/(k_2[A] + k_3) \tag{157}$$

so that

$$d[B]/dt = k_3k_1[A]^2/(k_2[A] + k_3). \tag{158}$$

From this enlarged point of view, then, in the general case of the unimolecular reaction corresponding to Eq. (146), the overall rate r is defined by Eq. (158), from which one can see that in general the reaction has no assignable order of kinetics; in particular it is of neither the first nor the second order. However, it seems to be agreed that there are two limiting cases in which the rate may be characterized by a definite kinetic order.

CASE 1 HIGH PRESSURE IN A GASEOUS REACTION OR HIGH CONCENTRATION [A]: In this case the condition $k_2[A] \gg k_3$ allows Eq. (158) to be well approximated by

$$d[B]/dt \approx (k_3k_1/k_2)[A]. \tag{159}$$

In these circumstances the reaction would be an "apparent" first-order reaction with first-order rate constant $k = k_3 k_1 / k_2$. Thus, in this case the order of molecularity and the order of kinetics would agree and, aside from decomposing k into its three factors, concentration–time courses would be analyzed in the same way as if the total mechanism were given by

$$A \xrightarrow{k} B + C, \tag{160}$$

the "apparent" mechanism (devoid of bimolecular activation–deactivation process) that would be perceived phenomenologically. Examples of this, that will be discussed later in the stochastic context of Section III.F, are sucrose inversion and the decomposition of di-*t*-butyl peroxide (or, more generally, di-*t*-alkyl peroxide decompositions).

CASE 2 LOW PRESSURE IN GASEOUS REACTION OR EXTREMELY LOW [A]: This conditions amounts to taking $k_2[A] \ll k_3$, in which case Eq. (158) is approximated by

$$d[B]/dt \approx k_1[A]^2. \tag{161}$$

This would go along with apparent second-order kinetics, phenomenologically perceived, in stoichiometric terms as the elementary bimolecular mechanism

$$2A \xrightarrow{k'} B + C, \tag{162}$$

where now the overall reaction rate constant k' would be judged as a second-order rate constant. Simply on intuitive grounds the results would be expected under the conditions specified because at low concentration of reactant the bimolecular collisional probability would be reduced; and thereby the possibility of deactivation by collision would be removed.

It is customary to rewrite Eq. (158) as

$$d[B]/dt = (k_1 k_3[A])/(k_2[A] + k_3) \cdot [A] \tag{163}$$

and to refer to the first factor on the right as k_{uni}, the unimolecular rate "constant," or the "pseudounimolecular rate constant,"

$$k_{\text{uni}} \equiv k_1 k_3[A]/(k_2[A] + k_3). \tag{164}$$

Thus, a rather anomalous situation obtains here kinetically; with changing conditions of concentration or pressure the order of kinetics may shift from first- to second-order kinetics, going through states in between with no perceivable order.

2. *A Generalization of the Lindemann–Hinshelwood Mechanism*

The Lindemann–Hinshelwood mechanism has been generalized to the extent of allowing for activation and deactivation by collisions with molecules other than the reactant A molecules. Letting M represent any other chemical species in mechanical interaction with A, the Lindemann–Hinselwood mechanism is considered to obtain in the form

$$A + M \underset{k_2}{\overset{k_1}{\rightleftharpoons}} A^* + M, \tag{165}$$

$$A^* \overset{k_3}{\longrightarrow} B + C. \tag{166}$$

Indeed, in the most recent work we find the dissociation step, Eq. (166), elaborated to

$$A^* \longrightarrow A^+, \tag{167}$$

$$A^+ \longrightarrow B + C, \tag{168}$$

where A* is an "interesting" [see Slater, 1959] or "energized" molecule, and A^+ is the dissociable activated molecule. This will be discussed further in Section II.D.

In this case, then Eq. (158), is replaced by

$$d[B]/dt = (k_1 k_3[M])/(k_3[M] + k_3)\cdot[A] \tag{169}$$

so that

$$k_{uni} \equiv k_1 k_3[M]/(k_2[M] + k_3). \tag{170}$$

In fact, Eq. (170), is usually "linearized" by writing

$$k_{uni}^{-1} = k_2/k_3 k_1 + k_1^{-1}[M]^{-1} \tag{171}$$

so that a plot of $[M]^{-1}$ versus k_{uni}^{-1} should yield a straight line. In practice, this is very often not the case, which is explainable [Hinshelwood, 1940] on the basis that k_3 is really not a constant, being subject to energy fluctuations of the activated molecules.

B. Possible Precursor to the Lindemann–Hinshelwood Hypothesis

An earlier approach to the problem of deducing first-order kinetics in a particular unimolecular system, namely, the sucrose inversion reaction, represented stoichiometrically by the chemical equation

$$S = G + F, \tag{172}$$

where the reactant chemical species is sucrose and the products are glucose and fructose was published by Arrhenius [1889]. This process has an interesting history of reappearance in the entire history of both chemical and enzyme kinetics at significant juncture points in their development—probably because of its inorganic and enzymatic catalyzability properties.

The work of Arrhenius in retrospect would certainly appear to have anticipated the later Lindemann–Hinshelwood and transition-state theories. However, the specific goal of his investigation was the provision of a theory to account for the observed temperature dependence on the rate of inversion (hydrolysis) of sucrose, an increase of roughly 12% in rate per degree having been noted. He accomplished this by distinguishing between two forms of the sucrose reactant, "inert" and "active," in equilibrium with each other. This provided a point of entrance and indication for the extension of the so-called "reaction isochore" due to van't Hoff [1884], referring to the classical thermodynamic deduction of the following expression of the temperature dependence of the equilibrium constant K in reversible reactions:

$$d(\log_e K)/dt = E/RT^2. \tag{173}$$

Substituting (k_1/k_2) for K into Eq. (173) gives

$$d(\log_e k_1)/dT - d(\log_e k_2)/dT = E/RT^2, \tag{174}$$

so that, heuristically, introducing the energies E_1 and E_2 associated with the forward and reverse reactions, respectively, by the relation

$$E \equiv E_1 - E_2, \tag{175}$$

opens the possibility that the following identifications may be valid:

$$d(\log_e k_1)/dT \equiv E_1/RT^2, \tag{176}$$

$$d(\log_e k_2)/dT \equiv E_2/RT^2. \tag{177}$$

These equations taken by themselves suggested that corresponding to the (equilibrium) reaction isochore is the temperature dependence for any rate constant k,

$$d(\log_e k)/dT = E/RT^2 \tag{178}$$

or

$$k = Ae^{-E/RT}. \tag{179}$$

Arrhenius was able to arive at this same result in the particular case of sucrose inversion by proposing the following mechanism [the correspondence of his result with Eq. (179), reinforcing the validity of both sets of ideas]:

$$\mathrm{S_i} \underset{k_2}{\overset{k_1}{\rightleftharpoons}} (\mathrm{S_a}) \xrightarrow{k_3} \mathrm{G} + \mathrm{F}, \tag{180}$$

where S_i is the inert species of sucrose molecules and S_a the active species and where the following additional assumptions are made:

$$\begin{aligned} &\text{A1.} \quad [\mathrm{S_a}] \ll [\mathrm{S_i}], \\ &\text{A2.} \quad k_3 \ll k_1, k_2, \\ &\text{A3.} \quad [\mathrm{S}] \approx [\mathrm{S_i}]. \end{aligned} \tag{181}$$

The details of deducing first-order kinetics from Eq. (180) and these assumptions are described in the earlier work (Bartholomay, 1962a, pp. 22–24).

It will be seen later that there is also close correspondence with the Arrhenius mechanism, Eq. (180), of the basic Michaelis–Menten mechanism for an enzyme catalyzed reaction. Its relation to the Lindemann–Hinshelwood mechanism, as described by Eqs. (148)–(151), is of particular interest in this section, and in retrospect to Section II, of this chapter.

C. Subsequent Analysis and Study of the Pseudounimolecular Rate Constant k_{uni}

1. *Hinshelwood's Approach*†

It was pointed out in Section III.A.2 that significant deviations from the linear relation prescribed by such relations as

$$k_{uni}^{-1} = k_2/k_3k_1 + k_1^{-1}[A]^{-1} \tag{182}$$

had been noted. It was apparently Hinshelwood who first observed that such discrepancies could be explainable on a statistical (or probabilistic) basis. A deficiency of applying ordinary bimolecular collision-theoretic reasoning to Step 1, is seen in his observation that bimolecular k_1 would be much larger than collision theory would predict, due to the neglect in that case of internal degrees of freedom, in particular, vibrations.

The background for invoking the probabilistic point of view of quantum theory in this was extrapolated from the earlier knowledge about the vibrations of diatomic molecules. Since they are, at least approximately simple harmonic, the molecules must gain or lose vibrational energy in integral multiples of $(h\nu)$, where ν is the natural vibration frequency. But, the vibration frequency of a simple harmonic vibration is given by

$$\nu = (1/2\pi)(R/m)^{1/2}, \tag{183}$$

where R is the restoring force per unit displacement and m is the mass. The more tightly bound the atoms (that is, the greater the restoring force), the greater the quantum—and thus, the smaller the probability that a given molecule will acquire it at a given temperature. Extremely stable diatomic molecules (such as H_2), one infers from this, are not set in vibration until high temperatures are reached. Extrapolating broadly, one would conclude, as did Hinshelwood, that in general, in chemical reactions molecules are not set in vibration unless the force of collisional impact exceeds a certain critical level. In other words, in recognizing the role of virbational degrees of freedom in bringing about dissociations, one is forced to discuss the mechanism in the probabilistic energetic context of quantum theory and to formulate the critical energy requirement in the framework of an energy-distribution theory.

†From "The Kinetics of Chemical Change" by C. N. Hinshelwood, pp. 37–39, 1940. By permission of the Clarendon Press, Oxford.

Accordingly, Hinshelwood [1940, pp. 37–40], derived what amounts to what may be called the following basic lemma and corollary for determining the probability that a reactant molecule will have a total energy in excess of a fixed amount in many degrees of freedom, without explicit attention to the apportionment of the energy among them.

Lemma.

$$\Pr\{\varepsilon < \text{Energy } E < \varepsilon + d\varepsilon\} \equiv f(\varepsilon)\, d\varepsilon = [\varepsilon^{-1/2} e^{-\varepsilon/RT} / (\pi k_B T)^{1/2}]\, d\varepsilon. \quad (184)$$

In later usage, we shall be indicating as the probability density function of total energy E the expression

$$f(E) = E^{-1/2} e^{-E/RT} / (\pi k_B T)^{1/2}. \quad (185)$$

The proof is omitted here, but it should be noted that the Maxwell–Boltzmann equilibrium assumption is made in the proof, namely, that the number n_i of the total set of n molecules in quantum state ε_i is given by

$$n_i = \frac{n \exp(-\varepsilon_i / k_B T)}{\sum_{i=1}^{M} \exp(-\varepsilon_i / k_B T)}, \quad (186)$$

employing our previous notation.

The applicability of this result is not restricted to any particular kind of energy. All that is required in the proof is that $\varepsilon_i = (2m)^{-1} p_i^2$ (where p_i is a momentum coordinate referring to any one of the possible types of motion); that is, that the energy is expressible as a "squared" or quadratic term. Since rotational, vibrational, and even the potential energy associated with simple harmonic reaction are of this form the result applies in all of these cases.

In applying the result to molecules of complex structure such as those subject to unimolecular decomposition, the total energy E may be apportioned in many ways. Assuming that each kind of potential and kinetic energy in each degree of freedom is representable by a square term considered as a variable governed by a probability density function of the form of Eq. (185), the following corollary is obtained.

Corollary. The joint probability density function $f(E_1, E_2, \ldots, E_n)$ corresponding to a molecule with total energy E partitioned into n square terms: $E = \sum_{i=1}^{n} E_i$ subject to the condition that the n random variables $\{E_i\}$ are pairwise, mutually independent is

$$f(E_1, E_2, \ldots, E_n) = (\pi RT)^{-n/2} \prod_{i=1}^{n} E_i^{-1/2} \exp\left(-\sum_{i=1}^{n} E_i / RT\right). \quad (187)$$

Expressing the nth component of E in the form of $E_n = E - \sum_{i=1}^{n-1} E_i$

allows the deduction of a marginal probability density function for E from Eq. (187)

$$g(E) = (\pi RT)^{-n/2} \int_{E_{n-1}=0}^{E} \int_{E_{n-2}=0}^{E} \cdots \int_{E_1=0}^{E} \left(\prod_{i=1}^{n-1} E_i^{-1/2}\right)\left[E - \sum_{i=1}^{n} E_i\right] \times e^{-E/RT}\, dE_1 \cdots dE_{n-1} \tag{188}$$

so that

$$\Pr\{E > \varepsilon\} = (\pi RT)^{-n/2} \int_{\varepsilon}^{\infty} g(E)\, dE \tag{189}$$

leads to

$$P_r\{E > \varepsilon\} = e^{-\varepsilon/RT}[(n/2 - 1)!^{-1}(\varepsilon/RT)^{n/2-1} + (n/2 - 2)!^{-1}(\varepsilon/RT)^{n/2-2} + \cdots + 1]. \tag{190}$$

When ε/RT is "large" (as in most chemical reactions), use of the first term only in the series of Eq. (190) leads to a decent approximation,

$$\Pr\{E > \varepsilon\} \approx \frac{e^{-\varepsilon/RT}(\varepsilon/RT)^{n/2-1}}{(n/2 - 1)!} \tag{191}$$

which, of course, is interpreted as the probability that a total energy E summed over all n terms exceeds a critical level ε, without any constraints on the distribution of E into the n components $\{E_i\}$.

The physical significance of this result applied to the Lindemann–Hinshelwood mechanism, according to Hinshelwood, is that if the total energy can be made up by contributions in any proportions, then, combinatorially speaking, the number of different ways corresponding to the n degrees of freedom in which the energy requirements can be stored in the complex molecule is obviously very great. Thus, in the (Step 1) activating collisions the molecule acquires the energy in any form and, indeed, goes on accumulating it until it obtains enough to break a particular bond. Even though there is no reaction except as a result of the surging of energy from one part of a molecule to another, there is a concentration in a particular bond or bonds, (the "interesting" molecules of Slater) and chemical reaction then becomes possible.

The postulated time lag of the Lindemann–Hinshelwood mechanism is therefore interpretable in terms of the time required for the stored energy to find its proper place. These deductions correlate with the more recent interpretations of Slater [1959], to be discussed below, in particular with the "random gap distribution" idea.

Equation (191) may now be utilized as a means of transforming the ordinary bimolecular rate constant into an expression for k_{uni} which incorporates the special features of reaction just discussed. We begin by collapsing the Lindemann–Hinshelwood mechanism into the ordinary bimolecular reaction

$$A + M \xrightarrow{k} B + C \tag{192}$$

for which the bimolecular collision-theoretic rate constant giving the rate of B and C formation is

$$k = Ze^{-E/RT}, \tag{193}$$

where Z is the collision frequency.

Intuitively, from Hinshelwood's observations and deductions one would expect that k_{uni}, the rate of formation of products according to the Lindemann–Hinshelwood mechanism, would be expressible as the product of ordinary collisional k from Eq. (193), by the proportion of colliding molecules which satisfy the high energy requirement. This latter factor is another way of calling for the probability that a collision results in satisfaction of the critical energy requirement, Eq. (191):

$$k^{uni} = \frac{Z(E/RT)^{n/2-1}}{(n/2-1)!} e^{-E/RT}. \tag{194}$$

Note that when $n = 2$ this reduces to $k = Ze^{-E/RT}$ but as n increases beyond 2, k_{uni} becomes much larger than k. One way of demonstrating the increase of k_{uni} over bimolecular k is to compare Eq. (194) with Eq. (178). In the case of Eq. (194), one obtains

$$\frac{d(\log_e k_{uni})}{dT} = \frac{E}{RT^2} - \frac{(n/2-1)}{T}. \tag{195}$$

Thus, the experimental estimates of E obtained from the former expression, Eq. (178), namely,

$$E = RT^2\, d(\log_e k)/dT \tag{196}$$

is smaller than that obtained from Eq. (195),

$$E = RT^2 d(\log_e k_{uni})/dT + (n/2-1)RT. \tag{197}$$

D. The Rice–Ramsperger–Kassel and Rice–Ramsperger–Kassel–Marcus Theories of the Unimolecular Reaction

1. *Analysis of the k_3 (k_a) Dissociation Rate Constant*

The excess energy requirement which is demonstrated by this comparison between the pseudounimolecular rate constant k_{uni} and its ordinary bimolecular approximation can be analyzed down to the k_3 (or k_a) rate constant corresponding to the final decomposition of the activated A^* (or, A^+) species. The elaboration of this idea which forms a most noteworthy chapter in the theory of unimolecular reactions is called the RRK (Rice, Ramsperger, Kassel) theory [see Rice and Ramsperger, 1927, 1928; Kassel, 1928]; and in later form, the RRKM theory, taking note of the contributions of Marcus [see, for example, Marcus, 1952, 1965, 1968; Wieder and Marcus, 1962]. General references to the RRK and RRKM theories are to be found in the books by Kassel [1928], Slater [1959], and in Rice [1962]. Its eminent position in the modern theory of chemical kinetics is evident from the numerous references to it in the many research papers in this area of kinetics appearing, for example, in the *Journal of Chemical Physics*, particularly during the past 15 years.

In the RRK theory will be found an elaboration of the time-lag assumption of the Lindemann–Hinshelwood mechanism; namely, the postulate that the energy contained in the internal degrees of freedom has to accumulate in one particular degree of freedom. In accord with this idea the rate constant k_3 (or k_a) is assumed to be a function of the actual energy possessed by the A* molecule in its vairous degrees of freedom: the larger this energy, the greater the probability that the requisite amount will pass into a given bond and hence the greater will be the "specific rate of decomposition" k_3.

An expression for k_3 which reflects these considerations [see Glasstone *et al.*, 1941, pp. 284–285], is derived by a further application of the combinatorics of the class of occupancy problems. The probability, call it $P(m|j)$, for a particular "oscillator" to have m quanta, given that the totality of S oscillators have j quanta is deduced on this basis to be

$$P(m|j) = (j - m + S - 1)!j!/(j - m)!(j + S - 1)!. \tag{198}$$

For j very large, this is said to be well approximated by

$$P(m|j) \approx [(j - m/j)]^{S-1}. \tag{199}$$

Then, assuming (1) that the rate at which the required energy gets into a particular degree of freedom is proportional to $P(m|j)$ and (2) that k_3 is proportional to this, calling k_3' the constant of proportionality, the expression for k_3 obtained is

$$k_3 = k_3'[(j - m)/j]^{S-1}. \tag{200}$$

Taking the total number of quanta j proportional to E, the total energy of the molecule: and m proportional to the minimum requisite energy E_0 for the bond prior to decomposition:

$$k_3 = k_3''[(E - E_0)/E]^{S-1}. \tag{201}$$

The final result for k_3 in the RRK theory is obtained by a process of "integrating" between E_0 and ∞ [see Section III.D.2, Eqs. (227)–(233)].

In the RRK theory, considering the time lag to be a function of the amount of energy surpassing the critical level, the number of degrees of freedom effective in transferring energy to the "critical oscillator" is taken as an adjustable parameter. This number is found experimentally by fitting data to plots of log (k_{uni}) versus log(pressure).

According to Marcus [1968], the RRKM theory gives a more satisfactory account of k_{uni} as a function of pressure, without resort to an adjustable parameter. In this later form the species A^+ mentioned earlier is treated as a normal molecular species. In this sense, for example, it is considered to have $(3n - 4)$ rotational and vibrational degrees of freedom (for a nonlinear polyatomic activated complex with n atoms) and one degree of internal translation.

2. *Slater's Contributions to the RRKM Theory*†

The penetrating work of Slater [1959], which brought to light the "random-gap assumption" of the Lindemann–Hinshelwood and RRKM theories, forms the basis for a modernized, more rigorous mathematical treatment of those theories. The insight into their critical assumptions which he provided inspired, for example, the very interesting and patient work of Bunker [1962, 1964a, b], on the testing of the basic random-gap assumption as a validation test of the RRKM theory, using the STRETCH computer at Los Alamos in highly detailed simulation studies, to be discussed in Section III.D.2. The results led Bunker to suggest an additional reformulation which, as we shall see, is particularly compatible with the modern stochastic point of view.

In the introduction to his book, Slater mentions that R. H. Fowler suggested that he develop the thesis that the individual bonds of a molecule vibrate in a complicated way which can be broken into a sum of so-called normal modes of vibrations, decomposition occurring when the normal modes come sufficiently into place to give the vital link or bond its critical disruptive stretch. (The consequent time lag to dissociation generates the random-gap assumption.) Accordingly, the central theme for Slater's work became that of finding a general formula for the average time interval required for the attainment of a critical high value by a sum of individual harmonic vibrations. As we have seen, this was in line with the program envisaged by Hinshelwood originally.

In his reformulation of the classical harmonic model at the basis of the earlier interpretations the molecule is represented by a dynamical system having n independent internal coordinates $q_1, q_2, \ldots, q_n$ such as the stretches of interatomic distances and angles. These coordinates are expressible as linear combinations of "normal coordinates" $Q_1, Q_2, \ldots, Q_n$ which vibrate harmonically with respective frequencies $\nu_1, \nu_2, \ldots, \nu_n$. The amplitudes are supposed to have been determined by the "last" collision of the molecule

†From Slater [1959, pp. 106–109].

and to remain unchanged until the next collision, unless dissociation should occur before this.

The physical basis presumed here for the random-gap assumption consists of the assumption that the molecule dissociates when it reaches one of a set of critical configurations, molecules in this state being referred to by him as "interesting molecules" (the A* molecules in the Lindemann–Hinshelwood mechanism). He further assumes that the internal coordinates may be chosen so that the critical configuration is fully specifiable by q_1, say, attaining a particular large value independently of the individual contributions of the other coordinates, $q_2, \ldots, q_n$ (note correspondence here with the ideas of Hinshelwood mentioned in Section III.C.1). Then at high concentrations it is assumed that the equilibrium distribution of energies obtains in all the normal modes of the molecule. In this case, the k_3 (or k_a) rate constant of the Lindemann–Hinshelwood mechanism, referred to as "the first-order rate constant k_∞ in the classical limit," is calculated by finding the probability per second that coordinate q_1 will reach the critical value q, which is then averaged over the energy distribution (referring to the integration process of the RRKM theory, again).

Thus, corresponding to a point $P(p_1, \ldots, p_n; q_1, \ldots, q_n)$ in $2n$-dimensional phase space, where $q_1, \ldots, q_n$ are the generalized coordinates specifying the internal state of the molecule at a given instant, and $p_1, \ldots, p_n$ are the momenta, as a simplification of the notion of potential energy surface in transition-state theory, he introduces a $(2n - 1)$-dimensional hypersurface HS by the equation

$$\eta(q_1, q_2, \ldots) = 0. \tag{202}$$

This surface is such that if P satisfies Eq. (202); that is, P is a point on HS, the molecule is just dissociating. Undissociated and dissociated states correspond to points P with $\eta < 0$ and $\eta > 0$, respectively. The separation of points is carried out in terms of a Hamiltonian $\mathcal{H}(q_1, \ldots, q_n)$ in the following sense: Starting at an interesting point ($\eta \geqslant 0$ always) one proceeds along a unique Hamiltonian path of motion, which after a definite time $s = s(q_1, \ldots, q_n)$ will for the first time cross HS. In other words, an interesting molecule initially at P will dissociate in time s, if it is not disturbed before this.

a. Slater's General Method for Calculating the First-Order Rate Constant. Letting Pr{collision of molecule in time $\Delta\tau$} $= \omega\,\Delta\tau$, where ω is the mean frequency of collision of the molecule, Slater introduced the probability

$$\Pr\{\text{no collision in time } (0, t)\} = e^{-\omega t}. \tag{203}$$

And on this basis he takes $e^{-\omega s(q)}$ as the probaiblity of dissociation:

$$\Pr\{\text{dissociation} \mid q_1, \ldots, q_n \text{ initially}\} = \exp(-\omega s(q_1, \ldots, q_n)). \tag{204}$$

Then where $f(q_1, \ldots, p_n)\, dq_1 \cdots dp_n$ is the probability of finding a molecule in state $(q, q+dq; p, p+dp)$, the number of molecules entering an interesting (q, p)-range per second through collision equals $[c\omega f(q_1, \ldots, p_n)\, dq_1, \ldots, dp_n]$. Considering $e^{-\omega s}$ as the probability of successful dissociation he obtains as the formula (and method) for the first-order rate constant k

$$k = \int\int \cdots \int \omega e^{-\omega s} f(q_1, \ldots, p_n)\, dq_1 \cdots dp_n, \tag{205}$$

the integration being over all internal states (q, p). In effect then, k is describable as the probability of dissociation of a molecule per second, and in this sense corresponds to the stochastic parameter μ is Bartholomay's stochastic model to be discussed in Section III.F.

b. The General Method Applied to the Harmonic Oscillator Model, the "Random Gap Assumption." The preceding paragraph is a highly condensed statement of Slater's general method for calculating the first-order rate constant, which he has also applied to the classical harmonic oscillator model. Assuming a molecule undergoes strictly harmonic vibrations, dissociating when $q_1 \geqslant q$, for mathematical convenience he uses initial normal mode energies $\varepsilon_1, \ldots, \varepsilon_n$ and phases $\psi_1, \ldots, \psi_n$ as the (q, p)-phase space realization. Having shown [Slater, 1959, Chapter 5.3] that the equilibrium, or high concentration proportion of molecules in an "interesting range" is

$$p = \exp(-E/k_{\mathrm{B}}T) \prod_{i=1}^{n} \{d\varepsilon_i\, d\psi_i/k_{\mathrm{B}}T)\}, \qquad 0 \leqslant \psi_i \leqslant 1, \tag{206}$$

where $E = \sum \varepsilon_i$, and that the time course for q_1 in an undisturbed molecule is

$$q_1 = \sum_{i=1}^{n} a_{1i} \sqrt{\varepsilon_i} \cos 2\pi(\nu_i \tau + \psi_i), \tag{207}$$

he states as the criterion for the molecule to be interesting that

$$\sum_{i=1}^{n} | a_{1i} | \sqrt{\varepsilon_i} \geqslant q. \tag{208}$$

Then where $s = s(\varepsilon_1, , \ldots, \psi_n)$ is the smallest time $t > 0$ at which q_1 reaches q, he obtains

$$k = \int\int \cdots \int \omega \exp(-\omega s) \exp(-E/k_{\mathrm{B}}T) \prod_{i=1}^{n} (d\varepsilon_i\, d\psi_i/k_{\mathrm{B}}T). \tag{209}$$

It is the integration process implied by Eq. (209) which leads to the random gap assumption in the following way. The phase integration may be separated from the right-hand side of Eq. (209) by defining

$$\phi(\omega) \equiv \int_0^1 \int_0^1 \cdots \int_0^1 \omega\, e^{-\omega s}\, d\psi_1 \cdots d\psi_n, \tag{210}$$

which is interpretable as the "phase average" of $\omega e^{-\omega s}$. And, for later convenience, identifying s with the time to the first "upzero" of $(q_1 - q)$ [referring to $\lim_{q_1 \uparrow q} (q_1 - q)$], he imposes an ergodicity condition by equating the phase average of Eq. (210) to the long-time average of the function $\omega \exp[-\omega s'(t)]$:

$$\phi(\omega) \equiv \lim_{T\to\infty} T^{-1} \int_0^T \omega \exp(-\omega s'(t)\, dt, \tag{211}$$

where $s' = s'(t)$ is defined as the time from a given value of t to the next upzero of the following auxiliary function:

$$f(t) = \left(\sum_{i=1}^{n} a_{1i} \sqrt{\varepsilon_i} \cos 2\pi v_i t\right) - q. \tag{212}$$

Then if $t_1, t_2, \ldots, t_r, \ldots, t_m$ are the times of successive upzeros of $f(t)$, by the definition of s', if $(t_r \leqslant t < t_{r+1})$, $s'(t) = t_{r+1} - t$, Eq. (211) becomes

$$\phi(\omega) = \lim_{m\to\infty} \left\{(1/t_m) \sum_{r=1}^{m} \int \omega \exp(\omega(t_{r+1} - t)\, dt\right\}. \tag{213}$$

And letting $\tau_r = t_{r+1} - t_r$ = "gap" between the rth and the $(r+1)$th upzero,

$$\phi(\omega) = \lim_{m\to\infty} (m/t_m)\left[1 - \lim_{m\to\infty}\left(\sum_{r=1}^{m} m^{-1} \exp(-\omega\tau_r)\right)\right]. \tag{214}$$

Introducing

$$L \equiv \lim_{m\to\infty} (m/t_m) \tag{215}$$

as the "asymptotic frequency of upzeros," and designating

$$\langle e^{-\omega\tau}\rangle \equiv \lim_{m\to\infty} \left\{m^{-1} \sum_{r=1}^{m} \exp(-\omega\tau_r)\right\} \tag{216}$$

as the "asymptotic average over the gaps" (see below), so that

$$\phi(\omega) = L(1 - \langle e^{-\omega\tau}\rangle), \tag{217}$$

Eq. (209) becomes

$$k = \int \cdots \int L(1 - \langle e^{-\omega\tau}\rangle) \exp(-E/k_B T)\, d\varepsilon_1 \cdots d\varepsilon_n/(k_B T)^n. \tag{218}$$

In the special case when ω (collision frequency) is very large, since $\langle e^{-\omega\tau}\rangle \approx 0$, he recovers the "classical limit rate constant"

$$k_\infty = \int \cdots \int L \exp(-E/k_B T)\, d\varepsilon_1 \cdots d\varepsilon_n/(k_B T)^n \equiv v \exp(-E_0/k_B T) \tag{219}$$

and this in turn will be seen to agree with Kramers' results to be given in Section III.E.2.

Slater's discussion has thus succeeded in isolating what he calls the random-gap assumption in the following way. The average of $e^{-\omega\tau}$ has been taken with respect to a uniform distribution $\{1/m\}$; that is, in technical probabilistic terms $\langle e^{-\omega\tau}\rangle$ connotes the asymptotic mean value of the random variable $e^{-\omega\tau}$ with respect to the discrete probability distribution $\{1/m\}$. On the other hand, in his interpretation of Eq. (218), he goes beyond the special meaning of $\langle e^{-\omega\tau}\rangle$ just indicated, when he says that to perform the integration of the rate constant we need an expression for $\langle e^{-\omega\tau}\rangle$ as a function of energies e_{ij}; i.e., we need the distribution function of the gaps τ_r. In other words, having noted that $\lim_{m\to\infty}\{m^{-1}\sum_{r=1}^{m} e^{-\omega\tau}$ amounts to the asymptotic mean of $\exp(-\omega\tau_r)$ with respect to $\{m^{-1}\}$, and accordingly, having replaced it by $\langle e^{-\omega\tau}\rangle$, he then ignores the origins of this quantity and thinks of it heuristically in a very general sense as a mean value of $e^{-\omega\tau}$ with respect to some unspecified probability function (distribution). And on this basis he proceeds next to consider what the result of different "random-gap distributions" will be on the integral, that is, on the ultimate value of the rate constant k. In effect then, he uses the derivation to arrive at the abstraction of a general form (corresponding to a general interpretation of $\langle e^{-\omega\tau}\rangle$) corresponding to which alternative distributions may be examined, as new possibilities.

In the same way, looking at the gap τ as a discrete random variable with values $\tau_1, \ldots, \tau_r, \ldots, \tau_m$ to begin with, defined by a probability distribution $\{1/r\}$, then

$$\bar{\tau} = \sum_{r=1}^{m} \tau_n/r \tag{220}$$

so that where $\langle\tau\rangle = \lim_{m\to\infty} \bar{\tau}$, the parameter $1/L$ may be identified with $\langle\tau\rangle$ and then extrapolated to mean the mean value of τ with respect to a random-gap distribution in general terms.

In his investigation of the effects of assuming different gap distributions, for convenience he considers τ as a continuous random variable with probability density $h(\tau)$, so that the general meaning he imparts to L and $\langle e^{-\omega\tau}\rangle$ are

$$L = \left(\int_0^\infty \tau\, h(\tau)\, d\tau\right)^{-1} \tag{221}$$

and

$$\langle e^{-\omega\tau}\rangle = \int_0^\infty e^{-\omega\tau}\, h(\tau)\, d\tau. \tag{222}$$

Of particular interest here is what he calls the "random distribution,"

$$h(\tau) = Le^{-L\tau}; \tag{223}$$

in this case,

$$\langle e^{-\omega\tau}\rangle = L/(\omega + L), \tag{224}$$

and using this in Eq. (209) gives

$$k = \int \cdots \int \omega\, L(\omega + L)^{-1} \exp(-E/k_B T)\, d\varepsilon_1 \cdots d\varepsilon_n/(k_B T)^n \tag{225}$$

which is precisely the RRKM rate constant [see Slater, 1959, Chapter 7.1]. He also worked out the details corresponding to a number of other forms for $h(\tau)$, including the "regular" or "uniform" distribution.

3. *Bunker's Monte Carlo Computer Studies*

a. The Random Lifetimes Distribution and Assumptions. Bunker [1964a], interpolates between the RRKM theory and Slater's method for calculating k_a to obtain a new way of expressing the random-gap assumption, beginning with the following restatement of Slater's results. Let τ be the time required for a certain molecule of energy E to reach dissociation by virtue of its natural motion in the absence of collisions. For an equilibrium distribution of molecular states at E there will be a probability density function of these natural lifetimes, call it now $n(\tau)$.

The probability that a molecule will not undergo collision in time τ is taken as $\omega e^{-\omega\tau}$ if ω is the average collision frequency. And the absolute rate of reaction at energy E is

$$k_a(E) = \omega \int_0^\infty e^{-\omega\tau}\, n(\tau)\, d\tau. \tag{226}$$

He then expresses the pseudounimolecular rate constant k_{uni} as

$$k_{uni} = \int_{E_0}^\infty k_a(E) s(E)\, dE, \tag{227}$$

where $s(E)$ is the "statistical weight of energy E." He compares this equation with that of Marcus [1952, Eq. (5)]:

$$k_{uni} = \int k_a \frac{d(k_1/k_2)}{1 + k_a/k_2 p}, \tag{228}$$

the integration being over all energies in excess of activation energy E_0. He observes that if $s(E)\, dE$ is identified with $d(k_1/k_2)$ and ω with $k_2 p$ (where k_1 and k_2 are the activation, deactivation constants of the Lindemann–Hinshelwood mechanisms and k_a the dissociation step rate constant), then the correspondence between Eq. (227) and Eq. (228) will be complete if

$$n(\tau) = k_a \exp(-k_a \tau), \tag{229}$$

which is seen as follows.

Substituting Eq. (226) into Eq. (227),

$$k_{uni} = \int_0^\infty \omega \left(\int_0^\infty e^{-\omega\tau}\, n(\tau)\, d\tau \right) s(E)\, dE. \tag{230}$$

If one then replaces ω by $k_2 p$ and $n(\tau)$ by Eq. (229), the result is

$$k_{\text{uni}} = \int_{E_0}^{\infty} k_2 p \left(\int_0^{\infty} \exp(-k_2 p \tau) k_a \exp(-k_a \tau)\, d\tau \right) d(k_1/k_2) \tag{231}$$

but

$$k_a \int_0^{\infty} \exp\{-(k_2 + k_a)\tau\}\, d\tau = k_a/(k_2 + k_a) \tag{232}$$

and therefore

$$k_{\text{uni}} = \int_{E_0}^{\infty} k_a [k_a p/(k_2 + k_a)] d(k_1/k_2). \tag{233}$$

In the context of Bunker's work, Eq. (229) would be called the "random lifetime distribution" [which, in a sense, is equivalent to Slater's "random-gap distribution" of Eq. (223)]. Whereas, the latter refers to the distribution of times to the attainment of the minimum critical q_1 value, the former refers to the distribution of time required for energized molecules A* to proceed to activated complexes A^+, the constant k_a in Eq. (229) being identified with the high-pressure rate constant associated with the unimolecular reaction step,

$$A^* \xrightarrow{k_a} A^+. \tag{234}$$

Bunker's original motivation was in equating the validation of the random lifetimes assumption with the validation of the RRKM theory itself. Bunker [1964a] states the case in the following way: "Consider two molecules with the same energy, enough to achieve dissociation, but with different internal atomic velocities and positions. Implicitly the RRKM theory assumes that there is no systematic way of predicting which will dissociate first—which constitutes the "random lifetime assumption." Note that in view of the preceding paragraphs, the word "random" has a specific technical meaning in terms of probability theory; namely, τ is a random variable with the specific probability density function given by Eq. (229). Moreover, this should be considered in juxtaposition with Slater's probability density function of Eq. (223), which specifies the "random-gap assumption," the two assumptions being equivalent in the sense that one implies the other.

Bunker's view was that, due to the centrality of the random lifetimes or gap assumption in the RRKM theory, the validation of that theory depends on the validation of the assumption in the form stated. Furthermore, due to the impossibility of performing real validation experiments; namely, the direct observation of real dissociating molecules, Bunker, with the assistance of Blais, decided to perform *in numero* experiments (the latter term coined by Bartholomay [1964a, 1968a, b]) on the STRETCH computer at Los Alamos. Two years of planning were involved, anticipating the availability of that computer. The guidelines for carrying out simulation of this kind on the computer were found in the earlier work of others. For example, Wall *et al.* [1958] published the first computer simulation of atomic trajectories during reaction. And, Thiele [1961] and Wilson [see Bunker, 1964a, p. 1952] had obtained some results specifically on unimolecular decompositions. According to Bunker, such studies illuminated the nature of molecular motion and confirmed a growing suspicion that the RRKM method of dealing with molecular virbrations was preferable to the Slater method which stressed the temporal "regularity" of the dissociation process (described in terms of the regular or uniform distribution of gaps). But since these studies dealt with isolated reaction events, they provided no satisfying evidence that the random lifetime assumption was tenable. The hope for this, Bunker felt, was in carrying out astronomically large numbers of calculations by Monte Carlo methods.

b. In Numero Studies Using the Monte Carlo Method on the* STRETCH *Computer. The plan for these studies, then, was to program the computer to select representative three-atom molecular configurations, to follow their internal motions by the solution of the Hamiltonian equations of reaction, and to determine the lengths of time required for the molecules to disrupt, Each molecule would be "interesting," that is, assumed to contain sufficient virbrational energy to break apart if most of its energy were concentrated in one of its bonds. In the first paper of the series [Bunker, 1962], for example, six triatomic molecular models were studied, all of which were considered to be nonrotating and described in terms of their bond lengths r_1 and r_2 and bond angle α. All models were taken to be bound by simplified valence-type potentials separable into terms involving only r_1, r_2, α and constants. Four were based on N_2O and two on O_2:

1. An N_2O-like linear molecule constrained to vibrate in the plane and bound by harmonic potentials.
2. Same as 1 but with bond-stretching potentials replaced by Morse functions.
3. A fictitious N_2O-like molecule which is the same as 2 except that the N_2O potential has been replaced by an ordinary Morse curve whose dissociation energy equals the energy separation between the maximum and minimum of the curves used in 2.
4. A harmonic model for N_2O.
5. An O_3-like bent molecule bound by harmonic potentials.
6. Same as 5 with the bond-stretching potentials replaced by Morse functions.

The Hamiltonian for a nonrotating, triatomic molecule in terms of their coordinates (r_1, r_2, α), their conjugate momenta (p_1, p_2, p_α) and their atomic masses (m_1, m_2, M) is given by

$$\begin{aligned} H = {} & (m_1 + M)p_1^2/2m_1M + (m_2 + M)p_2^2/2m_2M \\ & + p_1p_2 \cos\alpha/M + Ap_\alpha^2/2m_1m_2Mr_1^2r_2^2 - p_1p_\alpha \sin\alpha/Mr_1 \\ & - p_2p_\alpha \sin\alpha/Mr_1 + V(r_1) + V(r_2)V(\alpha), \end{aligned} \tag{235}$$

in which

$$A = m_1(m_2 + M)r_1^2 + m_2(m_1 + M)r_2^2 - 2m_1m_2r_1r_2 \cos\alpha. \tag{236}$$

The masses m_1 and M are separated by r_1, the masses m_2 and M, by r_2, and the V's are potential-energy expressions. The canonical equations of motion were given as

$$\begin{aligned} \dot{p}_1 &= [(m_1r_2 + Mr_2 - m_1r_1 \cos\alpha)p_\alpha^2/m_1Mr_1^3r_2] \\ &\quad - (p_2p_\alpha \sin\alpha/Mr_1^2) - [dV(r_1)/dr_1], \\ \dot{p}_2 &= [(m_2r_1 + Mr_1 - m_2r_2 \cos\alpha)p_\alpha^2/m_2Mr_1r_2^3] \\ &\quad - (p_1p_2 \sin\alpha/Mr_2^2) - [dV(r_2)/dr_2], \\ \dot{p}_\alpha &= [(r_1r_2p_1p_2 - p_\alpha^2) \sin\alpha + (r_1p_1 + r_2p_2)p_\alpha \cos\alpha \\ &\quad - Mr_1r_2\, dV(\alpha)/d\alpha]Mr_1r_2, \\ \dot{r}_1 &= [(m_1 + M)p_1/m_1M] + (p_2 \cos\alpha/M) - (p_\alpha \sin\alpha/Mr_2), \\ \dot{r}_2 &= [(m_2 + M)p_2/m_2M] + (p_1 \cos\alpha/M) - (p_\alpha \sin\alpha/Mr_1), \\ \dot{\alpha} &= -(p_1 \sin\alpha/Mr_2) - (p_2 \sin\alpha/Mr_1) \\ &\quad + (Ap_\alpha/m_1m_2Mr_1^2r_2^2). \end{aligned} \tag{237}$$

These equations were integrated by a conventional fourth-order Runge–Kutta procedure. All kinetic quantities for the six classes of molecular models were determined from the total energy H, the final data to consist of families of normalized distributions $n(\tau)$ of lifetimes τ for molecules belonging to an equilibrium distribution at each fixed energy H (as the parametric determinant of the member of the family). And so, it was necessary to prepare a computer program for choosing the members of an equilibrium set of molecular configurations consistent with a given H. Bunker saw this as equivalent to the problem of forming a uniform distribution of points in the six dimensional phase space of the molecule, in a region bounded by the hypersurfaces in which H and $H + dH$ are constant. Of the many different, possible ways of doing this he chose that which made the most efficient use of computer time, using Monte Carlo methods for determining the required maximal relative density of points and $S(H)$, the relative volume of phase space between ($H + dH$). A total of 90,026 initial configurations, that is, planar nonrotating trajectories, were integrated and 10,559 dissociations recorded. The distributions of lifetimes were fitted by least-square methods. Conversion of the distributions $n(\tau)$ into rate constants was effected by the Slater method.

Continuation of this work led to Bunker's publication [1964b] of conclusions based on these, plus 235,000 additional trajectory calculations. The combined Monte Carlo studies led to the confirmation of the random lifetime assumption [Eq. (229)] for a large number of cases. However, as Bunker himself emphasizes, extension of the results to other triatomic molecules requires judgment; to more complex molecules, extreme caution. Furthermore, the simulated molecules employed, of course, do not allow definite conclusions to be drawn for real molecules.

Some of the deeper implications of the study had to do with the separations of these results into those pertaining to molecules in which the modes of vibrations were harmonic and those which are anharmonic, both types being represented in the six classes of models chosen. In agreement with Slater's views the results seem to confirm that the assumption holds in the case of the anharmonic models: that a sufficient condition for the assumption is the occurrence of vibrations that differ greatly in frequency. However, Bunker inferred also that "badly" behaved lifetime distributions will occur for only a few molecules and these, only under very special experimental circumstances Thus, on the basis of his findings, he endorsed the general validity of the RRKM theory quite generally [that is, the random-lifetimes distribution of Eq. (229)].

The theoretical follow-up in this work consisted of constructing a new unimolecular theory that could be applied to a particular model to predict whether the random-lifetimes assumption will be valid.

c. Bunker's Unimolecular Reaction Model. On the basis of his findings in these *in numero* studies, Bunker constructed a model for the unimolecular reaction which makes the following assumptions:

A1. Rotation may be ignored.

A2. Vibrations are approximately separable.

A3. Energy may exchange freely between two normal modes at a rate equivalent to that with which it oscillates between them in the presence of harmonic coupling.

A4. The strength of the equivalent harmonic coupling is proportional to the maximum anharmonicity generated when the normal modes have their average amplitudes.

The details are worked out in Bunker [1964a, Part 2, theory]. The beginning point is the specification of the Hamiltonian in the following form, which is inferred from A1–A4;

$$H = \tfrac{1}{2} \sum_{i=1}^{S} (P_i^2 + \lambda_i Q_i^2), \tag{238}$$

where the Q's are the amplitudes of the S normal modes and the P's are conjugate momenta,

$$P_i = dQ_i/dt = \partial H/\partial P_i, \qquad i = 1, 2, \ldots, S. \tag{239}$$

The λ's are related to the normal mode frequencies by

$$\lambda_i = 4\pi^2 \nu_i^2, \qquad i = 1, 2, \ldots, S. \tag{240}$$

This type of Hamiltonian may be used to represent a variety of different dynamical systems, for example, a collection of independent oscillators, an equivalent linear nonbending molecule. It is the first which is utilized in his model. And, in this case the λ's carry the coupling conditions which it is necessary to impose to insure the exchange of energy between the oscillators of the ensemble.

d. Bunker's Reflections on the Computer Methodology Employed. In an informal summary of his work, Bunker makes some comments which may well apply to *in numero* studies generally and which are worthy of note in the context of the present volume:

> (1) An interesting aspect of these computational experiments . . . is the way in which they unexpectedly inject a human element into an experimental science that is becoming largely one of instrumentation. To anyone familiar with the way a computer problem is most often solved this may seem a preposterous statement. The usual procedure is to formulate the problem in an artificial language, which is then translated into a program by the machine itself, after which the machine solves the problem automatically in the absence of the formulator.
>
> Computational experiments. however. cannot be conducted in this manner . . . it is necessary to operate the computer under the direct and continuous supervision of someone who is familiar with both the physical process being simulated and the details of the computer program. . . . Any unusual or unexpected result . . . has to be detected promptly. translated into physics and used as the basis for a spot revision of the strategy of the calculation. . . . hundreds of hours of computing time were saved by this kind of man–machine interaction. Without it many of our results would have been out of reach.
>
> (2) Computational experiments, then, furnish us a useful and also an entertaining way of arriving at new physical insights. They have already demonstrated that they can supplement laboratory experiment and theory, and eventually they may equal them in importance [Bunker, 1964b, p. 108].†

Looked at from the insight provided by the work of Slater and the subsequent work of Bunker, the Lindemann–Hinshelwood and RRKM theories, added to the earlier collision theory and transition-state theory, all seem in retrospect to have been pointing in the direction of a stochastic theory of chemical kinetics. The beginning of the microscopic aspect of that theory may be seen particularly in the work of Kramers which is taken up next. The compatibility of Bunker's studies with the microscopic aspects of such a theory

†From "Computer Experiments in Chemistry" by D. L. Bunker.

will be obvious even in terms of the later discussion of macroscopic kinetic stochastic models.

E. Origins of the Stochastic Approach to Microscopic Chemical Kinetics

1. *The Christiansen Diffusion Model*

Christiansen [1936] began moving the general transition-state theory problem of characterizing mathematically the passage of the activated complex across the potential energy barrier, toward the domain of stochastic processes by constructing as a model for the kinetics of the following complex chemical reaction:

$$A_1 \rightleftharpoons X_2 \rightleftharpoons X_3 \rightleftharpoons \cdots \rightleftharpoons X_{n-1} \rightleftharpoons B_n, \tag{241}$$

a one-dimensional diffusion process. In this representation he drew on the analogy between the rate of diffusion of a solute through n cells linearly arranged and the chemical reaction of Eq. (241). In other words, the reaction was treated theoretically as a one-dimensional diffusion process of a molecule migrating between two planar surfaces at $x = 0$ and $x = l$ under the influence of a potential as a function of the distance x, where $0 \leqslant x \leqslant l$ In this way he produced an equation identical in form to the Einstein diffusion equation.

Christiansen's model formed the point of departure for Kramers [1940] whose, now classical, paper served as the real trigger, not only for all later stochastic, microscopic kinetic models but also for the reexamination and evaluation of the basic, underlying "equilibrium hypothesis" of the transition-state theory. The equilibrium hypothesis with the random-cap assumption discussed in Section III.D.1, 2 preceding, form the basic tenets of the validity of the whole theory. In the next section an examination of the nature of that hypothesis and its controversial position will be undertaken, as a prologue to the Kramers model.

2. *Testing and Validating the Equilibrium Hypothesis of the Transition-State Theory*

The classical theories of chemical kinetics are "equilibrium theories" in which a Maxwell–Boltzmann distribution of energy or momentum or internal co-ordinate is postulated to persist during a reaction. This arises whether we are dealing with Hinshelwood's collision theory or the modern absolute reaction rate theory proceeding from the transition-state or activated complex approach. It is inherently a component of all the faces of chemical kinetics. For example, in collision theory the number of energetic reaction-producing collisions is calculated under the assumption that molecular velocity components remain Maxwellian. And in Eyring's absolute rate theory, as we have seen, the equilibrium assumption is stated in terms of the assumption of a time-invariant Maxwellian distribution of reactant molecules in the reactant valley of the potential-energy surface, far removed from the potential-energy barrier.

Kramers' original motivation apparently was in the direction of providing a demonstration of the validity of the transition-state theory in this form. The equilibrium postulate had long been regarded as a somewhat questionable approximation to reality since any process naturally leads to disturbance of the initial equilibrium state. And a number of other investigators subsequently have subjected this point to highly sophisticated examination, mathematically and physicochemically speaking.

For example, as we shall presently note in greater detail, Montroll and Shuler [1957, 1958] extended the Kramers model, in the course of investigating this very point, thereby laying the foundations for the application of the highly mathematical aspects of the theory of stochastic processes to microscopic kinetics. The important bearing of investigations of the equilibrium hypothesis on modern microscpoic kinetics may in fact be perceived in their statement that microscopic kinetics refers to the adjustment of the energy distributions induced by chemical reactions.

On the assumption that the probability of an inelastic collision is small, Prigogine and Xhrouet [1949; see also Prigogine and Mahieu, 1950], set up an integro–differential equation which forms a natural extension of the Maxwell–Boltzmann expression and solved it by the Chapman–Enskog method of using a perturbed distribution function of the form

$$f = f_0(1 + \Phi), \tag{242}$$

where f_0 is the Maxwell–Boltzmann distribution, and Φ is the perturbation caused by chemical reaction. The Chapman–Enskog method had been used previously on the physical problem of calculating the Maxwell–Boltzmann perturbation due to a spatial gradient. An important result of the studies of Prigogine *et al*, was the deduction that the perturbation and corresponding deviation from the equilibrium is, in the final analysis, quite small when

$$E_{\text{act}}/k_B T \geqslant 5, \tag{243}$$

so that the transition-state predictions (and the equilibrium hypothesis) would appear, after all, to have a large range of validity. Other independent investigations seem to be in good agreement with this conclusion (see below).

Commenting on the Kramers model, Prigogine and Xhrouet [1949] offered the criticism: in spite of its great interest, Kramers' method only provides qualitative results. Specifically, they note that it contains a friction coefficient (η) measuring the coupling between the reacting particles and their surroundings, the value of which is still difficult to calculate. But one encounters this difficulty in all microscopic kinetic theories it seems, including Prigogine's, which contains the incalculable collisional parameter σ. The transmission coefficient of the transition-state theory, the integration constant of the Arrhenius equation ultimately elaborated—but again, incompletely in terms of collisional theory, the collisional frequency factor ν in the collision-theoretic rate expression (later on we shall see), certain transition probabilities W_{nm} in the Zwolinski–Eyring model, and corresponding quantities in the work of Montroll and Shuler, all are ultimately not experimentally measurable. Moreover, they are incalculable in the widest possible sense: neither their values nor their closed analytical forms can be determined from our present knowledge, say, of intramolecular forces. In fact, as the earlier work [Bartholomay, 1962a] has emphasized, chronologically each new theory of kinetics arose from the expansion or reinterpretation of the incalculable rate constant or some factor of it, contained in the corresponding rate expression of the immediately prior theory. In a sense then, Prigogine's observation may really attest to (but one example of) the ultimate incalculability of rates from first physical principles. These considerations indirectly commend to our attention stochastic approaches as a way around the difficulty, emphasizing that interpretation of probability theory that is sometimes referred to as "the mathematics of uncertainties."

3. *The Kramers Brownian Motion Model*

As the classical paper on stochastic processes in the physical sciences by Chandrasekhar [1943] demonstrates, the Brownian motion problem is identifiable with the random walk (flights) problems, which in turn is formulable as a diffusion equation. These early ideas which were already in the process of generating a mathematical theory of stochastic processes were also apparently shared by Kramers who took his inspiration for a Brownian motion formulation of chemical kinetics from Christiansen's diffusion model.

Specifically, in the context of transition-state theory, and in obvious relation to the Lindemann–Hinshelwood mechanism, Kramers assumed that reactant molecules become activated through collisions with other molecules (such as the M species of the generalized Lindemann–Hinshelwood mechanism) of the surrounding medium which acts as a "constant temperature bath." And, as in the transition-state theory, he assumes that after many collisional exchanges of energy some of the reactant molecules (or the A* species of the Lindemann–Hinshelwood mechanism) acquire sufficient energy (of activation) to diffuse across the potential energy barrier, the "rate" of the "diffusion current" constituting the rate of chemical reaction. He considered the interaction of the reactant molecules with the heat bath analogous to the Brownian motion of a particle in a viscous medium under action of a force whose potential is identifiable with the potential energy surface along the reaction coordinates. The interaction (coupling) of the molecules with the heat bath is expressed through a "viscosity coefficient η" such that a large value of η corresponds to a strong interaction between the molecules and the heat bath.

a. The Langevin Equation in Brownian Motion. The mathematical theory of Brownian motion of a free particle in a forceless field generally begins [Chandrasekhar, 1943] with Langevin's stochastic differential equation,

$$d\mathbf{u}/dt = -\beta\mathbf{u} + \mathbf{A}(t), \tag{244}$$

where $\mathbf{u}$ is the velocity vector of the particle. Equation (238) reflects the subdivision of the influence of the surrounding medium into two components: (a) a systematic component ($-\beta\mathbf{u}$) due to dynamical friction; and, (b) a random or fluctuating vector component $\mathbf{A}(t)$ which is characteristic of Brownian motion. By the solution of Eq. (238) is meant the specification of a probability distribution density $f(\mathbf{u}, t; \mathbf{u}_0)$, the probability density of $\mathbf{u}$ as a function of time t given that $\mathbf{u} = \mathbf{u}_0$ at $t = 0$.

b. Kramers' Generalization of the Langevin Equation to Phase Space.† Kramers begins with a generalization of the Langevin equation in the phase space of generalized (q, p) coordinates (q corresponding to the position $\mathbf{r}$ of a Brownian particle, and p to its velocity $\mathbf{u}$),

$$\dot{p} = K(q) + X(t), \qquad \dot{q} = p \tag{245}$$

for a particle of mass 1 in a one-dimensional domain, the particle being acted upon by an external force $K(q)$ and an irregular environmental force $X(t)$ [where $X(t)$ is a stochastic variable]. As in the treatment of the Brownian motion problem, utilizing Eq. (239), he employed the so-called "Einstein method" or "Einstein pattern" of solution. The method depends on the assumptions: (1) that there exists a set of time intervals Δt so small that the change in velocity [from $p(t)$ to $p(t) + \Delta t$ here] is expected to be very small, that is,

$$p(t + \Delta t) \approx p(t), \tag{246}$$

and (2) on the other hand, such Δt's must be large enough that the stochastic variables $X(t + \Delta t)$ and $X(t)$ are guaranteed to be statistically independent; that is,

$$\Pr\{X(t + \Delta t) = x_1 \mid X(t) = x_0\} = P\{X(t + \Delta t) = x_1\}. \tag{247}$$

Under these assumptions a new random variable $B(\Delta t)$ is introduced as

†From Kramers [1940, pp. 284–304].

a function of $X(t)$,

$$B(\Delta t) \equiv \int_t^{t+\Delta t} X(u)\, du \tag{248}$$

which is assumed to be stationary, that is, independent of t. Referring to the probability density function of $B(\Delta t)$ as $\phi_{\Delta t}$, it is allowed that

$$\phi_{\Delta t} = \phi_{\Delta t}(B_i;\,(q, p);\,T). \tag{249}$$

Some unusual mathematical conditions are placed on the momenta of $\phi_{\Delta t}$. For example, it is assumed that the momenta

$$\overline{B^n(\Delta t)} = \int_{-\infty}^{+\infty} B^n\, d\phi_{\Delta t} \tag{250}$$

depend on Δt in such a way that they can be represented by the first non-vanishing term of a Maclaurin series expansion in Δt. Einstein's original Brownian motion theory is characterized in these terms by requiring

$$\begin{aligned} \overline{B^1(\Delta t)} &= -\eta p \Delta t, \\ \overline{B^2(\Delta t)} &= \nu \Delta t + \cdots, \\ \overline{B^n(\Delta t)} &= o(\Delta t), \qquad n > 2, \end{aligned} \tag{251}$$

where the viscosity n and the constant ν may still depend on temperature and position and where

$$\nu \equiv 2\eta T \tag{252}$$

(defined so that $k_B = 1$ for convenience). Thus, letting

$$\mu_n = (\Delta t)^{-1}\overline{B^n(\Delta t)}, \tag{253}$$

Kramers derives a diffusion equation for an ensemble of particles having density $\rho(q, p)$ in (q, p) space, beginning with a justification of the Fokker–Planck-type partial differential equation,

$$\partial\rho/\partial t = -K(q)\, \partial\rho/\partial p - p\, \partial\rho/\partial q - \partial/\partial p(\mu_1 \rho) + \tfrac{1}{2}(\partial^2/\partial p^2)(\mu_2 \rho) - \cdots, \tag{254}$$

which he refers to as the "Gibbs equation completed with terms due to Brownian motion."

From physical considerations eventually he is able to deduce in this case the simple Einstein result of the form

$$\partial\rho/\partial t = -K(q)\, \partial\rho/\partial q - p\, \partial\rho/\partial q + \eta\, (\partial/\partial p)(p\rho + T\, \partial\rho/\partial p). \tag{255}$$

c. Calculation of the Rate Constant of the Transition-State Theory. He obtains final solutions only for the limiting cases of large and small viscosity. For example, in the former case the implication is that the effect of Brownian forces on the velocity of the particle is much larger than that of $K(q)$. Assuming that K does not change much over a "distance" $(T/\eta)^{1/2}$ one expects that, starting from an arbitrary initial ρ distribution, a Maxwellian velocity distribution will be established soon, that is, after a time lapse of the order of $(1/\eta)$ for every possible value of q. In this case the density function is factorable into $\sigma(q, t)$, the density distribution in the q coordinate and a negative exponential term in p,

$$\rho(q, p, t) \equiv \sigma(q, t) \exp(-p^2/2T) \tag{256}$$

(using the system in which k_B has been normalized to unity).

In this case, he obtains a reduction of Eq. (255) to the following Smoluchowski diffusion equation:

$$\partial\sigma/\partial t = -(\partial/\partial q)[(K/\eta)\sigma - (T/\eta)\,\partial\sigma/\partial q] \tag{257}$$

in which case the expression inside the brackets would correspond to a quantity, call it w, which amounts to the concentration gradient in simple Fickian diffusion theory. It is called by Kramers "the stationary diffusion current," noting that w may be obtained as the integration constant where $\partial\sigma/\partial t = 0$, that is, by first quadrature of the second-order differential equation

$$(\partial/\partial q)[(K/\eta)\sigma - (T/\eta)\,\partial\sigma/\partial q] = 0, \tag{258}$$

what obtains is the stationary diffusion current

$$w \equiv (K/\eta)\sigma - (T/\eta)\,\partial\sigma/\partial q. \tag{259}$$

But where U is the potential function and

$$K \equiv -\partial U/\partial q, \tag{260}$$

Eq. (259) may be rewritten as

$$w = -(T/\eta)e^{-U/T}(\partial/\partial q)(\sigma e^{U/T}). \tag{261}$$

Then where

$$\partial(\sigma e^{U/T}) = (w/T)\eta e^{U/T}\,\partial q, \tag{262}$$

integration with respect to the q variable, using the limits $q = q_a$ and $q = q_b$ gives

$$\sigma e^{U/T}\Big|_{q_a}^{q_b} = (w/T)\int_{q_a}^{q_b} \eta e^{U/T}\,dq, \tag{263}$$

so that we may write

$$w = T[\sigma e^{U/T}]_{q_b}^{q_a}\Big/\left(\int_{q_a}^{q_b} \eta e^{U/T}\,dq\right). \tag{264}$$

In the case, then, of a potential function having the form shown in Fig. 8, which corresponds to the situation in which a molecule originally caught at A, must escape by passing over the barrier of height Q at C to the hole at B for reaction, the calculation of reaction rate resembles Christiansen's diffusion model. If initially the number of particles caught at A is larger than would

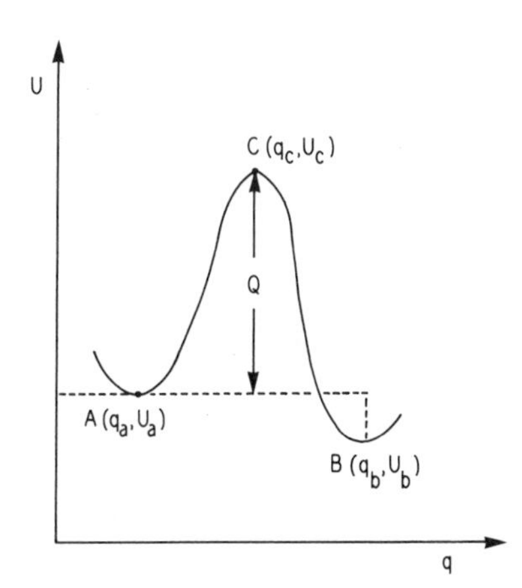

Fig. 8. Kramers' example of the potential function U, showing Q the height of the potential barrier. [From Kramers, 1940, p. 291].

correspond to thermal equilibrium with the number at B, a diffusion process begins, tending to establish equilibrium which for $Q \gg T$ can be compared with a stationary diffusion process. And he reasons that a Maxwellian distribution near A (also near B) will be established a long time before an appreciable number will have escaped, so that the quasistationary diffusion will correspond to a flow from a very large supply of Maxwellian-distributed particles at A to B.

Still for the case of large viscosity he then defines

$$\sigma_A \equiv \sigma e^{U/T} \qquad \text{(near } A\text{)}, \tag{265}$$

in which case Eq. (264) becomes

$$w = T\sigma_A \Big/ \left(\eta \int_{q_a}^{q_b} e^{U/T}\, dq\right). \tag{266}$$

Furthermore, assuming that U near A can be represented by $\frac{1}{2}(2\pi\nu)^2 q^2$ (analogous to case of an harmonic oscillator of frequency ν), he states that the number n_A becomes

$$n_A = \int_{-\infty}^{+\infty} \sigma_A \exp[-(2\pi\nu)^2 q^2/2T]\, dq = (\sigma_A/\nu)(T/2\pi)^{1/2} \tag{267}$$

(on the basis that the molecules at A follow the Maxwellian distribution).

Then, identifying the "rate constant" with the reaction velocity r, which he takes to be "the chance in unit time that a particle which was originally caught at A escapes to B," he sets

$$r \equiv w/n_A = (\nu/\eta)(2\pi T)^{1/2} \left\{ \int_{q_a}^{q_b} e^{U/T}\, dq \right\}^{-1}. \tag{268}$$

This would follow from the interpretation of the stationary diffusion current w in the numerator as the proportion of the total number n_A of activated complexes "near A" which diffuse to B.

Then assuming, as shown in Fig. 8, that U is well behaved near C, since the main contribution to the integral in Eq. (268) comes from a small region near C, he sets

$$U_{\text{near}\,C} \approx Q - \tfrac{1}{2}(2\pi v')^2(q - q_C)^2 \tag{269}$$

and obtains

$$r \approx (2\pi v v'/\eta)e^{-Q/T} \tag{270}$$

(for the large viscosity case, $\eta \gg 4\pi v'$). And, for small viscosity ($\eta < 4\pi v'$) he deduces

$$r \approx v e^{-Q/T}. \tag{271}$$

In the latter case may be seen most clearly the correspondence of his model and result with the transition-state method; for writing this in the usual form, r corresponds to the rate constant

$$k = v \exp(-E_0/k_B T), \tag{272}$$

where v is factored as shown in Section II.

d. Conclusions and Observations. 1. The reaction rate calculated by Kramers agrees with the rate calculated under the equilibrium assumption of the transition-state theory. In fact, as Montroll and Shuler [1958, p. 366] indicate, for $(E/k_B T) > 10$ the rate calculated from Kramers' model agrees with the latter to within about 10 % over a rather wise range of η. Thus, Prigogine's results quoted earlier are reinforced by these (Montroll and Shuler's in turn are also in agreement with them).

2. The activation energy Q (or E_0 earlier) should be the highest energy barrier (see Fig. 8) through which the system should have to pass; even if it chooses the most economic way, in terms of energy. And, referring to the corresponding point C in the diagram, that is, considering a concentration N_C of molecules in the transition state C, if these were in temperature equilibrium with the concentration N_A of molecules in state A, the temperature dependence of their relative concentration N_A/N_C would be given mainly by $e^{-Q/T}$ [that is, $\exp(-E_0/k_B T)$]. In this event, they would, according to Kramers, "correspond to the rarest of all intermediate products occurring in the reaction, and it is plausible that their concentration mainly determines the rate of reaction."

3. Kramers also deduced certain cases in which the transition-state method would give incorrect results; notably, autocatalytic reactions, failure in this particular case resulting from the fact that the necessary activation energy can be supplied by reaction products appearing between C and B. In other cases, incorrect values of r would be due to the fact that in some cases molecules in state C moving toward state B may not be compensated for by a backward flow; $B \longrightarrow C \longrightarrow A$, for example, a homogeneous gas reaction of very low

pressure, providing an insufficient number of activated molecules by collision. In yet other cases, the value of η, a measure of the intensity with which the molecules in the different states react with the surrounding medium, may be too small, yielding a reaction velocity below the transition-state prediction.

4. A detailed discussion of some of the strengths and weaknesses of the model is given by Kramers [1940, p. 302–204] with particular reference to a number of particular reactions: the slow racemization of an optically active substance in water; the decomposition of a polyatomic molecule with a great number of vibrational degrees of freedom (for example, N_2O_5).

5. Kramers refers to the treatment of Bohr and Wheeler [1939] of the fission of heavy nuclei, in which they considered the nucleus as a hot drop of homogeneously charged liquid. Such a drop has five modes of vibration capable of leading to fission. After justifying the applicability of the transition-state method to such a process, the probability of fission is treated by the transition-state method. As Kramers indicates, there is in fact a very close correspondence between the Bohr–Wheeler model and his own, his q corresponding to their "fission coordinate, and η corresponding to the resistance to which the vibration of the drop is subject as a consequence of the "viscosity" of the nuclear matter.

6. Finally, of particular interest to the stochastic theme is Kramers' general statement in reference to our Eq. (271): "The probability of escape represented by this formula corresponds exactly to the value which would be given by the transition-state method of calculating reaction velocities" [1900, p. 295]. Pursuing this correspondence further, it should be noted that this result in the form of Eq. (272) is in perfect agreement with the classical Arrhenius expression given in Eq. (179) and with the value of k_∞ deducible for the k_3 rate constant (dissociation step) of the Lindemann–Hinshelwood unimolecular reaction mechanism and recovered by Slater in the context of his general method [see Eq. (219) of Section III.D.1]. And, anticipating the macroscopic stochastic models to be presented in Section III.F, the basic probabilistic parameter μ interpretable in terms of Bunker's random-lifetimes model is also interpretable in the present context of Eqs. (271) and (272); particularly in view of Kramers' formulation of the rate r, in effect, as the probability that one of the n_A energized molecules near A "diffuses" to the state of dissociation.

4. *The Zwolinski–Eyring Model*

Kramers [1940, p. 301], addressing himself to the low-viscosity case, suggested that his model of the transition-state theory could be brought into even closer contact with reality by introducing discrete quantized energy states.

This suggestion is in the same spirit as Hinshelwood's early ideas about the unimolecular reaction, referring especially to the Lemma and Corollary in Section III.C.4. Kramers' suggestion seems in retrospect to have inspired the advent of the modern stochastic microscopic kinetics era in unimolecular reaction-rate theory when one considers the work of Zwolinski–Eyring [1947] and Montroll–Shuler [1958], the model due to the latter, in fact, being called by them "the discrete energy-level model for unimolecular reactions."

Zwolinski and Eyring [1947], began the construction of their microscopic, quantum mechanically oriented model by identifying the reactant molecules with one set of discrete quantum states and the product molecules by another set. However, in their theory these states are unspecific: the quantum levels are not necessarily thought of in terms of any particular degrees of freedom, so that from the quantum-theoretic point of view the treatment is left at a general level. It is simply assumed that chemical reaction consists of the transformation by collisions from the first set of states to the second. And the model applies to reactions which are of first-order molecularity with respect to each reactant species, so that it includes the unimolecular reaction. The essence of their model is contained in a system of linear differential equations with respect to time,

$$dx_n/dt = \sum_{m \neq n} \{W_{nm}x_m - W_{mn}x_n\}, \tag{273}$$

where $x_m(t)$ is the number of molecules in energy state m at time t and W_{nm} equals the probability per unit time of collisionally induced transitions out of energy state m into state n.

From the deterministic format of these equations it is seen that the W's have really the force of "rate constants." Thus, as in the classical theories. the probabilistic statistical, physical method is buried in a deterministic context. And, once again, like the viscosity constant of Kramers' model and the inelastic collisional cross-section parameter σ of the Prigogine model to which they are analogous, the W's are incalculable in the sense discussed earlier. One can say only that they are "in principle" calculable from a quantum mechanical interpretation of collision theory. In the absence of the requisite detailed analytical knowledge of the ultimate intermolecular forces and interactions involved, Zwolinski and Eyring assumed certain relations between the various W's and assigned plausible numerical values to expedite particular calculations.

The linear nature of the system of differential equations, Eq. (273),which constitutes the model guarantees its general solution in the form

$$x_n(t) = \sum_j B_{nj} \exp(-\lambda_j t), \tag{274}$$

where the λ's are the characteristic roots of the matrix (B_{nj}). They evaluated Eq. (274) numerically for a four-level model to obtain explicitly an expression for the $x_n(t)$'s. And the rate of reaction was obtained by computing appropriate products $W_{nm}x_m$ and summing these products over the reactant and product levels. Again, as in the work of Prigogine and Kramers, they turned to the main theme of such studies mentioned earlier, namely, the validation of the equilibrium hypothesis—this time in the form of the assumption that the concentration of reacting species is given at all times by equilibrium Maxwell–Boltzmann statistics. They reported that the rate calculated from this model was less than that predicted by the equilibrium assumption, the maximum deviation for a four-level model being 20%.

5. *The Nonequilibrium Discrete Energy-Level Stochastic Model of Montroll and Shuler*

Also motivated by the interest in testing the equilibrium assumption, Montroll and Shuler [see Montroll and Shuler, 1958; also Shuler, 1958, 1959] began with the viewpoint mentioned earlier that the study of the microscopic aspects of chemical reaction, in fact, consists of departures from the equilibrium state: "Microscopic chemical kinetics, or the adjustment of energy distributions induced by chemical reactions is the main theme of this paper" [1958, p. 362]. At the same time a related aim of the work was the evaluation of the harmonic oscillator model for unimolecular reactions.

a. Extension of the Zwolinski–Eyring Models. In this spirit the model of Zwolinski and Eyring and the Kramers' model led Montroll and Shuler to conceive of a model for the microscopic, nonequilibrium chemical kinetics of a unimolecular reaction as an extension of the Zwolinski–Eyring four-level model to one of $N > 4$ levels, enhanced by the mathematical theory of stochastic processes for treating the interlevel transition parameters as stochastic parameters in the classical random-walk problem, thereby pushing the formulation closer to a bonafide stochastic model. In that sense, a chemical reaction is conceived by them as the removal of reactant species from the reaction system, effected via the imposition of an "absorbing barrier" at level $(N + 1)$. And the rate of chemical reaction is then given in terms of the "mean first passage time," the average time required for a species to pass level N and reach the absorbing barrier at $(N + 1)$ the first time. Thus, this point of view is consistent with Slater's $s'(q_1, \ldots, q_n)$, the time to a first "up-zero," that is, with the details of the random-gap assumption, with the $n(\tau)$ distribution in Bunker, and, as we shall see, with the distribution of the times τ of individual molecular transformations implied in Bartholomay's macroscopic model.

b. Relation to the Kramers Model and the Lindemann–Hinshelwood Mechanism. Methodologically speaking, it is probably the closest to the

Kramers–Brownian motion model. Though, like Kramers' work, it was not specifically conceived for the special case of unimolecular reactions, it was interpreted by them in terms of the unimolecular Lindemann–Hinshelwood mechanism, specifically. The construction of the model is begun by considering an ensemble of reactant molecules with discrete quantized levels to be immersed in a large excess of (chemically) inert gas which is presumed to act as a constant-temperature heat bath throughout the reaction. The requirement of a constant temperature T of the bath implies that the concentration of reactant molecules is very small compared to the concentration of heat bath molecules. The reactant molecules are assumed to be initially in a Maxwell–Boltzmann distribution appropriate to a temperature $T_0 < T$. By collision with heat bath molecules the reactants are excited in a stepwise process into their higher energy levels until they reach level $(N + 1)$ when they are removed irreversibly from the system. The collisional transition probabilities W_{nm} governing the transfer of reactant molecules between levels with energies E_m and E_n are functions of the quantum numbers m and n. In terms of the Lindemann–Hinshelwood unimolecular mechanism they indicate that A* would refer to the species of Slater's interesting (energized) molecules with energy $> N$, that the species M in the generalized Lindemann–Hinshelwood mechanism corresponds here to the heat bath molecules. And the assumption that $[M] \gg [A]$ guarantees first-order kinetics: in other words, the differential equations governing the rate of activation will be linear with respect to [A].

c. Comments on the Harmonic Oscillator Model. The stochastic context for the W_{nm}'s, missing in the Zwolinski–Eyring model, is indicated by interpreting W_{nm} as the probability that a random "walker" will take a step $m \longrightarrow n$, this in turn being a function of the distance from the origin $(n = 0)$. (Of course, the closeness of the Brownian motion theory to the random walk problem offers also an obvious correlation of this formulation with the Kramers formulation). Whereas, these quantities are as difficult to calculate as the η and σ quantities in the other theories, Montroll and Shuler suggest that explicit calculations are in principle possible, following the method of Landau and Teller, provided that the reactant molecules can be treated as simple harmonic oscillators (see references to such assumption in the discussion of Bunker's work in Section II.D) and provided that only weak interactions occur between the oscillators and the heat bath molecules. The assumption of harmonicity is referred to in a note added in proof to their paper:

> In view of the failure of the harmonic oscillator model to account for the observed rate of activation in unimolecular dissociation reactions (the dissociation lag problem) these calculations have been repeated for a Morse anharmonic oscillator with

transitions between nearest and next-nearest neighbor levels [Montroll and Shuler, 1958, p. 392].

Then after referring to some unpublished results of Kim's and to some results by Montroll *et al.* (in press), on the relaxation of vibrational nonequilibrium distributions of a system of Morse anharmonic oscillations, they state: ". . . it seems clear, however, that the anharmonic oscillator model with weak interactions. . . does not constitute much of an improvement on the harmonic oscillator model in giving the observed rates of reaction."

d. The General Transport Equation.† On these assumptions they proceed by writing Zwolinski–Eyring Eq. (273), treated now as a "transport equation," in the form

$$dx_n/dt = \sum_{m \neq n} \{W_{nm}x_m - W_{mn}x_n\} = \sum_m A_{nm}x_m, \tag{275}$$

referring the reader to the paper by van Hove [1957], and to the work of Kohn and Luttinger (private communication to them) for critical discussions of the derivation and validity of transport equations such as Eq. (275), adding their intention to make a similar study of their validity in problems of chemical kinetics. Here $x_m(t)$ refers to the "fraction of molecules" in the mth energy state. It would appear in any case that the transport equation bears to microscopic chemical kinetics a relation similar to that which the general model composed of linear systems of differential equations bears to compartmental analysis in physiology.

In mathematically detailed Sections III and IV of their original paper, Montroll and Shuler, drawing heavily on matrix theory and the method of characteristic equations discuss general and restricted solutions of the general transport equation, Eq. (275), applying the results [Section VII, p. 382] to the case of the harmonic oscillator model of a diatomic molecule.

For example, beginning with the equilibrium distribution of x_n, treated as a random variable now in the probabilistic sense of statistical thermodynamics,

$$\text{(distribution of) } x_m(\infty) = \exp(-\beta E_m) \Big/ \sum_j \exp(-\beta E_j), \tag{276}$$

corresponding to which the condition of detailed balance at equilibrium is written

$$W_{nm} \exp(-\beta E_m) = W_{mn} \exp(-\beta E_n), \tag{277}$$

they eventually obtain for the solution of the transport equation:

$$x_n(t) = \exp(-\beta E_n) \Big/ \sum_i \exp(-\beta E_i) + \sum_{j>0} c_j \psi_j(n) \exp(\lambda_j t), \tag{278}$$

†From Montroll and Shuler [1958, pp. 361–399].

where the $\psi_j(n)$'s and λ_j's are respectively the normalized characteristic vectors and characteristic values of matrix $B = (B_{nm})$, where

$$B_{nm} = A_{nm} \exp[\tfrac{1}{2}(E_n - E_m)\beta] \tag{279}$$

and the c_j's are related to the initial level distribution.

This discussion is extended to include the possibility of chemical reaction; that is, the case in which attainment of the $(N + 1)$th level represents the completion of an individual reaction event. The reaction rate is then determined by the rate at which molecules in their "random walk" from level to level reach level $(N + 1)$ the very first time, noting that in the appropriate language of stochastic processes the mean time for reaching this level amounts to the "mean first passage time" for the Nth level. Utilizing the transport equation, the distribution of first passage times $P(t)$ is shown to be expressible in terms of $x_N(t)$ by

$$P(t) = W_{N+1,N} x_N(t), \tag{280}$$

whence the mean first passage time $\bar{t}$ becomes

$$\bar{t} = \int_0^\infty tP(t)\, dt = W_{N+1,N} \int_0^\infty t x_N(t)\, dt, \tag{281}$$

results which extend immediately to the case of transition to both nearest and next-nearest neighbor levels as

$$P(t) = \sum_{i=0}^{k} W_{N+1,N-i} x_{N-i}(t), \tag{282}$$

where transition can occur between the levels $n \pm 1,\ n \pm 2, \ldots, n \pm k$. And it is shown in the context of the general theory of mean first passage times (Section V) that

$$\bar{t} = \sum_{j=0}^{N} \sum_{i=1}^{k} W_{N+1,N-i} \exp(-\tfrac{1}{2}\beta E_{N-i})\, c_j \psi_j(N - i) \lambda_j^{-2}. \tag{283}$$

e. The Landau–Teller Calculations of Transition Probabilities. Results such as these, obtained on the basis of transport Eq. (275) are particularized to the pertinent example of the harmonic oscillator model of a diatomic molecule by recourse to the Landau–Teller treatment of the transition probabilities. Briefly, the idea is as follows. If the "transition probability per collision" P_{10} for the transition $0 \longrightarrow 1$ can be determined, Landau and Teller [see Montroll and Shuler, 1958, pp. 368–378] show that the linear perturbations of the diatomic molecules by the heat bath induce other transitions with probabilities per collision

$$P_{mn} = [(m + 1)\delta_{n-1,m} + m\delta_{n+1,m}]P_{10} = P_{nm}, \tag{284}$$

where δ_{mn} is the Kronecker delta (1 if $m = n$; 0, if $m \neq n$). Montroll and Shuler note that these are exactly the transition probabilities associated with transitions of a harmonic oscillator in a radiation bath, transitions only between adjacent levels being possible.

The relation between P_{mn}, the transition probabilities per collision, and W_{nm} the transition probabilities per unit time are given as

$$W_{n+1,m} = Z^*N^*e^{-\theta}P_{n,n+1}, \tag{285}$$

$$W_{n,n+1} = Z^*N^*P_{n+1,n}, \tag{286}$$

where Z^* is the collision number (in the case cited, the number of collisions per unit time by the oscillator when gas density is one molecule per unit volume). N^* is the total concentration of heat bath molecules, and $\theta = h\nu/k_BT$. Note that from Eqs. (285) and (286), there follows

$$W_{n,n+1} = e^{\theta}W_{n+1,n}, \tag{287}$$

as required by the principle of detailed balance.

f. The Harmonic Oscillator Transport Equation. The use of these transition probabilities in Eq. (275) particularizes it in the case of the harmonic oscillator model of a diatom to

$$dx_n/dt = c\{ne^{-\theta}x_{n-1} - [n + (n+1)e^{-\theta}]x_n + (n+1)x_{n+1}\}, \tag{288}$$

where c depends only on the coupling between molecules and heat bath. They solve this system subject to the top boundary condition

$$dx_n/dt = c\{Ne^{-\theta}x_{N-1} - [N + \alpha(N-1)e^{-\theta}]x_n\}, \tag{289}$$

where is α the absorption coefficient, eventually taken as 1. Using the trace of the corresponding determinant of the B^{-1} matrix, it is found that

$$\bar{t} = -c^{-1} \text{ trace } B^{-1}, \tag{290}$$

that is

$$\bar{t} = e^{(N+1)} \sum_{k=0}^{N} e^{k\theta}\{(N+1)^{-1} + N^{-1} + \cdots + (N+1-k)^{-1}\}, \tag{291}$$

corresponding to a totally absorbing barrier ($\alpha = 1$) at level $(N+1)$. And the rate of activation (the association step of the Lindemann–Hinshelwood mechanism) is proportional to the reciprocal of the mean first passage time, $(\bar{t})^{-1}$. The mean first passage time and rate of activation (association) ν_{act} deviate from their equilibrium values (obtained in the limit as $N \longrightarrow \infty$) by more than 10% when

$$N(1 - e^{-\theta}) < 10e^{-\theta}. \tag{292}$$

It is noted that in the high-temperature limit this corresponds approximately to

$$(Nh\nu/k_BT) = E_{act}/k_BT < 10, \tag{293}$$

which is in agreements with the Kramers and Prigogine results.

The complete characteristic vector analysis of the harmonic oscillator model is carried out using Gottlieb polynomials. In this case, the general solution of the transport equation Eq. (288) is

$$x_n(t) = \sum_{j=0}^{N} a_j l_n(\mu_j) \exp\{\mu_j(e^{-\theta} - 1)tc\}, \tag{294}$$

where the $l_n(\mu)$ are Gottlieb polynomials and $\lambda_k = \mu_k(e^{-\theta} - 1)$ are different characteristic values of the B-matrix, that is,

$$l_n(\mu) = e^{-n\theta} \sum_{\nu=0}^{N} (1 - e^{\theta})^{\nu}(1 - e^{\theta})^{\nu} \binom{n}{\nu}\binom{\mu}{\nu} \tag{295}$$

and

$$a_j = \left(\sum_{n=0}^{N} x_n(0) l_n(\mu_j) e^{n\theta}\right) \Big/ \left(\sum_{n=0} l_n^2(\mu_j) e^{n\theta}\right). \tag{296}$$

If, for example,

$$\text{initial distribution } x_n(0) = e^{-\theta_0}(1 - e^{-\theta_0}), \tag{297}$$

that is, a Boltzmann distribution with $\theta_0 \gg \theta$, it is found that

$$P(t) = -(d/dt)(\textstyle\sum x_n(t)) = W_{N+1,N} x_N(t), \tag{298}$$

$$P(t) = \alpha e^{-\theta}(N + 1)\, x_N(t). \tag{299}$$

In other words,

$$P(t) = c\alpha(N + 1)e^{-\theta} \sum_{j=0}^{N} a_j l_N(\mu_j) \exp\{-\mu_j(1 - e^{-\theta})ct\}, \tag{300}$$

and in the particular case when all molecules are in ground state,

$$P(t) = c\alpha(N + 1)e^{-\theta} \sum_{j=0}^{N} l_N(\mu_j) e^{\lambda_j t} \left(\sum_{n=0}^{N} l_n^2(\mu_j) e^{n\theta}\right)^{-1}. \tag{301}$$

Eventually, the result obtained for ν_{act} [defined, as above, as proportional to $(\bar{t})^{-1}$] is

$$\nu_{\text{act}} = (X/\bar{t}) = c(N + 1)X(1 - e^{-\theta})^2 e^{-(N+1)\theta}, \tag{302}$$

where X is the concentration of diatomic oscillator reactants in mole/cm^3. And, the rate constant k_{act} (corresponding to k_{uni}) is

$$k_{\text{act}} = (\nu_{\text{act}}/XN^*) = Z^* P_{10}(N + 1)(1 - e^{-\theta})^{-2} \exp(-\beta E_{\text{act}}). \tag{303}$$

For example, comparing this with the Arrhenius equation, written as

$$k_{\text{act}} = A \exp(-\beta E_{\text{act}}) \tag{304}$$

it is seen that the frequency factor of the model is given by

$$A = Z^* P_{10}(N + 1)(1 - e^{-\theta})^2. \tag{305}$$

g. Constrast with the Stochastic Model of Macroscopic Kinetics. Here again it is seen that the physicochemical and probabilistic details are contained in a detailed factorization of the unimolecular rate constant, not generally recognized or perceived in macroscopic kinetic measurements such as concentration–time data. Thus, the stochastic element is introduced at the microscopic quantum level in terms of expressions of the distributions of the number x_n of molecules at energy level n, and the results are not given in terms of the probability of attainment of a macroscopic concentration state as a function of time.

However, at about the same time that Montroll and Shuler were working on their model, the theory of stochastic processes was also being applied explicitly to obtain a theory of macroscopic kinetics which could provide such expressions. This was, in fact, visualized [see Bartholomay, 1957, 1959, 1964a] as a means of providing a more realistic framework for interpreting concentration–time data in experimental kinetic studies.

F. Macroscopic Stochastic Models of the Unimolecular Reaction

1. *Independent Origins*

The present author's interest in formulating macroscopic kinetics actually began in connection with studies of enzyme kinetic data. The possibility that established methods of high precision, accuracy, and reproducibility applied to purified systems could produce, on occasion, concentration–time courses with unaccountable fluctuations concerned him, for in this way possibly significant results on new systems might be discarded or misinterpreted.

It is interesting that apparently at about the same time Singer [1953], in England, was concerned over random fluctuations in complex processes such as nucleation processes and chain reactions. Singer emphasized similarly that, whereas ordinary chemical methods may not be sensitive generally to random fluctuations, there are some physicochemical processes in which one should expect macroscopic fluctuations. For example, he cited a class of chemical reactions in which the formation of one or a few reactive molecules is followed by a chain reaction which leads to the rapid (possibly explosive) completion of the reaction. In such a case, the formation of the reactive molecule which initiates the reaction being a rare event, the course of the reaction will be "irreproducible" (intrinsically). He considered also the case of chain mechanisms in which a slight change in concentration of reactive molecules, or so-called "chain carriers" can alter the reaction into a diverging process, pointing out that such a reaction would be irreproducible also, for example, if the critical change in the concentration of chain carriers resulted from a chance fluctuation. Finally, he offered the example of the formation of crystal nuclei in solution or melt or, more generally, nucleation in phase changes, pointing out that in such cases it should be considered probable that the stochastic behavior of the molecular aggregate representing the nucleus of a new phase would resemble that of a population for which, below a certain critical threshold, the probability of extinction exceeds the probability of survival (reminding one of the "threshold theorems" in epidemiology). The formation of a macroscopic nucleus would depend on the accidental growth of the population beyond the critical size. And if this were an event of sufficiently low probability, the time lapsing before the observation of the first nucleus would fluctuate randomly. Examples of such reactions are the oxidation of formic acid by potassium nitrate and the initial stages of certain polymerization reactions. To these cases can be added the extremely complex kinetics of DNA replication, which will be discussed in the last section of this chapter.

It is interesting that, while Singer's ideas and methods had been conceived of independently, there is a close correspondence conceptually and even methodologically with the model to be described in the next section. Also, in close agreement with the convictions expressed in the opening paragraph of this section, Singer had expressed the opinion that possibly a great deal of work in chemical kinetics hasd not been published or else had been abandoned becuase the results were unexplainably irreproducible, the possibility of a mode of analysis of inherent random fluctuations not having been recognized.

It should also be mentioned that as early as 1940, Delbrück had offered a statistical analysis of rate data in autocatalytic reactions that was also motivated by the appearance of random fluctuations in macroscopic kinetic data. The surveys by Bharucha-Reid [1960] and McQuarrie [1967] and the author's doctoral dissertation [Bartholomay, 1957] are sources of additional information on stochastic models of macroscopic chemical kinetics. And, in Section IV, stochastic models in enzyme kenetics will be discussed. In the remainder of this section we shall stress the foundational aspects in terms of the unimolecular reaction in particular.

In Singer's work and in Bartholomay's, can be seen the shift in macroscopic kinetics from the calculation of "rates of reaction" to the probabilities of the number of molecules present as functions of time. In both cases, the numbers, $n_1, n_2, \ldots, n_r$ of different molecular species $A_1, \ldots, A_r$ in a complex reaction system of constant volume are treated as discrete-valued random variables, as opposed, in the deterministic models, to the "concentrations" $[A_1], \ldots, [A_r]$ treated as ordinary analytic functions of time. Thus, whereas the culmination of deterministic models lay in the rate expressions $d[A_i]/dt$ $(i = 1, \ldots, r)$ giving the rates of change of such concentrations in terms of polynomial expressions with rate constants as coefficients, one corresponding to each reaction step, in these stochastic models the object has been to calculate the probabilities $p(n_1, \ldots, n_r; t)$ giving the distribution of the n's in time, together with the related measures of random variations. Bartholomay [1957, 1958] has shown that the stochastic models of this kind amount to Markov processes and he has invoked the machinery of this theory in subsequent analytical and experimental studies.

2. *Bartholomay's Stochastic Model of the Irreversible Unimolecular Reaction*

a. Relation to Microscopic Kinetics and the Random Lifetimes Assumption. While the stochastic approach was inspired originally by considerations of complex reactions, because of the apparently total lack of stochastic point of view and stochastic theory for dealing with macroscopic kinetics in the literature, the present author began with the unimolecular reaction as the logical first step. A general theory was visualized as obtainable by inductive generalization, that is, as a synthesis from stochastic formulations of basic elementary (indecomposable) reaction types [see Bartholomay, 1957, Introduction; 1962c, for a preliminary generalization]. Thus, the "unimolecular process" in his work is thought of as an elementary or unitary reaction step which contributes first-order kinetics to a complex reaction (of which it may be one step). It is therefore identifiable with the final dissociation step in the Lindemann–Hinshelwood mechanism or, possibly, with the whole mechanism in the cases in which first-order kinetics are produced. Examples of unimolecular decompositions of this latter kind which were, in fact, used to test the predictions of the model (see Section IV.F.2, 3) are: the classically studied sucrose inversion reaction mentioned in a number of other connections and the decomposition reaction of the di-*t*-alkyl peroxides. Similarly, radioactive decompositions and other "spontaneous" decompositions and simple isomerizations could be thought of as realizations of the unimolecular stochastic process.

Thus, in some cases the "stochastic parameter" of the unimolecular process [corresponding to the deterministic rate constant (see below)] may be related to the microscopic physicochemical aspects of a decomposition in a manner analogous to the relation of k_a to the final dissociation step of the Lindemann–Hinshelwood mechanism. Or, indeed, it may refer to a random event of apparently spontaneous origin. However, the intramolecular events involved in spontaneous unimolecular reactions which are analyzable down to the level of the deployment of the total energy of the unstable molecule into a particular degree of freedom to the extent necessary to break an interatomic bond call for the establishment of a mathematical relation between this parameter and the stochastic parameters of the stochastic models of microscopic kinetics.

The stochastic model of the unimolecular reaction discussed in this section is thus considered to apply either to individual nondecomposable steps in a complex reaction mechanism or to cases of one-step reactions satisfying simultaneously both the molecular and kinetic first-order criteria. Depending, then, on the particular realization of the model, a wide latitude of underlying microscopic physicochemical explanations is permitted.

b. The Purpose of the Unimolecular Stochastic Model and of a Stochastic Theory of Chemical Kinetics. With the long-range goal discussed in Section III.f.2.a, in mind, the next cases considered (see Bartholomay, [1957] and Bharucha-Reid [1960]) were the elementary irreversible bimolecular reaction and the simplest reversible reaction. This led to a method for the stochastic treatment of the basic enzyme-catalyzed reaction of biochemistry, which is, of course, a complex system composed of unimolecular and bimolecular steps resembling the Lindemann–Hinshelwood unimolecular mechanism. It is principally with such application in mind that the widest possible interpretation has been suggested for all of the stochastic models (and their parameters) of the elementary reaction types.

In any case the development of a stochastic theory of chemical kinetics should not be interpreted as aimed simply at providing statistical estimates of the corresponding rate constants. Rather, it calls attention to the existence of an inherent random component in observable fluctuations to be weighed against the random experimental error components, and, in this spirit, provides statistical characterizations of the former component, calling for reexamination of the whole concept of "rate constant." Nor were such studies undertaken with the contention, for example, that the unimolecular reaction generally is "irreproducible" in the sense implied in the more complex reactions cited in the previous section. It does happen, as we shall, see, that the statistical estimates of the stochastic parameters have been found to be indistinguishable from conventional estimates of the corresponding rate constants in a number of real cases. And, in the case of unimolecular reactions with small ranges of random fluctuations one may conclude from this that the stochastic parameter and its stochastic estimates, being simpler to carry out than a careful statistical estimate of the corresponding conventional rate constant, may be turned to by making the initial assumption that all the fluctuation is inherent as opposed to extraneous (or of experimental origin, the latter being the usual assumption). But this amounts to simply an empirical by-product of the theory, whose main ultimate goal is the provision of an appropriate and theoretical framework for studying the possible macroscopic implications of stochastic microscopic kinetics, including the possibility that occasional irregular sample curves may by generated. Such a theory could provide a formal and rational basis for estimating the "degree of irreproducibility" of a reaction or, otherwise said, for deducing conditions under which the variance or range of random fluctuations would indeed be below observable thresholds.

c. The Markovian Context of the Model. The unimolecular stochastic process is a member of the mathematical domain of Markov processes. In particular, it is describable as a Markov chain with finitely many states, continuous time parameter, and so-called" stationary transition probabilities." The chief characteristics of such a stochastic process make it a feasible one to propose as the basis for a large number of physical, chemical, and biological models: (1) the future state of the process depends on the present state, not on the entire past history; (2) the probability of a transition from an arbitrarily defined "state" to another depends on the length of time interval alone, independently of the time of attainment of the present state; (3) the transition occurs at any point in a time continuum. The "states" referred to in this general formulation correspond to the possible integral values of the number of molecules per constant volume of reaction mixture treated as a random variable, the "concentration random variable n."

All properties of the stochastic process are determinable from the matrix of transition probability–time functions, designated as $\{p_{ik}(t)\}$ where $p_{ik}(t)$ is defined as the probability of a transition from state i to state k in time t. In the case of the unimolecular reaction it is seen that where n_0 is the initial value of the concentration random variable, there are $n_0 + 1$ states possible, each state being identified with a value of n. Thus, one speaks of transition from concentration i to k, where $0 \leqslant k \leqslant i \leqslant n_0$.

Just as in the case of deterministic models wherein the concentration treated as an ordinary function of time is determined by integration of a differential equation of n, it has been shown [see Doob, 1953, Chapter 6] that the p_{ik}'s are determinable by integration of the following "forward system of differential equations":

$$p'_{ik}(t) = q_{kk} p_{ik}(t) + \sum_{j \neq k}^{n_0} q_{jk} p_{ij}(t), \qquad i, k = 0, 1, 2, \ldots, n_0, \tag{306}$$

subject to initial conditions

$$p_{ij}(0) = \delta_{ij} = \begin{cases} 1, & i = j, \\ 0, & i \neq j, \end{cases} \tag{307}$$

where the prime indicates differentiation with respect to time and the q coefficients sometimes referred to as "intensity numbers," comprise a Q matrix of elements defined by

$$\begin{aligned} q_{ij} &= p'_{ij} > 0, \qquad j \neq i, \\ q_{ii} &= (\sum_{j \neq i} q_{ij}) < 0. \end{aligned} \tag{308}$$

In utilizing the Markov chain in model construction, the Q matrix forms a convenient point of entrance, the individual elements being chosen so as to incorporate the leading stochastic features of the underlying system. And the forward system of Eqs. (306) is then arrived at by taking limits in a corresponding system of stochastic difference equations.

d. Probabilistic Implications of the Deterministic Model. † Bartholomay [1958] began the construction of the stochastic model for the unimolecular reaction A $\longrightarrow$ B by reinterpreting the deterministic model given by the ordinary differential equation

$$dn/dt = -kn. \tag{309}$$

Note that whereas n, the concentration of reactant in molecules per constant volume, say, is actually a discrete-valued variable with possible values $n = 0, 1, 2, \ldots, n_0$ (n_0 being its initial value) in Eq. (309) it is treated as a continuous variable. In fact, integration of this equation yields the familiar negative exponential form,

$$n = n_0 e^{-kt}, \tag{310}$$

giving the value of n corresponding to a particular time $t > 0$ in the reaction course. Within this framework, using infinitesimals, we may write

$$\Delta n = -kn\,\Delta t + o(\Delta t). \tag{311}$$

Taking the absolute value $|\Delta n|$ as an approximation to the integral number of individual decompositions taking place over the time interval $(t, t + \Delta t)$ and utilizing Eq. (311) we have

$$|\Delta n|/n = k\,\Delta t + o(\Delta t) \tag{312}$$

which shows that the relative frequency of occurrence of the decompositions in an infinitesimal time interval of length Δt is proportioned to the length of interval except for an infinitesimal of higher order.

Interpreting Eq. (312) heuristically in the context of the frequency theory of probability as the probability of occurrence of a decomposition, the same for each reactant molecule, which we now write as

$$p = \mu\,\Delta t + o(\Delta t), \tag{313}$$

the constant of proportionality μ becomes a stochastic parameter. Furthermore, considering that the set of n molecules present at time t undergo Bernoulli trials in time $(t, t + \Delta t)$ with probability p of success and $q = 1 - p$ of failure, the binomial distribution which governs such events yields as the expression for the probability of exactly k successes in n trials

$$b(n, k; p) = \binom{n}{k}[\mu\Delta t + o(\Delta t)]^k[1 - (\mu\Delta t + o(\Delta t)]^{n-k}, \tag{314}$$

from which it follows that

$$b(n, 1; p) = \mu n\,\Delta t + o(\Delta t), \qquad k = 1, \tag{315}$$

$$b(n, k; p) = 0 + o(\Delta t), \qquad k > 1. \tag{316}$$

†From Bartholomay [1958, pp. 175–190; 1959, pp. 363–373].

Thus, the probability of exactly one decomposition in the interval $(t, t + \Delta t)$ from the n molecules present at time t is directly porportional to the length of interval and to n. But the probability of more than one such decomposition in this time is an infinitesimal of higher order that Δt and hence approximately 0.

e. The Stochastic Model as a Markov Process. The observations and considerations of the preceding section were incorporated into an axiomatic, probabilistically oriented description of the unimolecular reaction as a process consisting of a collection of individual decomposition events applied to an initial population of n_0 molecules and governed by the following conditions:

A1. If exactly $(n_0 - n)$ decompositions have occurred in the time interval $(0, t)$, in the subsequent infinitesimal interval $(t, t + \Delta t)$ the probability of a single decomposition from the n molecules present at time t equals $[\mu n \, \Delta t + o(\Delta t)]$ And the probability of more than one such event in the time equals $o(\Delta t)$.

A2. The probability parameter μ has the dimensions and properties of the ordinary deterministic first-order rate constant k.

A3. Reactant species A and product species B coexist without significant interactions of any kind.

A4. The reverse reaction $B \longrightarrow A$ is impossible.

This formulation in turn was realized in the mathematical domain of the theory of stochastic processes as a Markov chain with $(n_0 + 1)$ concentration states, continuous time parameter, and stationary transition probabilities $p_{ik}(t)$ determinable from an intensity matrix $Q = (q_{ij})$ specified as follows: According to the Maclaurin expansion, the p_{ij}s may be expressed as

$$p_{ij}(\Delta t) = p_{ij}(0) + p'_{ij}(0) \, \Delta t + o(\Delta t), \qquad j \neq i. \tag{317}$$

Equations (308) reduce this to

$$p_{ij}(\Delta t) = q_{ij} \, \Delta t + o(\Delta t), \tag{318}$$

the identification of which with the content of A1–A4 gives

$$\begin{aligned} p_{ij}(\Delta t) &= \mu i \, \Delta t + o(\Delta t), \qquad && j = i - 1, \\ &= o(\Delta t), && j < i - 1, \\ &= 0, && j > i. \end{aligned} \tag{319}$$

In other words, the Q-matrix elements are chosen so that

$$q_{ij} = \begin{cases} 0, & j < i, \quad j > i, \\ \mu i, & j = i - 1, \\ -\mu i, & j = i. \end{cases} \qquad j = 0, 1, 2, \ldots, n_0, \tag{320}$$

Substitution into the system of Eqs. (306) yields the following forward system of equations:

$$p'_{ik} = -\mu k p_{ik}(t) + \mu(k+1)p_{i,k+1}(t) \tag{321}$$

for given i, where $0 \leqslant i \leqslant n_0$ and for all k's such that $k \leqslant i$. A customary and convenient way of solving Eq. (321) is to use the "probability generating function" (pgf),

$$\text{pgf}\{p_{ik}(t)\} = \phi_i(s, t) = \sum_{k=0}^{n_0} s^k p_{ik}(t), \tag{322}$$

in which s is an arbitrary mathematical variable such that $|s| < 1$ of no physical significance. Differentiating $\phi_i(s, t)$ partially with respect to s and t separately and making appropriate substitutions from Eq. (312) leads to the following homogeneous, first-order, partial differential equation:

$$\partial\phi_i/\partial t = \mu(1-s)\,\partial\phi_i/\partial s, \qquad i = 1, 2, \ldots, n_0 \tag{323}$$

with boundary conditions

$$\phi_i(s, 0) = s^i, \qquad \phi_i(1, t) = 1 \tag{324}$$

which is solved by using Lagrange's method of auxiliary differential equations, and gives

$$\phi_i(s, t) = [(-1 + e^{\mu t}) + s]^i e^{-\mu i t}, \qquad i = 1, 2, \ldots, n_0 \tag{325}$$

Expanding the right-hand side, collecting powers of s, and comparing with Eq. (322) yields the following expressions for the transition probabilities, completely defining the process:

$$p_{ik}(t) = \binom{i}{k} e^{-\mu i t}(-1 + e^{\mu t})^{i-k}, \qquad 0 \leqslant k \leqslant i \leqslant n_0. \tag{326}$$

The mean value and variance functions of k considered as the random variable governed by the probability distribution $\{p_{ik}(t)\}$ $(k = 0, 1, 2, \ldots, i)$ are, respectively,

$$E_{ik}(t) = ie^{-\mu i t}, \tag{327}$$

$$\sigma^2_{ik}(t) = i(e^{-\mu t} - e^{-2\mu t}). \tag{328}$$

A convenient reinterpretation of the p_{ik}s is obtained by rewriting Eq. (326) as

$$p_{ik}(t) = \binom{i}{k}(e^{-\mu t})^k(1 - e^{-\mu t})^{i-k}. \tag{329}$$

Thus, for fixed t, this distribution may be interpreted as $b_i(k, i;\, e^{-\mu t})$, a binomial distribution with $e^{-\mu t}$ taken as the probability p of success. Note, however, that from Eq. (329),

$$p_{11}(t) = e^{-\mu t}, \tag{330}$$

$$p_{10}(t) = 1 - e^{-\mu t}, \tag{331}$$

so that the "probability of success" would actually correspond to the probability that a molecule will not decompose and the "probability of failure" to the probability that it will.

f. The Related Prior Model. In later computer studies utilizing this model, Eqs. (326) or (329) were specified as defining the "full Markov model" (or the "posterior" or "sequential" model). In connection with comparing the predictions of this model with the classical deterministic theory which assumes that the initial state n_0 determines the entire future course of the reaction, [reflected in the use of Eq. (310) or a linearization of it by logarithmic transformation, in analyzing experimental data], it is convenient to restrict the model to the specification only of the transition probabilities $p_{n_0k}(t)$; that is, to the probability of attainment of state $k < n_0$ only from the initial state n_0. In other words, of the full set of Markovian probabilities $\{p_{ik}(t)\}$ associated with state k, only the case $i = n_0$ is considered. Subject to this restriction the model is referred to as the "Prior model."

Thus, in this case we would be interested only in determining the probability of attaining state $k < n_0$ (replacing k by n now) t time units after the initiation of the reaction. And in this sense the model is compared with the deterministic model by noting that just as the latter fives n as a function of time t after initiation of the reaction, the Prior model gives the prior probability of attainment of any value of n in this same time t.

On the other hand, as will be demonstrated presently, the full Markovian model, that is, the full set of transition probabilities $\{p_{ik}(t)\}$, prompts one to make detailed *in numero* experimental studies of the transitions from any state i to any subsequent state k, corresponding to a time interval of length τ, where now τ is used to denote that the time interval is not referred to $t = 0$ as its initial point. Formally, the Prior model is obtained by restricting i to the value n_0 in Eq. (326) and writing

$$p_n(t) = \binom{n_0}{n} e^{-\mu n_0 t}(-1 + e^{\mu t})^{n_0 - n} \tag{332}$$

or, in binomial form,

$$p_n(t) = \binom{n_0}{n}(e^{-\mu t})^n(1 - e^{-\mu t})^{n_0 - n}, \tag{333}$$

in which case the mean value and standard deviation functions become

$$E(n, t) = n_0 e^{-\mu t}, \tag{334}$$

$$\sigma(n, t) = [n_0 e^{-\mu t}(1 - e^{-\mu t})]^{1/2}. \tag{335}$$

Note that identifying μ with the first-order rate constant k makes Eq. (334) identical with Eq. (310), the predicted time course of the deterministic model, in which case Prior stochastic model is spoken of [Bartholomay, 1957, 1958] as "consistent in the mean" with the deterministic model. It is in this precise sense that the progress of a unimolecular reaction in terms of the macroscopic deterministic theory, holds "on the average," the degree of random dispersion about this being given by Eq. (335).

For large values of n_0 and of n, the probability of state n at time t can be approximated by the probability density function of the concentration variable n treated, as in the deterministic case, as a continuous random variable, call it now X, with initial value x_0 in accord with the central limit theorem which allows the approximation of the binomial distribution of Eq. (333) by a Gaussian distribution

$$p(x, t) = (2\pi)^{-1/2}\sigma^{-1}(n, t) \exp\{-\tfrac{1}{2}\sigma^{-2}(n, t)[X - E(n, t)]^2\}. \tag{336}$$

Relative to this approximation, a region defined by the boundaries $[E(n, t) \pm 2\sigma(n, t)]$ may be constructed in which the points (x, t) obtained in repeated sample runs would be expected to fall with roughly 97% probability. A typical sample curve of the model, corresponding to an idealized run, free of experimental error, is predicted by the Markov theory to be a curve, fluctuating randomly about $E(n, t)$ but lying mostly inside the region.

(1) THE STOCHASTIC REACTION SURFACE. The relationship between the (Prior) stochastic model and the deterministic model is visualized graphically in Fig. 9, the so-called "stochastic reaction surface for the unimolecular reaction." In effect, this shows the reaction probability force p as a function of n and t. Here, the profile in the family of planes $t = K =$ constant have been sketched using the continuous Gaussian probability densities $p(x, K)$ determined from Eq. (336). Since the maxima of this family of surface curves occur at the mean values of the approximating Gaussian probability densities $p(x, K)$, they correspond to the points $(n_0 e^{-K\mu}, K)$ in the (n, t)-plane which lie on the deterministic curve $n = n_0 e^{-\mu t}$ shown in the (n, t)-plane (taking $\mu \equiv k$). Thus, the (n, t)-plane is the plane of the mean value and ensemble of sample curves.

In the family of planes $n = c =$ constant, using the actual probability function Eq. (333), the profiles have maxima located over points on this same curve; that is, $[n = c,\ t = -\mu^{-1} \log_e (c/n_0),\ 0 \leqslant c \leqslant n_0]$. Consequently, there is a sharp crest on the surface, defined by the maxima of these two orthogonal families of profiles, the orthogonal projection of this crest in the (n, t)-plane being the deterministic curve. In these terms then, the deterministic model is obtained by contracting all the profiles to zero width and height one directly above the deterministic curve in the plane. And, according to the stochastic theory (Prior model) at time $t > 0$ the concentration n has any value between 0 and n_0 corresponding to $t =$ constant with probability read off as the p coordinate.

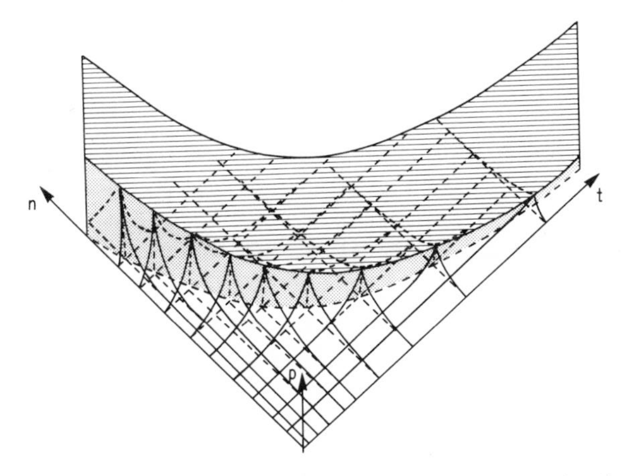

Fig. 9. Stochastic reaction surface for the Prior unimolecular stochastic process. [From Bartholomay, 1958, Fig. 1, p. 183.]

As the surface shows, beginning with initial concentration n_0, with decreasing n and/or increasing t there is a corresponding contraction of probable values. It is possible to have an increase in the value of n as t increases, according to the Prior model interpretation. But, the full Markovian model does not allow this, the theory guaranteeing that each sample curve is a stepwise constant, nondecreasing curve. This latter property follows from the fact that $p_{ik}(\tau) \equiv 0$ for $k > i$.

It may be shown that

$$\lim_{t\to 0} p_n(t) = \begin{cases} 1, & n = n_0, \\ 0, & n \neq n_0; \end{cases} \tag{337}$$

$$\lim_{t\to \infty} p_n(t) = \begin{cases} 0, & n \neq 0, \\ 1, & n = 0. \end{cases} \tag{338}$$

(2) Measure of Inherent Variability. The degree of inherent random variability, or natural fluctuation of the unimolecular reaction called for by the stochastic theory is contained in the $\sigma(n, t)$ function of Eq. (335). In particular, from this it is seen that

$$\lim_{t\to\infty} \sigma(n, t) = \lim_{t\to 0} \sigma(n, t) = 0, \tag{339}$$

$$\sigma_{\max}(n, t) = \tfrac{1}{2} n_0^{1/2}, \tag{340}$$

the maximal value occurring at time $t = \mu^{-1} \log_e 2$. corresponding to the reaction half-time, so that beginning with zero variability as the reaction proceeds, by half-time it grows to its maximum when (when n_0 has been reduced to $n_0/2$) it begins receding and reaches zero by the end of the reaction. It is thus possible to quantify the threshold of observable or macroscopic inherent variability of the process, considering the precision of a method of determination. Clearly such fluctuation would be relatively large in dilute reactions. The result is in good agreement with the "$n^{1/2}$" law by Schroedinger [1945].

The coefficient of variation $\mathrm{CV}(t) = \sigma(t)/E(n, t)$ is a measure of relative inherent variability, becoming in this case

$$\mathrm{CV}(t) = [(e^{\mu t} - 1)/n_0]^{1/2}, \tag{341}$$

so that

$$\lim_{t \to 0} \mathrm{CV}(t) = 0, \tag{342}$$

$$\lim_{t \to \infty} \mathrm{CV}(t) = \infty. \tag{343}$$

The latter limit shows that while the absolute inherent variability may be declining; this declines much less rapidly than the mean value. The relation between these two measures of variability is pictured in Fig. 10.

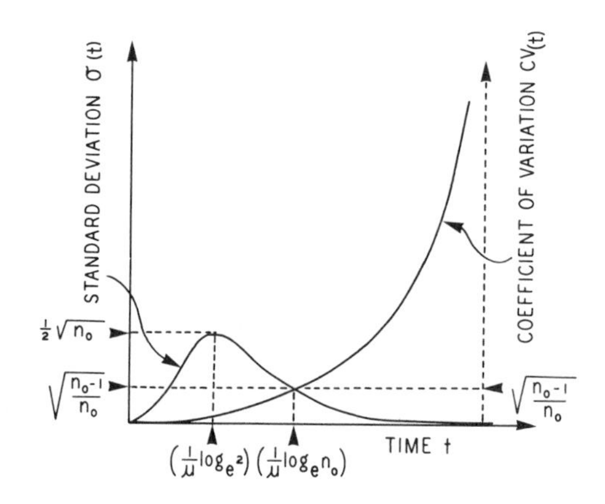

Fig. 10. Theoretical standard deviation and coefficient of variation curves for the Prior unimolecular stochastic process.

(3) Probability Measure of Completion Time. The "apparent" completion time of unimolecular reactions can be characterized in this theory, also on a realistic probabilistic basis. If $\varepsilon > 0$ is an arbitrarily small positive number, then the time T_ε required to obtain completion with probability equal to $(1 - \varepsilon)$ is determined by solving the equation

$$p_0(t) = (1 - e^{-\mu t})^{n_0} = 1 - \varepsilon, \tag{344}$$

the result being

$$T_\varepsilon = -\mu^{-1} \log_e[1 - (1 - \varepsilon)^{1/n_0}], \tag{345}$$

where $(1 - \varepsilon)^{1/n_0}$ is the principal n_0th root of the probability number $1 - \varepsilon$.

g. Estimation of the Stochastic Parameter. The deterministic model in macroscopic kinetics cannot be applied directly to the analysis of real data because of observable random fluctuations which produce in actual runs an irregular concentration–time course (with negative exponential trend $n = n_0 e^{-kt}$). It has been emphasized that the macroscopic stochastic model predicts the presence of an inherent random component, which in some cases could become macroscopic, particularly in the later, dilute stages of reaction. The deterministic theory does not recognize this factor, or, indeed, any random component. However, in practice recognition is given to the existence of a random error component, generally attributed to possibly controllable extraneous experimental error. And on this basis, statistical procedures such as empirical curve fitting and more sophisticated regression theory are employed to estimate the rate constant. Usually this is done in terms of the linear form,

$$\log x = \log x_0 - kt, \tag{346}$$

where x is considered a continuous variable corresponding to relative concentration, or percentage initial concentration (see below). In such cases [see Bartholomay, 1959, 1968b] the original model is replaced (implicitly) by a semistochastic or statistical model, in the following sense, which is most simply indicated in terms of the original form $x = x_0 e^{-kt}$. The variable x becomes stochastic when this equation is extended to the form

$$x = x_0 e^{-kt} + \varepsilon(t), \tag{347}$$

where $\varepsilon(t)$ is considered to be a Gaussian process. An assumption of this kind is implicit in the statistical estimation employed (see Section III.F.4.c [10]).

The opposite idealization is implied in the basic stochastic theory: namely, reaction runs are assumed to be free of experimental errors. Under this assumption a convenient method is possible for estimating the stochastic parameter μ; as mentioned earlier, the following results have been derived [Bartholomay, 1959] in this case.

CASE 1 ENSEMBLE OF R SAMPLE CURVES WITH RESPECT TO THE MARKOV MODEL: Suppose in executing a unimolecular reaction run, serial determinations of the concentration random variable n are made at $(S + 1)$ equispaced times $t_0 = 0$, $t_1 = \tau$, $t_2 = 2\tau, \ldots, t_S = S\tau = T$ over an interval $[0, T]$ of the time axis. If the reaction is repeated R times so that an ensemble of R such sample curves is obtained, let m_{rs} be the value of n in the rth run at time $s\tau$, where $r = 1, 2, \ldots, R$ and $s = 0, 1, \ldots, S$. These data then make up an $R \times (S + 1)$ "observation matrix" $M = (m_{rs})$. In this case the following maximum likelihood estimate $\hat{\mu}$ for μ was derived:

$$\hat{\mu} = T^{-1}\left[\log_e\left(\sum_{r=1}^{R}\sum_{s=1}^{S} m_{r,s-1}\right) - \log_e\left(\sum_{r=1}^{R}\sum_{s=1}^{S} m_{r,s}\right)\right]. \tag{348}$$

CASE 2 $R = 1$, A SINGLE SAMPLE CURVE: In this case dropping the R subscript and the corresponding summation in Eq. (348) gives

$$\hat{\mu} = T^{-1}\left[\log_e\left(\sum_{s=1}^{S} m_{s-1}\right) - \log_e\left(\sum_{s=1}^{S} m_s\right)\right]. \tag{349}$$

This latter result may be compared with the case of applying regression theory on the statistically enlarged deterministic basis [using now Eq. (346) as the underlying deterministic model]

$$\hat{k} = \sum_{s=0}^{S} (t_s - \bar{t})(\log_e m_s - \overline{\log m}) \Big/ \sum_{s=0}^{S} (t_s - \bar{t})^2. \tag{350}$$

Reference has already been made in Section III.F.b to comparisons made between k and μ in an early study [Bartholomay, 1959]. The published data of Pennycuick [1926], on sucrose inversion and those of Raley *et al.* [1948], on di-*t*-butyl peroxide (first-order kinetics having been validated in both cases), were employed. For example, in the case of Pennycuick's data, application of Eq. (350) yielded the result

$$\hat{k} = 3.732 \times 10^{-3} \quad \text{min}^{-1}. \tag{351}$$

Pennycuick had used a method of successive averaging to obtain

$$\bar{k} = 3.686 \times 10^{-3} \quad \text{min}^{-1}. \tag{352}$$

The 99% confidence interval for the result in Eq. (351) was found to be

$$[3.732 \pm 0.048] \times 10^{-3} \quad \text{min}^{-1}. \tag{353}$$

As an exercise in the application of Eq. (349), making the opposite assumption that the random fluctuations are entirely inherent in origin, the result obtained was

$$\hat{\mu} = 3.729 \times 10^{-3} \quad \text{min}^{-1}. \tag{354}$$

In the case of the unimolecular gaseous decomposition of di-*t*-butyl peroxide shown by Raley *et al.* (1948) to satisfy first-order kinetics,

$$(CH_3)_3COOC(CH_3)_5 \longrightarrow 2(CH_3)_2CO + 2C_2H_6, \tag{355}$$

the data at two different temperatures had been given, on the basis of which the following averaged rate constants had been reported:

at 154.6°C:

$$\bar{k} = 3.22 \times 10^{-4} \quad \text{sec}^{-1} = 1.932 \times 10^{-2} \quad \text{min}^{-1}, \tag{356}$$

at 147.2°C:

$$\bar{k} = 1.43 \times 10^{-4} \quad \text{sec}^{-1} = 8.580 \times 10^{-3} \quad \text{min}^{-1}. \tag{357}$$

Using larger intervals than they had employed and Eq. (350), Bartholomay [1959, p. 371] reported

at 154.6°C:

$$\hat{k} = 3.26 \times 10^{-4} \quad \text{sec}^{-1} = 1.956 \times 10^{-2} \quad \text{min}^{-1}, \tag{358}$$

at 147.2°C:

$$\hat{k} = 1.43 \times 10^{-4} \quad \text{sec}^{-1} = 8.580 \times 10^{-3} \quad \text{min}^{-1}. \tag{359}$$

Using Eq. (349),

at 154.6°C:

$$\hat{\mu} = 3.28 \times 10^{-4} \quad \text{sec}^{-1} = 1.968 \times 10^{-2} \quad \text{min}^{-1}, \tag{360}$$

at 147.2°C:

$$\hat{\mu} = 1.42 \times 10^{-4} \quad \text{sec}^{-1} = 8.580 \times 10^{-3} \quad \text{min}^{-1}. \tag{361}$$

It was suggested that as a by-product of the stochastic theory, because of the ease of calculation of $\hat{\mu}$ as opposed to $\hat{k}$, the possibility of applying these formulas should be kept in mind. However, they were applied in the spirit of testing preliminarily the consistency of the stochastic model with macroscopic classical deterministic kinetics. In this sense, this was the first attempt at experimental testing of the stochastic model, to be followed by the *in numero* studies to be discussed next.

3. *Preliminary Considerations for the in Numero Studies*

Because of the impossibility of error-free real reaction runs, it is not possible to obtain direct verification of the experimental implications of the stochastic model. However, attempts are currently being made (see Section III.F.4.c) to simulate on the computer random experimental error. If this can be accomplished then, by addition of this random component, the simulations (see below) can be optimized. With this restriction in mind *in numero* studies based on computer simulations of the sucrose and peroxide reactions were carried out with encouraging results.

a. Sample Size and Scaling Considerations. An immediate problem which has to be faced generally in computer simulation studies of chemical kinetics is the sample size, that is, initial reactant concentration. In the case of the unimolecular reaction one has to decide on a reasonable value for n_0, which technically is the integral number of molecules per constant volume. Considering that the starting number of reactant molecules in an ordinary chemical reaction is a multiple of 10^{23} molecules generally, obviously such a value is out of the question in such studies. It happens, however, as the dimensions (sec^{-1} or min^{-1}) of the first-order rate constant in the deterministic model show, that only relative numbers for n are required in kinetic rate determination studies, that is, only the ratio n/n_0 is considered. In fact, in the preceding section the various calculations, including the original sources by Pennycuick and by Raley *et al.* were all given in terms of percentage initial concentration. In other words, for comparisons with deterministically treated real data, direct values of n may be avoided, since

$$dn/dt = -kn \tag{362}$$

can be replaced by

$$d(n/n_0)/dt = -k(n/n_0), \tag{363}$$

so that if $x = (n/n_0 \times 100)\%$, the basic formula is still of the form $dx/dt = -kx$. It happens conveniently, that the stochastic formulas for estimating the corresponding stochastic parameter can be normalized similarly, for division of each m_{rs} value in Eqs. (348) and (349) by n_0 still yields the same results, the terms $\log n_0$ and $-\log n_0$ canceling each other.

However, in applying the set of transition probability functions $\{p_{ik}(\tau)\}$ of the defining stochastic model as the basis for simulations, absolute numbers cannot be avoided, so that the final results only can be normalized to n_0. Consequently, a value for n_0 must be assigned *a priori*. It is not too important whether n_0 be chosen as very large, say, 10^4 as opposed to 10^3 or 10^2, for compared to realistic initial concentrations all of these numbers are indeed minuscule. Of course, the intrinsic rate, or stochastic parameter would still be computable on this basis and moreover, the plotting of results on a relative percentage basis would still be comparable with the results of real experiments. To increase the meaningfulness of the comparison one could compare the simulated results with the lower parts of the real concentration–time courses. Or, in general, one can take the statistical point of view that the value of n as a concentration random variable per unit volume is taken with respect to an extremely small unit volume, in effect, discussing a simulated sample run in the context of sampling theory. This would be enhanced by guaranteeing that corresponding real experiments to be used in such comparisons are of uniform density with respect to the reactant-containing mixture. Of course, if one is applying the stochastic model to the Lindemann–Hinshelwood mechanism, supposing it to be the final dissociation of energized molecules; or, more generally, in simulations of spontaneous first-order disruptions (in the most extreme case), the spatial distribution aspects of the problem disappear and the simulated result obtained on a small initial sample basis is realistically interpretable. There are then a number of interpretations in which the restrictions in magnitudes imposed by simulation studies can still produce meaningful comparisons in the unimolecular process. It should also be noted that if one wished to simulate extremely dilute reactions such as may take place intracellularly, then the extra time spent in a simulation beginning with a very high n_0 value would be justified as a means of estimating the magnitudes of statistical fluctuations *in vivo*.

In cases for example, where the real data to be simulated go from a relative concentration of 100% initially to, say, 10% over a total observation interval

[0, T] one can estimate from the deterministic theory an appropriate comparative value of n_0 for the simulation study. Thus, from Eq. (363), it follows that

$$\int_1^{n'/n_0} d(n/n_0)/(n/n_0) = -k \int_0^T dt, \tag{364}$$

so that the time T required to reach a relative percent concentration of $(n'/n_0 \times 100)$ is

$$T = k^{-1}[-\log_e(n'/n_0)]. \tag{365}$$

So the time required for 90% of the total decomposition of $n_0 = 10^3$ molecules is the same as that required for 90% decomposition of $n_0 = 10^2$ (n' being 10^2 in the first case and 10 in the second). On the other hand, it would take twice as long for the reduction of 10^3 molecules to 10, as it would for the corresponding reduction of 10^2 molecules to 10.

In the case of the sucrose reaction studied, Pennycuick [1926], followed and reported the reaction down to roughly 10% initial concentration. If in the simulation study one begins with 10^3 molecules, using Eq. (365), with $k = 3.73 \times 10^{-3}$ min^{-1}, one obtains $T = 617$ min, which is close to 625 min, the actual time of Pennycuick's runs. Another estimate of the time required for a reaction run might be obtained, for example, by using Eq. (345) of the stochastic theory which leads to

$$T_{0.05} \approx \mu^{-1} \approx 270 \quad \text{min} \tag{366}$$

as the 95% probability measure of completion time. (We shall see presently that this quantity μ^{-1} arises in a much more significant theoretical manner.) Finally, from Eq. (365) one obtains the reaction half-time $T_{1/2}$ corresponding to $(n/n_0) = 0.5$,

$$T_{1/2} = 0.69/k \approx 185 \quad \text{min}. \tag{367}$$

But knowing that it takes, for example, 617 min for 90 of 100 molecules to decompose according to the deterministic kinetics, or that according to the stochastic theory the reaction half-time is 185 min does not tell us how long it requires for a single molecule, say, on the average, to dissociate. For example, the time for each molecule could be close to 185 min. The stochastic theory, in fact, allows great variability in these times, as we shall see in the next paragraph. Still, some safe estimates of the minimum time per individual decomposition is required in simulations based on equispaced sampling intervals. A method of arriving at this brings the macroscopic model into an interesting coincidence with the Slater and Bunker random lifetimes theory, as will be demonstrated in Section b following.

For the studies reported here, it was decided to select $\tau = 1$, or $\tau = 2$ min in most cases as the minimum time for individual decomposition. This is far below the mean value time computed on the basis of the consideration of Section b following. It was decided originally on the empirical basis that if we were to apportion, equally, say 270 min to 180 molecules (the numbers attempted in pilot experiments) assuming the extreme case of successive nonoverlapping decomposition events, then this could be a safe estimate. It turned out that in all of the studies made, which varied between $n_0 = 40$ and $n_0 = 10^3$, values less than $\tau = 2$ revealed no great differences. The question of characterizing τ, the actual time for an individual decomposition, as a random variable can be approached as follows.

b. The Random Distribution of Times Required for Individual Unimolecular Decompositions. Equation (331) gives $p_{10}(t) = 1 - e^{-\mu t}$ the probability of decomposition of a single molecule in time τ. Let this expression be used to introduce the new function $P(t)$,

$$P(t) \equiv 1 - e^{-\mu t}, \tag{368}$$

thought of as the cumulative probability function for the time interval τ, regarded now as a random variable giving the time for an individual decomposition, that is,

$$P(t) = \Pr\{0 < \tau \leqslant t\} = 1 - e^{-\mu t}. \tag{369}$$

Then, differentiating with respect to t,

$$dP/dt \equiv p(t) = \mu e^{-\mu t} \tag{370}$$

determines $p(t)$ as the probability density function. We may also write

$$p(t)\,dt = \mu e^{-\mu t}\,dt = \Pr\{t < (t) < t + dt\}. \tag{371}$$

That this is a bonafide probability density function may be seen from the fact that

$$\int_0^\infty p(t)\,dt = \mu \int_0^\infty e^{-\mu t}\,dt = 1. \tag{372}$$

Moreover, it is seen that this distribution has the following properites:

$$\bar{t} = \mu^{-1}, \qquad \text{mean value,} \tag{373}$$

$$\sigma = \mu^{-1}, \qquad \text{standard deviation.} \tag{374}$$

This probability density is of the Poisson type and skewed to the right of $\tau = 0$ with roughly 60% of the values of τ lying between 0 and $1/\mu$, the mean value, (with 100% coefficient of variation). Noting also that $\Pr\{0 < \tau < 2 \text{ min}\} < 0.001$ for $\mu = 3.73 \times 10^{-3}\ \text{min}^{-1}$ the choice of $\tau = 2$ referred to in the preceding parapragh would appear to be small enough to guard against missing individual decompositions in an equispaced time sampling process.

Of particular interest here is the fact that if we identify, as indicated earlier, the parameter μ with the rate constant $k_a = k_3$ of the dissociation process of the Lindemann–Hinshelwood mechanism, then the distribution given by

Eq. (370) is identical with the "random gap" of Slater's or the "random life-times distribution" of Bunker [see Eq. (229)]. The correspondence would be complete by reinterpreting τ, which in this general case represents the time between individual decompositions, as the time required for the dissociation of an energized molecule, that is, as τ as defined in Bunker's distribution—thus identifying the macroscopic stochastic model with the microscopic models. In this sense, we might also consider the macroscopic stochastic model as another indication of what Slater has called the basic "random-gap assumption" present in all unimolecular theories. It is also because of this association that Bunker's experiments may be interpreted as feasibility experiments for the macroscopic model.

4. *In Numero Computer Studies*

There are two obvious alternatives in experimental studies aimed at simulating stochastic models such as the unimolecular decomposition model. One can either fix a sampling interval and incorporate the state-to-state changing transition probabilities into a random number-theoretic Monte Carlo procedure for selecting successive states of a sample curve or "simulated reaction run," or one can consider τ as a random variable (as in the preceding paragraph) and devise a Monte Carlo procedure consisting of algorithms for selecting specific values of τ and then testing to see whether this value exceeds a critical value necessary for the accomplishment of a decomposition. The details of the Monte Carlo procedure and a few samples of the results of using the first method will be presented in Section III.F.4.b, below. While the second method was never implemented by the author, independently Jacobs [1963; see also McQuarrie, 1963] employed this viewpoint in his computer studies of this process. Thus, the method discussed next is similar to that implemented by Jacobs. It may be regarded as an extension of the point of view embodied in deriving Eq. (370) of the immediately preceding section.

a. A Monte Carlo Procedure Using Random Time Intervals. For the class of Markovian processes, Kendall [1950] was apparently the first to attempt computer simulations, specifically in connection with his generalized birth and death processes. His method was extended by Shea and Bartholomay [1965] in *in numero* studies based on a stochastic model for hepatocyte proliferation in toxic cirrhosis. This same method is adapted here to the unimolecular process.

(1) Indexing and Redesignation of States. For purposes of the *in numero* sampling and simulation studies the conventions employed in the general theory of Markov chains and in the previous sections here, namely, of identifying the number of molecules j with state j of the system is no longer suitable.

Instead, the jth state will refer to the jth computation step, that is, to the jth sampling interval. Accordingly, the number of molecules present at the jth state will now be designated as x_j. Thus, if it is found that at state j, x_j has the value n, at the next state, that is, the $(j + 1)$th state x_{j+1} can have the value $(n - 1)$, or indeed, $(n - k)$. (And in the case of the prior model, it could have the value $[n + k]$ where $k \geqslant 1$). Thus, x_j is now defined as the number of molecules present at the jth sampling state, where $0 \leqslant x_j \leqslant n_0$ and $j = 0, 1, 2, \ldots, n_0, \ldots$, (that is, j can exceed n_0 now). This new convention will be followed in both this section and Section b.

(2) The Derivation of the Probability Distribution of Times Required for at Least One of the x_j Molecules Present in Computation State j to Decompose. To determine the probability density function for the decomposition of at least one of x_j molecules, it is convenient to begin by obtaining the probability that in a time interval of length t no decomposition of the x_j present at the beginning of the interval can occur. By substituting $x_j = i = k$ into Eq. (326), one obtains

$$P_{x_j, x_j}(t) = \exp(-\mu x_j t), \qquad j = 0, 1, 2, \ldots. \tag{375}$$

This same result may be arrived at as in Kendall [1950] or Shea and Bartholomay [1965], by defining the intensity q numbers so as to forbid the transitions $x_j \longrightarrow x_{j\pm k}$ $(k \geqslant 2)$. In this way it can be guaranteed for the sampling process considered here that the complement, $1 - p_{x_j, x_j}(t)$, would be the probability of exactly one decomposition in time t and no more than one. Thus, as in the case of the derivation of Eq. (368), one may interpret

$$P_{x_j}(t) = 1 - \exp(-\mu x_j t) \tag{376}$$

as the cumulative distribution function $\Pr\{0 < \tau \leqslant t\}$ that the value of the new random variable τ corresponding to exactly one decomposition is $\leqslant t$. And again, from this, by differentiation,

$$dP_{x_j}(t)/dt = p_{x_j}(t) = \mu x_j \exp(-\mu x_j t), \tag{377}$$

that is,

$$p_{x_j}(t)\, dt = \mu x_j \exp(-\mu x_j t)\, dt \tag{378}$$

is the probability that the time τ required for the next decomposition lies between t and $(t + dt)$; that is, $p_{x_j}(t)$ is the probability density function of the random variable τ.

For the Monte Carlo procedure it is convenient to introduce first of all a new random variable U_j related to τ by the equation

$$U_j = x_j \tau, \qquad j = 1, 2, \ldots, \tag{379}$$

where now x_j is considered simply as a constant of proportionality defining the direct variation of U_j with τ at the jth computational step. Then, where the value u_j of U_j corresponds to the value t of random variable τ, so that $u_j = x_j t$, one obtains

$$du_j = x_j \, dt. \tag{380}$$

And so, with this substitution the corresponding probability density function $f(u_j)$ is determined by

$$f(u_j)du_j = p_{x_j}(u_j/x_j) \, |dt/du_j| \, du_j. \tag{381}$$

In other words,

$$f(u_j) \, du_j = \mu \exp(-\mu u_j) \, du_j, \qquad j = 0, 1, 2, \ldots, \tag{382}$$

which corresponds precisely to the distribution $p(t)$ in Eq. (370). The random variable U_j which has been introduced thereby (and the earlier τ of Section III.F.3.b) is a type of χ^2 distribution with two degrees of freedom. Specifically, U_j is distributed as $(2\mu)^{-1}\chi^2$.

Given a (pseudo) random number generating procedure, implementable on a computer, which is programmed to generate random numbers fitting the rectangular or regular distribution $R[0, 1]$ on the unit interval, let it be assumed that a sequence of such numbers $\{r_1, r_2, \ldots\}$ is available. These rs may in turn be transformed into random samples of the random variable U_j by taking advantage of the transformation of the distribution $R[0, 1]$ into the χ^2 distribution just discussed. Thus, consider a new random variable R_j whose values r_j relate to the values of u_j of U_j by the transformation

$$r_j = \phi_j(u_j) = \exp(-\mu u_j), \qquad j = 0, 1, 2, \ldots. \tag{383}$$

In this case, then, the probability density function $g(r)$ corresponding to $f(u)$ follows from the relation

$$g_j(r_j) \, dr_j = f_j(u_j = \phi_j^{-1}(r_j)) \, |du_j/dr_j| \, dr_j, \tag{384}$$

so that where

$$u_j = -\mu^{-1} \log_e r_j = \phi_j^{-1}(r_j), \qquad j = 1, 2, \ldots \tag{385}$$

and

$$(du_j/dr_j) = -(\mu r_j)^{-1} \, dr_j, \tag{386}$$

$$g_j(r_j) \, dr_j = 1 \cdot dr_j. \tag{387}$$

In other words, R_j has the rectangular $R[0, 1]$ distribution so that via the relations between r_j and u_j in Eq. (383) a randomly generated number r_j becomes transformed into a choice u_j from the random distribution of U_j. Note that as the R_j interval $[0, 1]$ corresponds to the U_j interval $[0, \infty]$, according to the transformation of Eq. (383), a long sequence of generated rectangularly distributed numbers $\{r_1, r_2, \ldots, r_j, \ldots, r_N\}$ becomes transformed into a pertinent random sequence of Us, $\{u_1, u_2, \ldots, u_j, \ldots, u_N\}$ where $u_j = -(\log r_j)/\mu$.

Suppose now that the sampling process has progressed to the point where x_j molecules of species A are present at time t_j. Then, if t_{j+1} is the time to the next decomposition, so that $(t_{j+1} - t_j)$ is the value of the random variable τ (corresponding to the time of the next upzero in the Slater formulation, for example), since, by definition $u_j = x_j\tau$, $(j = 1, 2, \ldots)$, the value of u_j corresponding to the value $(t_{j+1} - t_j)$ of τ is

$$u_j = x_j(t_{j+1} - t_j), \tag{388}$$

so that

$$t_{j+1} = t_j + u_j/x_j \tag{389}$$

and

$$x_{j+1} = x_j - 1. \tag{390}$$

In this way a sequence of cooridnate pairs would be determined which determines the sample curve of the process. Sampling would terminate as soon as $x_{j+1} = 0$ or $t_{j+1} > T$ where $[0, T]$ is a predetermined interval and the resulting sample curve should be a stepwise constant function defined by

$$x(t) = x_j, \qquad t_j \leqslant t < t_{j+1}. \tag{391}$$

Note that this process corresponds to the full Markov model inasmuch as each solution depends on the immediately prior state.

b. A Monte Carlo Procedure Using Equispaced Time Intervals for Sampling. As in the previous case let x_j be the number of molecules present at the jth sampling state, occurring in this case at predetermined time $t_j = j\tau$ where τ is the length of a prechosen constant sampling interval (such as $\tau = 2$, as indicated in Section III.F.3.c). Then, corresponding to the next sampling time $t_{j+1} = (j + 1)\tau$ let X_{j+1}, the number of molecules present, be treated as a new random variable whose (discrete) probability distribution $p_{j+1}(x_{j+1})$ is defined in terms of the transition probabilities $p_{ik}(t)$ of Eq. (326) or (329) by

$$p_{j+1}(x_{j+1}) \equiv p_{x_j, x_{j+1}}(\tau), \qquad 0 \leqslant x_{j+1} \leqslant x_j \leqslant n_0, \tag{392}$$

so that the cumulative distribution function of X_{j+1}, call it $P_{j+1}(x)$, is

$$P_{j+1}(x) = \sum_{x_{j+1}=0}^{x} p_{j+1}(x_{j+1}). \tag{393}$$

Now, given a random number r_{j+1} from the rectangular distribution $R[0, 1]$ (see preceding section), the value x_{j+1} of X_{j+1} is determined by solving the equation for x

$$P_{j+1}(x) = \sum_{x_{j+1}=0}^{x} p_{j+1}(x_{j+1}) = r_{j+1} \tag{394}$$

(note that $0 \leqslant r_{j+1} \leqslant 1$). In other words, a value x_{j+1} (of x) is sought, such that

$$P_{j+1}(x_{j+1}) = r_{j+1}. \tag{395}$$

The solution of this $(j+1)$th equation is recorded as x_{j+1}, the number of molecules present at time t_{j+1}. This procedure is repeated over and over again until the solution x attains the value 0; that is, $x_N = 0$ at $t = N\tau$.

This procedure, which was carried out originally on an IBM 1620 digital computer, begins by setting n_0 equal to an arbitrary value and μ equal to the rate constant of the particular unimolecular reaction to be simulated. Then for $x_0 = n_0$ at $t = 0$, and a fixed sampling interval of length τ, the value x_1 at time $t = \tau$ is determined by calling first for a random number r_1, followed by calling for a solution to the equation

$$P_1(x) = \sum_{x_1=0}^{x} p_1(x_1) = r_1, \qquad j = 0, \tag{396}$$

where $p_1(x_1) = p_{x_0,x_1}(\tau)$. The solution x becomes the value x_1 associated with $t = t_1$. Then setting $j = 1$ and using $p_2(x_2) = p_{x_1,x_2}(\tau)$ together with another random number, the procedure is repeated, and so on. This describes the overall procedure. In practice the following additional considerations have been incorporated.

1. The random number generation was effected in the IBM 1620 computer using a pseudo-random number generating subroutine written by D. L. Fink of IBM and based on the customary "power residue method."

2. Closed analytical solutions of the equations corresponding to Eq. (394) are not obtainable, so that a number of alternative methods of successive approximations were devised and incorporated into the final computer program. All of these alternatives are based on the trial of all possible (monotonically increasing from zero) integral values of x, until a value, call it x'_{j+1} is found such that $P_{j+1}(x'_{j+1}) \geqslant r_{j+1}$ for the first time. Then x_{j+1} is assigned the value x'_{j+1}. The alternatives refer to the following forms of the transition probabilities $p_{j+1}(x_{j+1}) = p_{x_j,x_{j+1}}(\tau)$.

FACTORIAL METHOD I (THE INDIRECT METHOD): Consider the new random variable Y_{j+1} defined by

$$Y_{j+1} \equiv X_j - X_{j+1}, \qquad j = 1, 2, \ldots, N-1, \tag{397}$$

so that Y_{j+1} corresponds to the number of molecules decomposed in the time interval $(t_j, t_{j+1} = t_j + \tau)$. Since, in this case,

$$x_{j+1} = x_j - y_{j+1}, \qquad j = 1, 2, \ldots, N-1, \tag{398}$$

the probability function $p_{j+1}(y_{j+1})$ of the random variable Y_{j+1} is obtained by substitution into the expression for $p_{x_j,x_{j+1}}(\tau)$,

$$p_{j+1}(y_{j+1}) = \binom{x_j}{x_j - y_{j+1}} \exp(-\mu x_j \tau)(-1 + e^{\mu\tau})^{y_{j+1}}, \qquad j = 1, 2, \ldots, N-1. \tag{399}$$

Again, the value y_{j+1} corresponding to time t_{j+1} is obtained as the solution

$y = y_{j+1}$ of the equation

$$P_{j+1}(y) = \sum_{Y_{j+1}=0}^{y} p_{j+1}(y_{j+1}) = r_{j+1}, \tag{400}$$

and Eq. (398) transforms this into the value x_{j+1} corresponding to $t = t_{j+1}$.

FACTORIAL METHOD II (THE DIRECT OR BINOMIAL METHOD): An alternative to the preceding is the use of the regular binomial form from Eq. (329), that is,

$$p_{j+1}(x_{j+1}) = b_{j+1}(x_{j+1}, x_j; e^{-\mu\tau})$$

or (401)

$$p_{j+1}(x_{j+1}) = \binom{x_j}{x_{j+1}}(e^{-\mu\tau})^{x_{j+1}}(1 - e^{-\mu\tau})^{x_j - x_{j+1}}, \qquad j = 1, 2, \ldots, N-1$$

so that solutions of the equation

$$B_{j+1}(x) = \sum_{x_{j+1}}^{x} b_{j+1}(x_{j+1}, x_j; e^{-\mu\tau}) = r_{j+1}, \qquad j = 1, 2, \ldots, N-1 \tag{402}$$

are required.

METHOD III (THE GAUSSIAN APPROXIMATION METHOD): Because of the excessively large numbers generated by the factorial quantities involved in the binomial coefficients of Methods I and II, for values encountered in setting $n_0 > 40$, say, the Gaussian approximation of the binomial transition probabilities utilized in ealier Eq. (336) of the prior model is introduced here in relation to Eq. (401). In other words, b_{j+1} is approximated by the probability density function $g_{j+1}(x_{j+1})$ with x_{j+1} treated now as the value of a continuous random variable X_{j+1},

$$g_{j+1}(x_{j+1}; \tau) = N_{j+1}[E_j(\tau), \sigma_j(\tau)] \equiv (\sqrt{2\pi}\,\sigma_j)^{-1} \exp\{(-2\sigma_j^2)^{-1}(x_{j+1} - E_j)^2\},$$
$$j = 1, 2, \ldots, N-1, \tag{403}$$

where

$$E_j(\tau) = x_j e^{-\mu\tau},$$
$$\sigma_j^2(\tau) = x_j e^{-\mu\tau}(1 - e^{-\mu\tau}), \qquad j = 1, 2, \ldots, N-1. \tag{404}$$

In this case, x_{j+1} is determined as the solution of the equation

$$G_{j+1}(x) = \int_{-\infty}^{x} g_{j+1}(x_{j+1}; \tau)\, dx_{j+1} = r_{j+1}. \tag{405}$$

In some cases Method III has been combined with Method I or II, by using the Gaussian method for values of x_j down to 40 and then generating the remainder of the sample curve by one of the other methods.

STUDIES USING THE PRIOR MODEL. The preceding methods were also applied in the very earliest studies [see Bartholomay, 1964a] to the Prior model. In that case, instead of using the full set of transition probabilities $p_{x_j, x_{j+1}}(\tau)$, for τ fixed throughout and varying x_j from instant to instant, the prior transition probabilities in the form $p_{n_0, x_j}(j\tau)$ were used, varying the time interval by units of τ in determining the appropriate probabilities; that is,

$$p_j(x_j) \equiv p_{n_0, x_j}(j\tau) = \binom{n_0}{x_j}(\exp(-\mu jt))^{x_j}(1 - \exp(-\mu j\tau))^{n_0 - x_j}. \qquad (406)$$

The process begins by solving the equation

$$P_1(x) = \sum_{x_1=0}^{x} p_1(x_1) = r_1, \qquad (407)$$

where

$$p_1(x_1) = p_{n_0, x_1}(\tau),$$

the solution of the equation being assigned as the value x_1 at time $t = \tau$. Next, the equation

$$P_2(x) = \sum_{x_2=0}^{x} p_2(x_2) = r_2, \qquad (408)$$

is solved, where

$$p_2(x_2) = p_{n_0, x_2}(2\tau),$$

repeating the process until finally $x_N = 0$.

c. Some Results. Some samples of the results of these studies obtained in 1963 and 1965 are discussed next.

1. Using the Prior model, the first simulation attempted specified the fixed values: $n_0 = 10^3$, $\tau = 2$ min and $\mu = 3.732 \times 10^{-3}$ min^{-1} to test the predictions of that model against Pennycuick's original data. Method III, was used and the absolute results (x_j, t_j) plotted as shown in Fig. 11, 750 sampling instants having been used. The curve was drawn as a polygonal curve, joining the successive data points. The negative exponential trend is apparent and fluctuations of relatively small magnitudes, the whole appearance suggesting strongly the real reaction run. Using the stochastic estimate $\hat{\mu}$ in Eq. (349), the result obtained was $\hat{\mu} = 3.726 \times 10^{-3}$. Note also, that roughly 90 % conversion has been obtained in 650 minutes, which also closely parallels Pennycuick's real reaction run.
2. Figure 12 shows the closeness of Pennycuick's data, the stochastic sample points (taken from Fig. 11) and the theoretical deterministic curve $n = n_0 e^{-\mu t}$ compared on the basis of percent initial concentration versus time, at 25-min intervals.
3. Figure 13 demonstrates three different sample curves obtained using the three different methods and a starting value of $n_0 = 40$, only intervals of 50 min having been plotted.

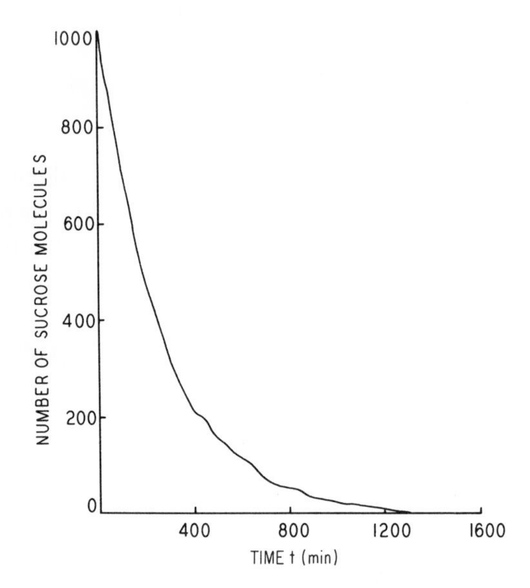

Fig. 11. A sample curve generated by Monte Carlo method III applied to the Prior model, as a simulation of (Pennycuick's) sucrose inversion reaction. Rate constant had the value 3.732×10^{-3} min^{-1}. The stochastic estimate made from the sample curve data was $\hat{\mu} = 3.726 \times 10^{-3}$ min^{-1}. Initial value $n_0 = 10^3$ molecules.

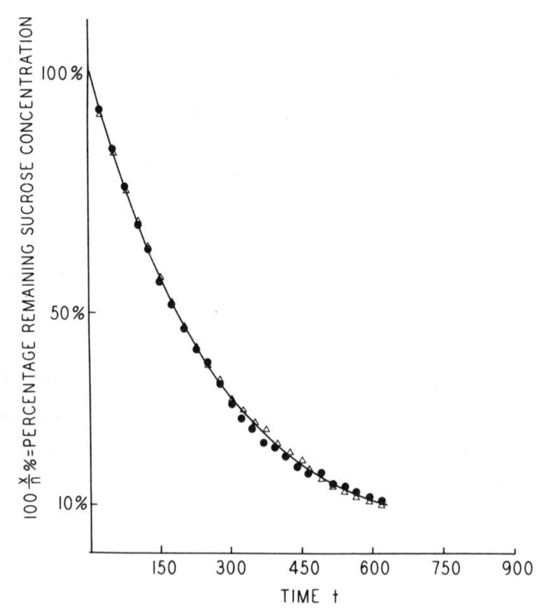

Fig. 12. Comparison of Pennycuick's data (△), the Monte Carlo simulation of Fig. 11 (●), and the deterministic prediction (——).

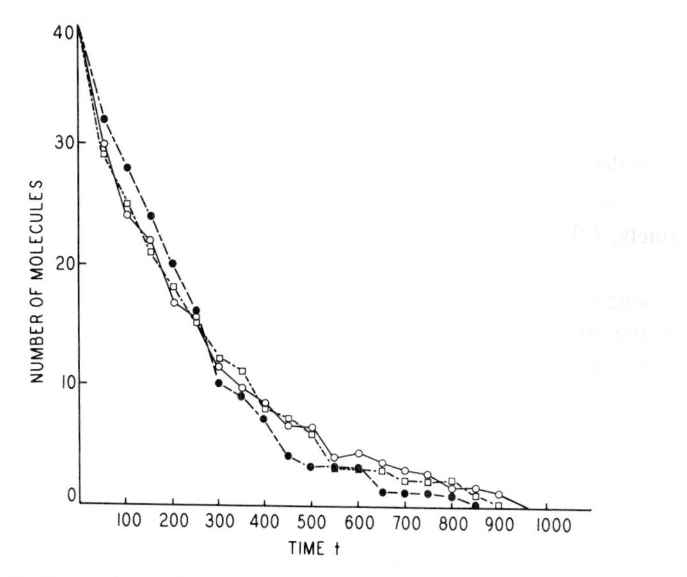

Fig. 13. Comparison of the three Monte Carlo methods applied to the simulation of sucrose inversion, using the Prior model, with μ set at 3.732×10^{-3} and $n_0 = 40$ molecules: (●- — -) factorial method I; (□- - -) factorial method II; (○——) Gaussian approximation method III.

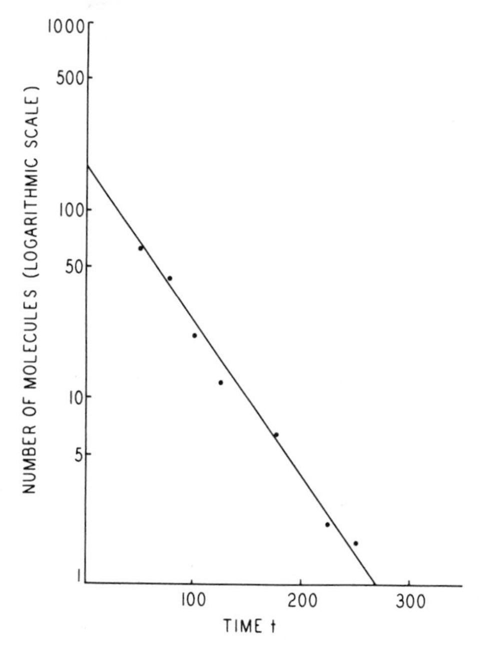

Fig. 14. Prior model simulation of decomposition of di-*t*-butyl peroxide, with reference to reaction run of Raley *et al.* [1948] for absolute $T = 154.6°$ and unimolecular rate of 1.932×10^{-2} min^{-1}. For the simulation n_0 was set at 170 molecules. A semilogarithmic plot for which the stochastic parameter estimate was $\hat{\mu} = 1.939 \times 10^{-2}$.

4. Figure 14 shows a semilogarithmic absolute plot of the results of using the Prior model corresponding to $\mu = 1.932 \times 10^{-2}\ \text{min}^{-1}$, $T = 154.6°$, and $n_0 = 170$ molecules to simulate the reaction run of Raley *et al.* [1948] for the decomposition of di-*t*-butyl peroxide. The estimate $\hat{\mu}$ of μ obtained from the data, namely, $1.939 \times 10^{-2}\ \text{min}^{-1}$ is almost identical with the starting value of μ.

The closeness of the estimated $\hat{\mu}$'s to the starting values μ, together with the closeness the simulated runs to the corresponding theorefical deterministic negative exponential trends, and the similarities (between the simulated and real reaction runs) in the appearances of fluctuations about the trends attest to the feasibility of such stochastic models for the unimolecular reaction.

5. To obtain some idea of the extent of random fluctuations contained in the sample curves of the Prior model, it is necessary to generate a whole ensemble of such samples. Table I gives the results of statistically estimating the variances contained in an ensemble of 25 runs, using only the values at intervals of 50 min, starting with $n_0 = 40$ and using Monte Carlo method I. Comparisons of the statistical estimates s^2 with the theoretical predictions σ^2, were obtained using Eq. (336) of the Prior model. In all cases, it is seen that the theoretical variances lie within the 95% confidence intervals of the estimates.

The following figures and tables are examples of some of the detailed studies made of the sucrose inversion reaction using the full Markov model.

6. Figure 15 shows the result of statistical estimations of the mean value and least squares curves and standard deviations based on an ensemble of 25 sample curves of the full Markov model using Monte Carlo Method I, with $n_0 = 40$ molecules. The mean value curve is the polygonal curve joining the points corresponding to the means of all 25 x_j values corresponding to $t = t_j = j \cdot 50$ ($j = 1, 2, \ldots, 12$) only the data at 50 min. intervals having been utilized. The points above and below the points on the mean-value curve are located at $(t_j, \bar{x}_j \pm s_j)$, where s_j is the standard deviation at t_j. The three different estimates of μ reported are all in good agreement with Pennycuick's rate constant.

7. The same data and statistical analyses of the 25 simulations represented in Fig. 15 were utilized to obtain Fig. 16. The relation of the standard deviation and coefficient of variation curves to the mean value curve can be seen to fit the qualitative expectations of the process.

8. The 25 simulations each of the Prior (see Table I) and the full Markov models, corresponding to $n_0 = 40$ were utilized to compare the extents of variation generated by the Prior model with those generated by the full Markov model as shown in Fig. 17. The excess of Prior standard deviations over the Markov standard deviation may be explainable on the basis that

TABLE I

Prior Model: 95% CONFIDENCE INTERVALS OF ESTIMATED VARIANCE s^2 AT 50-MIN INTERVALS[a]
(based on 25 sample curves, $n_0 = 40$)

τ	Estimated s^2	$\frac{24s^2}{39.4151}$	$\frac{24s^2}{12.4001}$	Theoretical σ^2
50	3.8228	2.3272	7.3973	5.6496
100	5.3324	3.2467	10.3199	8.5784
150	6.1385	3.7374	11.8799	9.7963
200	6.1385	3.7374	11.8799	9.9729
250	6.1663	3.7545	11.9341	9.5450
300	5.3749	3.2722	10.4012	8.7948
350	5.0266	3.0603	9.7277	7.8995
400	4.3932	2.6749	8.5025	6.9691
450	3.2761	1.9948	6.3406	6.0683
500	3.0723	1.8705	5.9457	5.2317
550	2.8224	1.7183	5.4619	4.4766
600	2.4161	1.4711	4.6761	3.8076

[a]The probability is 0.95 that the actual value of σ^2 will be in the interval. $24s^2/39.1451 < \sigma^2 < 24s^2/12.4001$.

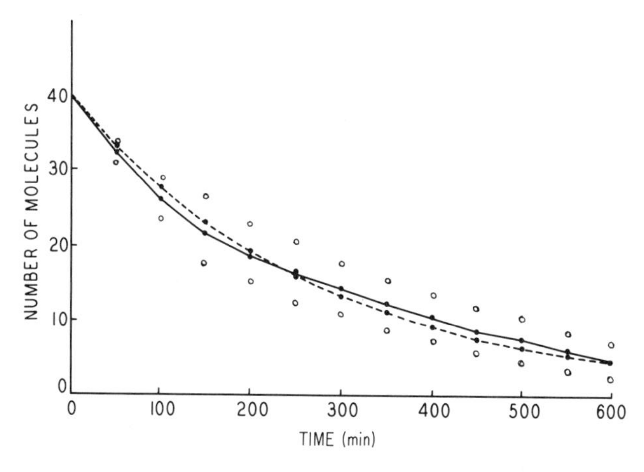

Fig. 15. Mean-value and least-squares curves and standard deviations estimated from 25 simulations using the Markov model with $n_0 = 40$ molecules and $\mu = 3.732 \times 10^{-3}$, corresponding to Pennycuick's rate constant. Stochastic parameter estimates: 3.543×10^{-3} from mean-value curve; 3.644×10^{-3} from least-squares curve; 3.524×10^{-3} from Eq. (348). Curves: (——) mean-value curve; (– –) least-squares curve. Open circles represent one standard deviation above and below the mean value.

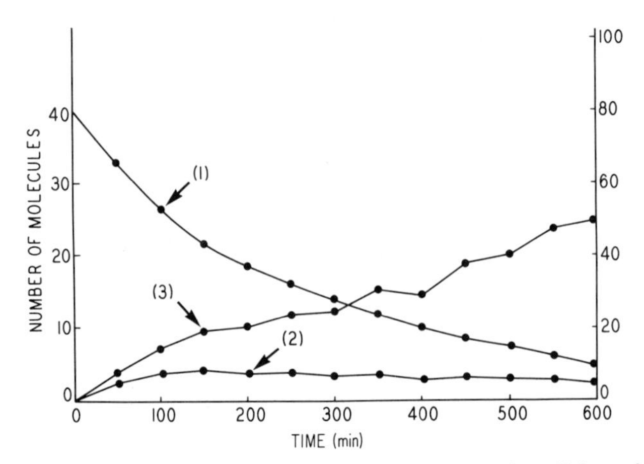

Fig. 16. Mean-value curve (1), standard-deviation curve (2), and coefficient-of-variation curve (3), plotted from data represented in Fig. 15. Axis on right-hand side for curve (3).

the two variations have different base lines and are really not measuring the same thing. In other words, in the case of the Markov simulation, having attained a value x_j at t_j, the probability measure imposed on the selection of an x_{j+1} at $t_{j+1} = t_j + \tau$, is obtained from the transition probability giving the probability of going from x_j to some value of the next random variable X_{j+1}. But in the Prior model simulation, having attained the

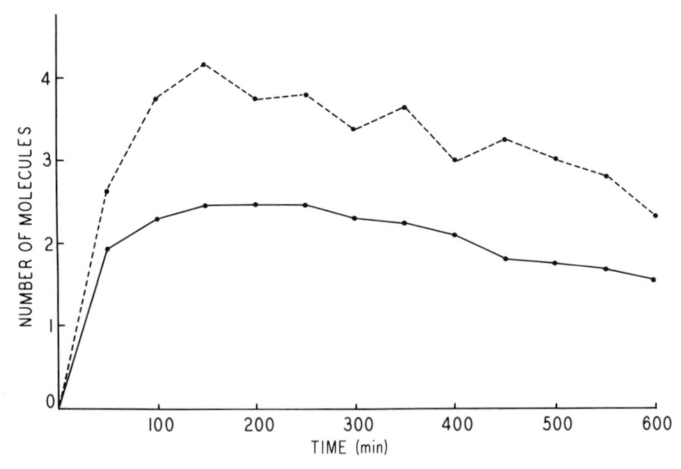

Fig. 17. Comparison of standard-deviation curves obtained from 25 simulations each of the Prior and Markov models corresponding to sucrose inversion reaction with rate of 3.732×10^{-3} and $n_0 = 40$ molecules. Estimate of μ made from the mean-value curve of the Prior model was 3.921×10^{-3}; that from the Markov model, 3.543×10^{-3}. Curves: (——) Prior standard deviation curve, (– – –) Markov standard deviation curve.

value x_j, the chice of x_{j+1} is made by using as probability measure the function $p_{n_0, x_{j+1}}(t_j + \tau)$ always starting from $x_0 = n_0$, independently of the present state x_j, (except as to the time t_j required for its attainment). This means that each new selection x_{j+1} can be in the full range $0 \leqslant x_{j+1} \leqslant n_0$, as opposed to the range $0 \leqslant x_{j+1} \leqslant x_j$ in the Markov case. At first sight simply from the differences in ranges of variation one would expect a larger standard deviation at each time in the Prior case. This is a matter which has received some attention, as the results accompanying Fig. 18 and Table II show. These latter analyses pertain to the Markov model, similar analyses of the Prior model not having been made.

Finally, the rationale for the comparisons made in Fig. 17 should be mentioned. With the reservations of the preceding paragraph in mind, it was felt, nevertheless, that the two ensembles of sample curves, once obtained could be treated on a real experimental, empirical basis, simply as sucrose reaction runs, without regard to their *in numero* origins.

9. The first detailed pilot study of the variability associated with Markovian transitions out of a given state produced the data utilized in the preparation of Table II. For this purpose, a new ensemble of 10 sample curves was generated. A new variable $\xi_{x_j}(10)$ defined as the number of molecules of A present 10 min after the attainment of concentration state x_j, was introduced. On a given sample curve showing x_j as a function of t_j, the value of $\xi_{x_j}(10)$ is the ordinate corresponding to the time $(t_j + 10)$. The totality of 10 such values, one from each of the 10 sample curves, describes the distribution of values of the concentration 10 min, after the attainment of concentration level x_j. In this study in which again $n_0 = 40$, x_j was taken arbitrarily to be 40, 35, . . . , 5 molecules.

In effect, the distributions so produced represent estimates of the transition probability $p_{x_j, x}(10)$, so that this latter is the theroretical probability function of the new random variable X corresponding to the values $\xi_{x_j}(10)$. The quantities in the table were calculated from the following formulas (superscripts referring to the number of the sample curve):

$$\bar{\xi}_{x_j}(10) = \tfrac{1}{10} \sum_{i=1}^{10} \xi_{x_j}^{(i)}(10) \qquad \text{(estimated mean value),} \tag{409}$$

$$E_{x_j}(10) = x_j e^{-10\mu} \qquad \text{(theoretical mean value),} \tag{410}$$

$$s_{x_j}(10) = \tfrac{1}{3}\left[\sum_{i=1}^{10} (\xi_{x_j}^{(i)}(10) - \bar{\xi}_{x_j}(10))^2\right]^{1/2} \qquad \text{(estimated standard deviation),} \tag{411}$$

$$\sigma_{x_j}(10) = [x_j e^{-10\mu}(1 - e^{-10\mu})]^{1/2} \qquad \text{(theoretical standard deviation).} \tag{412}$$

It can be seen that as the initial x_j values drop from 40 to 5, the standard

TABLE II

Markov Model: COMPARISON OF ESTIMATED AND THEORETICAL MEAN VALUES, STANDARD DEVIATIONS, AND COEFFICIENTS OF VARIATION (10 sample curves, $\tau = 10$ min)

Molecular state x_j	Mean value		Standard deviation		% Coefficient of variation	
	Est. $\bar{\xi}_{x_j}(10)$	Theor. $E_{x_j}(10)$	Est. $s_{x_j}(10)$	Theor. $\sigma_{x_j}(10)$	Est. $s/\bar{\xi}$	Theor. σ/E
40	37.805	38.534	1.070	1.188	2.830	3.083
35	32.985	33.717	0.539	1.111	1.635	3.296
30	28.865	28.901	0.494	1.028	1.712	3.560
25	23.535	24.084	0.549	0.939	2.333	3.900
20	19.090	19.267	0.730	0.840	3.827	4.360
15	14.485	14.450	0.417	0.727	2.879	5.034
10	9.415	9.633	0.367	0.594	3.902	6.166
5	4.673	4.816	0.346	0.426	7.418	8.720

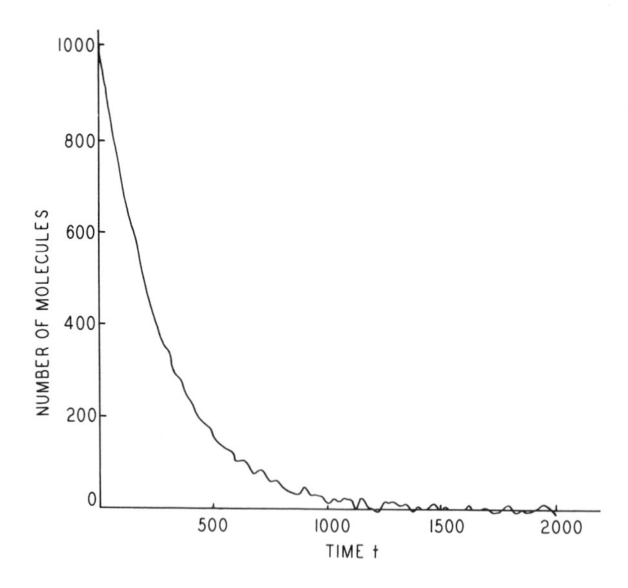

Fig. 18. Simulation of an experimental run according to deterministic theory, approximating random experimental error component $\varepsilon(t)$ by a normal probability density function $N[0, 0.01n_0]$.

deviations drop very abruptly to a much lower set of values than the corresponding theoretical values. However, aside from the values corresponding to $x_j = 25$ and 20, a steady decrease continues until at $x_j = 5$, the experimental and theoretical standard deviations grow close, just as they were initially. These features are, of course, reflected in the coefficients of variation. A decrease in standard deviation with decreasing x_j values would be predicted on the basis of the rationale expounded in the first paragraph of Item 8, Of course, the smallness of the samples and the fact that transitions over only one arbitrary time interval of length (10) have been chosen, preclude a general interpretation. These particular experiments are discussed here mainly to indicate a way of analyzing the finergrained predictions of the full Markov model, which should be pursued much further.

10. Sample curves for the stochastic extension of the deterministic model: Figure 18 shows the results of another pilot experiment indicating yet another interesting investigative direction into which the macroscopic model leads. The motivation for this experiment came from the interest mentioned earlier in discovering some way of characterizing the difference between experimental fluctuations and the inherent random fluctuations predicted by the stochastic model. It has already been emphasized that actual chemical experiments cannot be carried out which can do this, since they would require an unattainably high degree of experimental precision. The Monte Carlo experiments just discussed do give some idea of the fluctuation component contained in the chemical process itself.

It was reasoned that similar experiments might be devised which eliminate the inherent component itself by beginning with the deterministic curve and adding to the precise deterministic prediction $x = x_0 e^{-\mu t}$ a random component $\varepsilon(t)$ such as that indicated in Eq. (347). Different forms of stochastic error processes $\varepsilon(t)$ could be tried to see which determinations most resemble actual chemical runs. In other words, what is suggested here is a set of *in numero* studies based on simulations of the stochastically augmented deterministic model. For example, beginning with the Gaussian approximation to the Prior model, one can represent the Prior unimolecular stochastic model by

$$x(t) = x_0 e^{-\mu t} + \varepsilon'(t) \tag{413}$$

where now $\varepsilon'(t)$ as a stochastic variable is characterized as a Gaussian probability density

$$N[0, [x_0 e^{-\mu t}(1 - e^{-\mu t})]^{1/2}]. \tag{414}$$

Then, the total experimental, or real process would have a form

$$x(t) = x_0 e^{-\mu t} + \varepsilon(t) + \varepsilon'(t), \tag{415}$$

indicating both components of random fluctuations (on an assumption of stochastic independence between them). Results of this kind could then be used to assess the relative magnitudes to the two sources of random fluctuations.

The first attempt in this direction is shown in Fig. 18. A Monte Carlo procedure similar to that of Method III was devised for generating samples corresponding to different times t in the deterministic time course from the Gaussian probability density $N[0, 0.01x_0]$. In other words, simply as a sample of different hypotheses that could be tested, it was assumed here that the experimental error process is describable independently of the progress of the reaction by a Gaussian distribution of 0 mean and fixed standard deviation equal to 10^{-2} the initial value x_0. In this case, then, corresponding to Eq. (414), the stochastically augmented deterministic model may be thought of itself as a stochastic process of the form $N[x_0e^{-\mu t}, 0.01x_0]$.

The *in numero* experiments discussed in this section, like Bunker's microscopic kinetic studies, indicate the use of the computer in theoretically based studies of chemical kinetics. With reference to the biological and medical sciences they are examples at the molecular level of a new kind of numerical experimentation that may help fill in the gaps of ordinary experimentation which is limited by practically unattainable conditions.

5. *A Generalization of the Unimolecular Stochastic Process*†

Darvey and Staff [1966] have obtained a generalization of the stochastic model of the elementary unimolecular reaction in the case of a system of n interconverting chemical species $\{A_i\}$ $(i = 1, 2, \ldots, n)$ with first-order interconversion kinetics assumed throughout the system, in which the initial stage of reaction is defined by assigning zero concentration to all except the mth species. Such a system may be regarded as a first-order (or linear) approximation to a complex multicomponent system such as might occur in cellular biology. At the same time such a system might arise in connection with industrial petroleum processes where one encounters a system consisting of an enormous number of chemical species which for theoretical treatments must be conveniently "lumped" into a few classes.

The Darvey–Staff model has the following axiomatic description: Given the set $\{X_i(t)\}$ of random variables corresponding to the numbers of molecules of $\{A_i\}$ $(i = 1, 2, \ldots, n)$ present at time t with initial conditions

$$\begin{aligned} X_m(0) &= N, \\ X_j(0) &= 0, \qquad j = 1, 2, \ldots, m-1, m+1, \ldots, n, \end{aligned} \tag{416}$$

the following axioms are assumed to hold for the system.

†From Darvey and Staff [1966, pp. 990–997].

***Axiom* 1.** The probability of a single intermolecular conversion $A_i \longrightarrow A_j$ in time $(t, t + \Delta t)$, given that x_i molecules of A_i are present at time t, is taken to be

$$\mu_{ij} x_i \, \Delta t + o(\Delta t), \qquad i, j = 1, 2, \ldots, n; \; i \neq j.$$

***Axiom* 2.** The probability of more than one conversion in $(t, t + \Delta t)$ equals $o(\Delta t)$.

***Axiom* 3.** The processes are all homogeneous in time, and future changes are independent of the past (the stationary Markovian property).

The stochastic model for this system is described in terms of $P(x, t) = P(x_1, x_2, \ldots, x_n; t)$, the probability that at time t there are x_i molecules of A_i $(i = 1, 2, \ldots, n)$ present. As in Bartholomay's treatment, the function $P(x, t)$ is determined by a differential equation in t obtained from the difference equation for $[P(x, t + \Delta t) - P(x, t)]$ by dividing by Δt and passing to the limit as $\Delta t \longrightarrow 0$,

$$\begin{aligned} dP(x, t)/dt = {} & \sum_{i=1}^{n} \sum_{\substack{j=1 \\ (j \neq i)}}^{n} \mu_{ij}(x_i + 1) P(x_1, x_2, \ldots, x_i + 1, \ldots, x_j - 1, \ldots, x_n; t) \\ & - \sum_{i=1}^{n} \sum_{\substack{j=1 \\ (j \neq i)}}^{n} \mu_{ij} x_i P(x, t), \end{aligned} \tag{417}$$

the initial condition being

$$P(x, 0) = 1 \qquad \text{for} \quad x_m = N \quad \text{and} \quad x_i = 0, \quad i \neq m. \tag{418}$$

The corresponding multivariate probability-generating function is given by

$$G(s_1, s_2, \ldots, s_n, t) = \sum_{x_1} \sum_{x_2} \cdots \sum_{x_n} P(x, t) \prod_{i=1}^{n} s_i^{x_i}, \tag{419}$$

noting that $\sum_{i=1}^{n} x_i = N$, where $0 \leqslant x_i \leqslant N$ and that the s_i are arbitrary variables (with $|s_i| \leqslant 1$). Multiplying Eq. (417) by $\prod_{i=1}^{n} (s_i^{x_i})$ and summing over all values of x_i, a first-order partial differential equation for $G(s, t)$ is obtained,

$$\partial G/\partial t = \sum_{i=1}^{n} \sum_{\substack{j=1 \\ (j \neq i)}}^{n} \mu_{ij}(s_j - s_i) \, \partial G/\partial s_i, \tag{420}$$

where $G(s, 0) = s_m{}^N$. And, again, as in the model of the previous section, a solution of this first-order partial differential equation is obtained by using the Langrange method of auxiliary ordinary differential equations. The generalization expected from the binomial form of the unimolecular reaction (Prior model) probability function, namely, a multinomial probability function obtains

$$P(x, t) = \left(N! \Big/ \prod_{i=1}^{n} x_i!\right) \prod_{j=1}^{n} (w_i(t))^{xi}, \tag{421}$$

where the $w_i(t)$ are time functions determined from the coefficients of an Nth-order polynomial expansion of $G(s, t)$.

It is not perfectly general that all of the current stochastic models in kinetics are "consistent in the mean" with the corresponding deterministic models, a point which will be demonstrated in Section IV.E. However, like the earlier model, the Darvey–Staff Model does satisfy the criterion. In other words, where the deterministic model, derivable as in the simple unimolecular case from the law of mass action, consists of the system of ordinary differential equations

$$dx_i/dt = \sum_{\substack{j=1 \\ (j\neq i)}}^{n} k_{ji}x_j - \sum_{\substack{j=1 \\ (j\neq i)}}^{n} k_{ij}x_i, \qquad i = 1, 2, \ldots, n, \tag{422}$$

and where the k_{ij}'s are identified with the μ_{ij}s, as before, the substitution of the mean values of the various X_i's treated as random variables, satisfy the same system. In this sense, the class of stochastic models satisfying the mean-consistency property suggest a stochastic reformulation of the law of mass action.

The mean-value functions $\mu_i(t)$ of X_i, the variance functions σ_i^2 of X_i, and the convariance functions $\sigma_{ij}(t) = \mathrm{cov}(X_i, X_j)$ $(i, j = 1, 2, \ldots, n; i \neq j)$ have been obtained as functions of the $w(t)$ functions, using the derivatives of the probability generating function, as

$$\mu_i = Nw_i(t), \tag{423}$$

$$\sigma_i^2(t) = Nw_i(t)[1 - w_i(t)], \tag{424}$$

$$\sigma_{ij}(t) = -Nw_i(t)\, w_j(t). \tag{425}$$

Note that where $i = 1$ and $w_i(t) = e^{-\mu t}$ Eqs. (423) and (424) reduce to the mean value and variance function of the Prior model of the simple unimolecular process.

6. *The Krieger–Gans Analog in Microscopic Kinetics of the Unimolecular Stochastic Model in Macroscopic Kinetics*†

Stochastic formulations very similar to those employed in macroscopic kinetics have been derived for microscopic chemical kinetics, so that at the methodological or metatheoretical level of abstraction an interesting connection between macroscpoic and microscopic kinetics can be utilized as a possible mode of junction between these two sides of chemical kinetics. A case in point is the stochastic model in microscopic kinetics of Krieger and Gans [1960], which bears the same relation to the detailed microspoic stochastic model of Montroll–Shuler for the unimolecular reaction that the

†From Krieger and Gans [1960, pp. 247–250].

Darvey–Staff model in macroscopic kinetics bears to the stochastic model of the elementary unimolecular decomposition process. In this sense, and also in its details, the Krieger–Gans model corresponds closely to the Darvey–Staff model.

Krieger and Gans studied a closed system of n noninteracting molecules (particles), which, however, may interact with a heat bath of other molecules. They considered the accessible states of the molecules as a set or r cells in a space, each of which is accessible by some path from every other cell. The state of the system is then assumed to be specifiable in terms of a vector $n(n_1, \ldots, n_i, \ldots, n_r)$ of "occupation numbers," n_i being the number of molecules in the ith cell. And, $p_{ij}(n, t)$ is defined as the probability per unit time that a molecule makes the transition from state i to state j.

A first-order process is defined by the requirements (1) that for all i and j,

$$p_{ij}(n, t) = \alpha_{ij} n_i, \tag{426}$$

where the "transition probabilities" α_{ij} correspond to the μ_{ij} parameters in the Darvey–Staff model, (2) implicitly, that

$$p_{ij}\,\Delta t = \alpha_{ij} n_i\,\Delta t + o(\Delta t) \tag{427}$$

is the probability that a single transition $i \longrightarrow j$ takes place in an infinitesimal time interval of length Δt; (3) that the probability of more than one such transition is $o(\Delta t)$; and (4) explicitly that

$$p_{ii} = -\sum_{j \neq i} p_{ij}\,\Delta t \tag{428}$$

is the probability that a molecule in stage i undergoes no change in this same interval Δt.

Aside from different boundary conditions (see below) and their different underlying systems—this earlier model being based on a microscopic chemical kinetic system and Darvey and Staff's later model decribing an underlying macroscopic system—the Krieger–Gans and the Darvey–Staff mathematical models are identical. Thus, the defining differential equation for the present model is the same as our Eq. (417) of the latter model if we identify α_{ij} with μ_{ij} and n_j with x_j. Similarly, the probability generating function of the present model is the same as our Eq. (420) in the Darvey–Staff model. This probability-generating function in this case had been simplified to the form

$$\partial G/\partial t = \sum_{j=1}^{r} \sum_{i=1}^{r} \alpha_{ij}\, s_j\, \partial G/\partial s_i, \tag{429}$$

by defining arbitrarily

$$\alpha_{ii} \equiv -\sum_{j \neq i}^{r} \alpha_{ij} \tag{430}$$

(as in the case of the intensity matrix of the Markov theory).

Using the general boundary values $\{n_{i0}\}$ of $\{n_i\}$ $(i = 1, 2, \ldots, r)$, the general solution of Eq. (429) was obtained, using the method of characteristics, as

$$G = \prod_{i=1}^{r} [B^{-1} \sum_k B^{ki} \sum_j b_{kj} s_j e^{\lambda_k t}]^{n_{i0}}, \tag{431}$$

where $B = \det(b_{kj})$, B^{kj} is the cofactor of b_{kj}, and $\{b_{kj}\}$ is a set of arbitrary multipliers subject to the constraint

$$\sum_i b_{ki} \alpha_{ij} = b_{kj} \lambda_k. \tag{432}$$

The latter is, of course, equivalent to the familiar eigenvector problem implied by the equation written in the form

$$\sum_i b_{ki}(\alpha_{ij} - \lambda_k \delta_{ij}) = 0, \tag{433}$$

δ_{ij} being the Kronecker delta, and $\det(\alpha_{ij} - \lambda_k \delta_{ij})$ the characteristic determinant.

The solution of the differential equation for $\partial P(n; t)/\partial t$ has been obtained for the case of the infinite time limit, starting with

$$\lim_{t \to \infty} G(s_1, \ldots, s_r; t) = G_\infty = \prod_i \Big[\sum_j b_{rj} s_j \Big/ \sum_j b_{rj} \Big]^{n_{i0}}, \tag{434}$$

which is rewritten as

$$G_\infty = \Big[\sum_j b_{rj} s_j \Big/ \sum_j b_{rj} \Big]^N, \tag{435}$$

where $N = \sum_{i=1}^{r} n_i$.

Thus, the infinite time limit is independent of the initial population distribution. But Eq. (435) is the probability generating function of the multinomial distribution; that is,

$$P(n_1, n_2, \ldots, n_r; \infty) = \Big(N! \Big/ \prod_{i=1}^{r} n_i!\Big) \prod_{i=1}^{r} \Big[b_{ri} \Big/ \sum_{j=1}^{r} b_{rj} \Big]^{n_i}. \tag{436}$$

a. The Transition-State Theoretic Interpretation. Krieger and Gans observed the following interesting connection with the transition-state theory. First of all, note that the mean values of the n_i components "at equilibrium," that is, as $t \longrightarrow \infty$, are obtainable directly from Eq. (435) as

$$\langle n_i \rangle = [\partial \log_e G_\infty / \partial \log_e s_i] = \Big(N b_{ri} \Big/ \sum_{j=1}^{r} b_{rj}\Big), \qquad \text{all} \quad s_j = 1. \tag{437}$$

Then, setting $\{b_{ri}\} = \{g_i \exp(-\varepsilon_i / k_B T\}$ and dividing both sides by N transforms Eq. (437) into

$$\langle n_i \rangle / N = g_i \exp(-\varepsilon_i / k_B T) \Big/ \sum_{j=1}^{r} g_j \exp(-\varepsilon_j / k_B T). \tag{438}$$

But with $\langle n_i \rangle / N$ interpreted in a probability frequency-theoretic sense as

$p(\varepsilon_i)$, this is identical with the Maxwell–Boltzmann distribution of Eq. (54). So, in the statistical thermodynamic context of transition-state theory, Eq. (436) is interpretable as the distribution function for (the "canonical ensemble"):

$$P(n, \infty) = Z^{-1} \exp(-E/k_B T), \tag{439}$$

where $E = \sum_{i=1}^{r} n_i \varepsilon_i$ is the total energy and

$$Z = \left(\prod_{i=1}^{r} n_i!/N!\right)\left(\sum_{j=1}^{r} g_j \exp(-\varepsilon_j/k_B T)\right)^N \tag{440}$$

is the partition function for the system.

b. Relaxation of the System from an Equilibrium State. Assuming that the system is initially at equilibrium, with initial values $\alpha_{ij}(0)$, they examined the time effects of disturbing it by adjusting the initial α's to the new α_{ij}. They deduced that the system in this case still maintains a multinomial equilibrium distribution. In other words, a system relaxing by first-order processes from one equilibrium distribution to another maintains at all times a multinomial distribution. As they point out, this result constitutes the generalization of a theorem of Montroll and Shuler on the relaxation of a system of harmonic oscillators.

7. *Conclusion*

In this section we have seen some examples of different levels of elaboration of the simplest chemical reaction, the unimolecular reaction, and the ways in which different approaches can come together from apparently independently conceived different starting points; also, the extra and prolonged effort required to explore in depth a particular type of process. We have here, also, examples of the necessary arbitrariness of both the underlying physical models and the overlying mathematical models. As difficult as it may at times appear to resolve the different points of view reflected in these different choices, an integration of these views is necessary before a completely acceptable interpretation, or model, or theory can be synthesized from the various findings. Such a theory will have to perform an integration between the classical deterministic and the stochastic deductions, between the microscopic and the macroscpoic aspects. In this section, an effort has been made to emphasize a few such interconnections which are already discernible. Hopefully, a continuation of this process may lead to a clearer elucidation of the unimolecular reaction, and to its incorporation into the general theory of chemical kinetics. This is certainly desirable, for there still remains a vagueness of definition of some of the most basic ideas referred to in this Section III. It is the author's conviction that where such vagueness perisists, the weakness can be overcome, at least in part, by insistence on clearer mathematical exposition of ideas and, in part, by leaving the beaten path in favor of trying some reformulations. The same comment holds for experimental studies. It is for this reason that the few existing and pertinent, computer-assisted *in numero* studies have been discussed here in some detail.

So much of the general microscopic theory of chemical kinetics still rests on inferences drawn from detailed studies of the simplest and smallest molecules and simplified models of these, that the time appears quite distant when a fairly complex reaction can be discussed in great detail both inter- and intramolecularly. Empirical, or semiempirical/semitheoretic studies of such systems must be turned to in such cases with the realization that, as in the total history of chemical kinetics, it may be necessary to penetrate from the outside in—to pass from macroscopic studies and interpretations to microscopic studies, counting on an ultimate resynthesis in the opposite direction. In fact, it is such studies that provided the incentive to expand the domain of chemical kinetics.

In the final section of this chapter we consider, as an important example of this, the relatively new area of enzyme kinetics whose growth and development reflects the strong insistent force of the rapidly expanding area of biochemistry. The main reaction to be discussed is the basic enzyme-catalyzed reaction consisting of an enzyme species and a substrate. An effort will be made to delineate the directly applicable aspects of chemical kinetics and those which require departures and outbranchings from the classical theory. Again, both deterministic and stochastic approaches to questions at the microscopic and macroscpic levels will be discussed. It will be seen that the complexity of the molecules involved has caused the microscopic development to lag behind the macroscpic and experimental findings.

IV. Enzyme Kinetics

A. The Enzyme–Substrate System

1. *In Vivo versus in Vitro Characterizations*

Enzyme molecules, like all protein moelcules, are found only within or closely associated with living systems, which means that the final tests of feasibility of proposed modes of action and kinetic hypotheses would have to be carried out experimentally at the *in vivo* level, idealistically speaking. On the other hand, all of our current detailed information on enzymes has had to come from the interpretation of the results of isolated *in vitro* experiments, inferences drawn from the body of existing chemical knowledge and the study of physicochemical analogs, and mathematical models and idealizations. In addition, there is the new *in numero* approach requiring the use of large electronic computers, still in its infancy, but with some exciting potential in terms of the simulation of intact biochemical activities, unapproacheable by real experimentation.

Certain special features of enzyme catalyzed reactions make obvious the necessity for an enlargement of the purely chemical point of view. For example, as has been emphasized also by Schroedinger, they may occur in extremely dilute concentrations in the cells, so that random fluctuations may be relatively macroscopic. There is also the question of the relation of the structure of the enzyme molecule to its chemical activity and similarly the question of the effects of environment and compartmentalization on the mechanism of reaction. Neither can the heterogeneous nature of such reactions *in vivo* be overlooked, involved as they may be in the coupling of diffusion, transport, and bioelectric processes nececssary for the maintenance and control of life processes at all levels of biological organization. Finally, the question of relative specificities of the mixtures of substrates and enzymes in the same natural biological locale must be considered in translating *in vitro* results to biological reality. Clearly then, in this case the minimum requirement for the ultimate elucidation of the kinetics of enzyme catalyzed reactions is the purely chemical one; and, the maximum requirement is the integration of our knowledge of the chemical aspects into the appropriate context of a biological system $\mathcal{B}(S, C, E, F, \tau)$ whose environment E contains a great variety of other chemical, biological, and physical systems. In other words, it is necessary in studies of enzyme kinetics to keep in mind the dual nature of such a system which must at the same time be considered as a chemical and as a full biological system.

In this chapter it has been necessary to declare many of these questions beyond the intended scope of discussion and to confine the discussion mainly to some of the basic aspects (including the stochastic aspects) of the most fundamental and elementary enzyme caralyzed reaction, the so-called Michaelis–Menten enzyme–substrate reaction. It will be seen that even in this case, there is far from a complete theory. Some idea of why this is so may be inferred simply from considering the highly complex chemical nature of enzyme molecules. And, in this very complexity may be seen also some reasons for expecting that the probabilistic approach may have to be featured over the classical deterministic approach.

2. *Enzyme Molecules as Proteins*

Proteins (of which enzyme form a subclass) like nearly all biological macromolecules are polymers of high molecular weight beyond 5×10^3, the term polypeptide being reserved for substances of molecular weight lower than this level but of similar unitary composition. Carbon, hydrogen, oxygen, and nitrogen atoms form their basic atomic-level units, and the amino acids their higher level compositional or monomeric units. The immense variety of possible protein species is evident from the fact that 20 different amino acids form the basic protein alphabet, a given protein molecule containing 100 or more of these units.

As early as 1900, Hofmeister and Fischer independently suggested that proteins were assembled through the formation of secondary amide linkages between the α-carbonyl and α-amino groups of adjacent amino acids—bonds which have come to be called "peptide bonds." It is from this that conjugated amino acids are referred to as peptides, the individual amino acid compositional units being referred to in this context as "residues." The coplanarity of the six atoms of a peptide unit $C\alpha^{(1)}$—CO—NH—$C\alpha^{(2)}$ is an important factor in the maintenance of the structure of the molecule. The multivalent electrolytic nature of a protein molecule imparts additional special properties to such systems.

A source of great difficulty in experimental studies derives from the contamination of protein molecules which require elaborate purification procedures. Empirical criteria have had to be set up for deciding how far to take the purification process in order to guarantee the significance of experimental findings.

Four different levels or dimensions of structural description have evolved, further attesting to the complexity of such molecules.

a. Primary (One-Dimensional) Structure. "Primary structure" refers simply to the protein molecule thought of as a sequence of covalentaly bonded amino acid residues without any reference to its topology, in simplest terms then, as a linear array. Questions of an information-theoretic kind generally refer to this simplest aspect of structure inasmuch as knowledge of the sequence allows its identification as a word unit of protein language. It goes one step further than a knolwedge simply of the component amino acid residues allows. Though many such primary structures are now settled and methods for accomplishing this have been worked out to the point where amino acid analyzers have been constructed for automation of the procedure, this is a relatively recent development. The first such determination was not accomplished until 1955, roughly, when F. Sanger, at Cambridge, succeeded in determining the primary structure of insulin. Next came the determination of the primary structure of the enzyme ribonuclease, shown to consist of 124 amino acid residues. Others since identified include chymotrypsinogen, trypsinogen, papain, myoglobin, hemoglobin, cytochrome C, and lysozyme. The latter will be discussed in some detail presently. Currently evidence seems to be accumulating through the synthesis of proteina anologs that the primary structure may in fact determine the entire structural conformation of the protein molecule.

b. Secondary (Two-Dimensional) Structure. Secondary structure is generated by the formation of hydrogen bonds between the components of the peptide linkage. It refers to the helical conformation of the amino acid chain maintained by these intramolecular hydrogen bonds and also to so-called sheet structures derived from the formation of intermolecular hydrogen bonds. The work of Pauling *et al.* [1951; see also Pauling, 1939, pp. 498–500; and Pauling and Corey, 1951, 1953] utilizing x-ray diffraction studies made on crystalline low molecular weight amides led to the establishment of a set of criteria including the following four, for the most stable secondary structures.

1. The peptide group must be planar and have bond lengths and angles identical to those found in crystals of secondary amides,
2. every carbonyl oxygen and amide nitrogen must be involved in hydrogen-bond formation.
3. the hydrogen-bonded hydrogens lie close to a line joining the oxygen and nitrogen atoms.
4. translations and rotations of residues relative to a central axis must be uniform on a residue-to-residue basis.

Among the many possible helical structures considered, only one was found to meet the specifications for maximum stability: the α helix, which covers 5.4 Å per turn, a complete turn being made every 3.6 residues, the side-chains extending away from the helix axis. The hydrogen bonds between the components of the peptide linkages are approximately parallel to the longitudinal axis of the helix. The sheet structures that meet the stability requirements are of two main types: the "parallel" and "antiparallel pleated sheets," the former being composed of a series of chains in parallel, the latter having every other chain oriented in the opposite direction. In contrast to the α helix, the hydrogen bonds are nearly perpendicular to the longitudinal axis of polypeptide chains. The nearly completely extended polypeptide chains form hydrogen bonds with the neighboring chains.

Inferences from infrared spectroscopy and x-ray diffraction calculations are utilized in settling secondary structure. For example, the position and number of bonds are determined from infrared spectra, and the transition from a randomly coiled molecule to an α helix or sheet structure is accompanied by chracteristic alterations in the spectra. The method of optical rotatory dispersion has revealed that most proteins have helical regions interspersed with random coil regions.

The method of x-ray diffraction [see, for example, Mahler and Cordes, 1966, pp. 101–103; also Crick and Kendrew, 1957; and Low, 1961] employed in the determination of secondary and tertiary structure is a highly involved computational procedure requiring many iterations of multidimensional Fourier analysis to such an extent that the procedure would be impracticable to propose were it not for the existence of large-scale digital computers. The methods rest on the Bragg equation for describing the monochromatic beam of x-rays by scattering atomic centers. From the general features of x-ray diffraction patterns the dimensions of the so-called "unit cell" of a given molecule, the number of molecules per unit cell, the positions of atoms in the unit cell, and an estimate of molecular weight are obtainable.

Very briefly, the positions of atoms within the cell are established by x-ray diffraction in the following way: Let $\rho(x, y, z)$ be a continuous function giving the electron density at point $P(x, y, z)$ of the crystal. Connection with the atomic structure is made via the Fourier representation of $\rho(x, y, z)$,

$$\rho(x, y, z) = V^{-1} \sum_{h=-\infty}^{+\infty} \sum_{k=-\infty}^{+\infty} \sum_{l=-\infty}^{+\infty} F_{h,k,l} \exp[-2\pi i(hx + ky + lz)], \tag{441}$$

where V is the volume of the cell, and the Fourier coefficients $F_{h,k,l}$ depend on the atomic scattering factors for each atom responsible for a particular maximum. The h, k, l so-called "Miller indices" define a particular plane in the unit cell and are associable with each individual diffraction maximum. Information on the structural characteristics of the molecule are thus contained in the Fourier coefficients. These, in turn, are obtained directly from the intensities of the diffraction maxima, that is, from the darkness of spots on the photographic plate.

The resolving power of the technique is in the neighborhood of 1 Å; interatomic distances for most covalent bonds being in excess of this limit, it is possible to identify each of the atoms of the crystal. For example, the work of Kendrew on the structure of myoglobin involved the calculation of more than 10^4 components in this way. In such cases a three-dimensional graphic representation of the density function and hence of the molecular structure is produced by tracing out two-dimensional electron maps (as profiles of the surface) on lucite sheets for a number of planes of the crystal and superimposing the plots. The shape of the surface is then reconstructed from this family of profiles. An extremely large number of maps of increasing resolution power is required in an asymptotic approach to the actual structure. Lysozyme (to be discussed in Section IV.D.3) was the first enzyme structure to be detailed in this way.

c. Teritary (Three-Dimensional) Structure. Viscosity measurements, frictional coefficients, and light-scattering studies have provided ample evidence of significant variability in the overall topography of protein molecules, providing preliminary evidence that in the native state the protein forms a rather tight, compact structure. It is therefore considered that the secondary helical structure or pleated-sheet structure is modified by a highly specific folding property producing a variety of different shapes. "Tertiary structure" refers to this folded conformation of the molecule. It too is enlightened by x-ray diffraction studies.

Speaking more generally, now, in polymeric biological macromolecules (including proteins as well as the nucliec acids) the helix is natural to propose as the final (secondary) structure that would be compatible with groups such as—CO—NH— which repeat over and over again, the helical configuration being maintained by the secondary bonds in compliance with stability requirements. But in case certain sets of these secondary bonds should be much stronger than others, then the helical form no longer supports a stable configuration. The attachment of irregular side groups to the backbones of the molecules contribute to this change and to the attainment of an identifiable tertiary structure. A tertiary arrangement that is energetically very satisfactory for the backbone of the protein or nucleic acid may produce very unsatisfactory bonding of the side groups. Thus, the tertiary structure, it may be regarded, is arrived at as a compromise between the tendency of the regular backbone to form a regular helix and that of the side groups to twist the backbone into a configuration that maximizes the strength of the secondary bonds formed by the side groups.

Four different types of interaction are considered to be principally responsible for the maintenance of secondary and tertiary structures: interpeptide hydrogen bonds, side-chain hydrogen bonds, ionic bonds, and apolar or hydrophobic bonds, the latter being associated with the tendency of nonpolar molecules to associate in solution. Thus, while some evidence exists that the higher order structure is determined by factors in the primary structure. it is far from simple to guess the correct shape of a macromolecule from its covalent linear structure alone. Combinatorially, the number of possibilities on this basis is immense, certainly a problem that requires use of the computer. Another drawback to such guesswork is that, in actual fact, our knowledge of the nature of weak bonds is very incomplete. In many cases we are not sure of either the exact bond energies or of the possible bond angles of the formation. Only x-ray diffraction studies can enlighten the problem of higher order structure determination. Such experimental studies may allow general algorithms to be set up on a semiempirical basis eventually.

Finally, it should also be emphasized that these higher-order structural aspects may not be maintainable because of the weakness of the forces involved so that possibly more than 50% of a protein molecule may at a given time assume a variety of random conformations, a result simply of thermal movements, for example. On the other hand, most of these conformations are themselves thermodynamically unstable, so that eventually thermal movements can bring together groups that form "good" weak bonds. And, it is conejctured that these groups tend to stay together because more free energy is lost when they form than can be regained by their breakup.

d. Quaternary (Four-Dimensional) Structure. It is also known that many proteins are composed of several independent polypeptide chains—independent in the sense of not being covalently bonded to each other. For example, hemoglobin contains four chains and no disulfide bridges. Relatedly, ferritin of molecular weight 4.8×10^4 and 4×10^3 amino acid residues is really an agrregate of 20 almost identical smaller polypeptide chains of about 200 amino acid residues each. Similarly, the protein component of tabacco mosaic virus was originally thought to have the molecular weight of 36×10^6. But, it was later learned to consist of 2.15×10^3 smaller protein molecules each containing 158 amino acid residues. Such loose collections or aggregates of polymeric units are referred to as the quaternary structure, so that quaternary structure is to protein structure what population biology is to organismic biology, in a sense. The situation can become exceedingly complex. For example, the enzyme glutamic dehydrogenase is considered to have three identifiable levels of structural organization at the quaternary level. On experimental evidence, enzymic activity is generally thought to be retained by the isolated subunits of such aggregates.

3. *Enzyme Molecules as Catalysts*

The enzyme molecule does not satisfy all aspects of the original Ostwald definition of chemical catalyst or accelerator of reactions [see Baldwin, 1963, Chapter 1, for an excellent discussion of this point]. But there are also a number of important features which inorganic and enzyme catalysts share. For example, enzyme catalyzed reactions like other catalyzable reactions are reactions which can proceed in the absence of catalysts, though at an extremely slow rate; they may also in some cases be catalyzed by inorganic catalysts (for example, sucrose hydrolysis or inversion). The essence of enzyme activity, however, lies in the proclivity of these complex biological macromolecules to accelerate a reaction's velocity by breaking specific covalent bonds. Their highly complex structure apparently accounts for their specific affinities for certain molecules, though as we shall emphasize, the catalytic action is probably associated with a relatively small portion of the molecule referred to as the "active site" or "active center," to be discussed in Section e, following, and in greater detail in Section IV.C. Their exclusiveness is of such a high degree that it appears that there is a unique one-to-one correspondence between many enzymes and their catalyzed reactions,

a. Activation of the Substrate. In the absence of enzymes most of the covalent bonds of biological molecules or substrates whose transformations they catalyze are stable and disrupt only under high temperatures. Enzymes may act by somehow lowering the stability threshold of a given bond. The manner in which this is accomplished, that is, the "activation" of a substrate molecule may be regarded as a resultant of its total complex structure acting through its specific site of action. It is assumed that secondary forces are the basis by which the enzyme and its substrate combine in the formation of an enzyme-substrate complex, which may be likened to the activated complex of the transition-state theory. However, an enzyme molecule is not itself transformed by this interaction, apparently returning to its normal state after catalyzing the transformation of a given substrate molecule, ready to "activate" the next substrate molecule. On a biological time scale enzymes work very fast, some being able to catalyze as many as 10^6 individual transformations per minute (referred to as the "enzyme turnover number"). It is this property particularly which suggested the characterization of enzyme catalysis in the language of queueing theory (see Section IV.F.2). Often, of course, no transformations of substrates will occur in a minute in the absence of specific enzymes.

b. Energy Considerations. The free energy of bonding ΔF, derivable from the "equilibrium constant" usually hints at which types of bonds may be responsible for the enzyme–substrate (E–S) interaction, an important aspect of the study of the mechanism and kinetics of enzyme catalysis. For example, van der Waals bonds carry energies of 1–2 kcal $mole^{-1}$, only slightly higher than the kinetic energy of heat motion. Hydrogen and ionic bonds have energies between 3 and 7 kcal $mole^{-1}$. Thus, ΔF values between 5 and 10 kcal/mole suggest that several secondary bonds are involved in E–S interactions. This would imply that only weak bonds attach enzymes to substrates. Enzyme–substrate complexes can thus be made and broken apart as a result of thermal motion, a clue to the large turnover numbers (or speeds of action).

In the transition-state theory, as we have seen, the activated complex intermediate species, though not a normal molecule has been treated thermodynamically with some success, including, for example, the determination of the rate constant of the reaction as a function of the equilibrium constant associated with the equilibrium between reactants and active species. The E–S intermediate complex presents obviously many more difficulties. However, one can hypothesize that the E–S complex is sufficiently like the activated complex to be considered as such, and then hope to proceed to discuss the kinetic rate constant as a function of the equilibrium constant between the E–S complex and the E and S species considered as ordinary reactants. The special features of biological macromolecules would render this a crude approximation based on our present knowledge, if it could be done. But the enzymetic equilibrium constant is not experimentally available or determinable. The interesting situation in biochemistry is that the kinetic rate constant is, in fact, used to determine the equilibrium constant, or thermodynamic quantities such as the free energy of activation related to this constant [see, for example, Bray and White, 1966, Chapter 6].

Further complications of interpretation in energy considerations arise from the fact that the E–S complex may be a generic term for a whole sequence of E–S postures, or activated states, $(ES)_i^*$, corresponding to which the potential energy curve would show a sequence of relative maxima, until the critical level is attained by one of them; that is, until the energy barrier is surmounted by the true activated form of E–S complex. Evidence in favor of this interpretation is found in the fact that generally, the energy level of the E–S complex is lower than would be expected form "normal" complexes.

c. Temperature Characteristics. Like most chemical reactions, enzyme-catalyzed reactions are temperature dependent, the reaction velocity increasing exponentially with increasing temperature. However, in *in vitro* experiments, forcing the temperature beyond a certain point destroys the catalytic properties of the catalyst. Thus, a unique problem in enzyme kinetics is the determination of an optimum temperature. Most enzymes show an optimum temperature in the range 30°–40°C. According to Baldwin [1963, p. 17], "This led to the suggestion that when animals became homoiothermic they settled on a body temperature of the order of 37°C because their enzymes would "work better" in that neighborhood. . . . "

d. pH Characteristics. Furthermore, catalytic activity is highly pH sensitive. As in the case of temperature there is also an optimum pH which is characteristic of a given enzyme. For example, a given carbohydrase has always the same optimum pH, even when paired with different substrates. A typical pH-versus-activity curve resembles that obtained by plotting the degree of ionization of a simple ampholyte such as glycine versus pH. It is pertinent to point out here that most of the physical properties of solutions of ampholytes, such as osmotic pressure, solubility, viscosity, in fact pass through a maximum or minimum when plotted against pH, the pH at which these extrema occur being referred to as the "isoelectric pH." Similar changes are attributable to the ionic states of the enzymes considered as ampholytes; that is, being multipolar, any given protein or enzyme can eixst in a number of different ionic forms. Enzymes are most stable in the neighborhood of optimum pH, the optimum pH being therefore considered even more a characteristic feature than optimum temperature. Most enzymes have their optimum pH close to neutrality or physiological pH, in the range between pH 5 and pH 7. Extreme pH values lead to denaturation of the enzyme molecule, so that pH is another critical parameter in experimental studies.

e. The Active-Site Concept. Almost from the beginning of biochemistry it has been realized that not the whole surface of an enzyme is involved in its catalytic activity. Discussions of this point through the years have spanned the scientific method, all the way from philosophy and teleolgy to deliberate scientific experiment and sometimes combinations of all of these aspects. At times, the so-called active site has been considered more as an operational unverifiable concept than a biological reality. Evidence seems now to be accumulating in favor of the view that the active site is a chemically characterizable, though small portion of the enzyme molecule, a local geometrical property as opposed to an (overall) topological one. It is taken to include those side chains and peptide bonds which come into direct physical contact with the substrate and other side chains or peptide bonds which, although not in direct contact with the substrate, perform a direct function in the catalytic process. Thus, it is now mandatory to investigate the details of a structure of the enzyme molecule at the active site or center including the bonds and orientations of the chians. Clearly, the elucidation of the site must take into account the four dimensions of structure spoken of earlier; this explains why complete active site characterizations have only just begun to appear in the literature.

(1) FISCHER'S LOCK–KEY HYPOTHESIS. In the late nineteenth century, Emil Fischer, in effect, equated enzyme specificity to the active site by suggesting that the catalytically active portion of the enzyme molecule is constructed and rigidly maintained so as to complement the structure of its substrate. This hypothesis has come to be called the "lock–key" hypothesis.

(2) KOSHLAND'S INDUCED-FIT HYPOTHESIS. More recently, as an alternative, Koshland [see Boyer *et al.*, 1959, Chapter 7] proposed the so-called "induced-fit" hypothesis which ascribes a flexibility to the active-site local geometry, allowing it to conform to the approaching substrate molecule. In other words, he hypothesizes that the mere presence of the substrate induces an orientation in the groups constituting the active site.

(3) VALLEE–WILLIAMS' "ENTATIC STATE." Specifically with reference to "metalloenzymes" (see Section IV.C.2), Vallee and Williams [1968] very recently have proposed a new concept, complementary to both Fischer's and Koshland's hypotheses, aimed at explaining for the case of a metalloenzyme why a substrate is attracted to the active site and how, once there, it fits into the site or conversely the site conforms to it. In discussing the possible roles of a metal atom at an active site they postulate the existence of an "entatic state," or, "state of entasis" (Greek: *ϵνταασισ*, a stretched state or state of tension): ". . . the existence in an enzyme of an area with energy, closer to that of a unimolecular transition state than to that of a conventional stable

molecule, thereby constituting an energetically poised domain." Thus, this region of the enzyme molecule would be characterized as existing in a state corresponding to the "energized state" or the Slater "interesting molecule" of the unimolecular rate theory of Section III, or the activated complex of the general transition-state theory. The kinetic implications of this hypothesis may be seen in analogy with the transition-state theory if one considers that in the latter case the rate of reaction depends on the ease with which the transition state of the activated complex can be reached. Apparently in this proposal it is assumed that the maintenance of this state is related to the metal–enzyme association, in which case one would expect that addition or replacement of the metal by other metal ions would result in some effect on the enzymetic activity, and this seems to be the case. This hypothesis also implies the necessity to reconsider the notion of stability of an enzyme configuration along the lines mentioned in Section IV.A.3.a, preceding.

Vallee and Williams therefore caution against making inferences from the study of simplified analogs of metalloenzymes in the form of metal complexes, since the latter relax to stable configurations. They emphasize that the difference between the properties of metalloenzymes and simple metal complexes may reflect the role of secondary and tertiary structure on animo acid side chains serving as their metal ligands; that, in fact, the cooperation of a metal and its ligand, like amino acid chains, would appear to define the entatic state of a metalloenzyme which would then be thought of as a characterizing property of the active site in such a case.

From the viewpoint of enzyme kinetics the situation is complicated further by the fact that in some cases more than one active site may exist in an enzyme molecule in which case the molecule is sometimes referred to as a "multivalent enzyme." Further details on active sites are reserved for Section IV.C and some mathematical models for dealing with the kinetics of multivalent enzymes will be discussed in Section IV.D. In the Section IV.B, we shall briefly review the classical deterministic mathematical model of the basic Michaelis–Menten mechanism and examine its principle assumption.

First, it should be emphasized that we have limited these discussions mainly to the "elementary" Michaelis–Menten mechanisms involving simply an enzyme and a substrate species. In actual fact, it has been found that most enzyme catalyzed reactions involve an additional species such as metal ion or coenzyme as cofactors which cooperate with the enzyme in the activation process. And in the case of a coenzyme, for example, corresponding to the active site would be an additional site for the coenzyme–enzyme association. The question of a mechanism for explaining this cooperation is unsettled in the literature, the preferred view being that it acts as an enzyme modifier, perhaps exerting its influence indirectly in terms of stability effects on the molecule, or mediating additional energy transfer to the active site. The treatments of the kinetics of the enzyme–modifier system and even of the enzyme–inhibitor systems, however, follow the mathematical pattern of the kinetics of the Michaelis–Menten mechanism, with obvious extensions, all based on additional complexation steps such as enzyme–modifier or enzyme–inhibitor resembling in principle the E–S complex. The details of these more complicated enzyme-catalyzed systems will be found in some of the texts cited in the references [for example, Laidler, 1958; Boyer *et al.*, 1959; Mahler and Cordes, 1966; Bray and White, 1966; see also London, 1968].

B. The Classical Mathematical Model of the Elementary Enzyme-Catalyzed Reaction

1. *The Michaelis–Menten Mechanism and the Original Equilibrium Model for Its Kinetics*

In the earlier work, [Bartholomay, 1962a, pp. 60–69] the origins and details of the formulation of the hypothesis concerning the mode of action of an enzyme molecule upon its substrate advanced by Michaelis and Menten [1913], originally in terms of the enzyme (saccharase)-catalyzed sucrose hydrolysis reaction (which has already been mentioned in this chapter in a number of other pivotal studies) were given. With an empirical equation expressing the initial rate v of inversion as a function of initial substrate concentration [S] [each cooridnate pair ([S], v) obtained from the initial phase of a new reaction run] as guide, and in keeping with the activated or active sucrose species hypothesis advanced by Arrhenius (see Section III.B) and perhaps in the presence of (or contributing to) the preliminary thinking that gave rise to the transition-state theory, Michaelis and Menten proposed that the explanation to the equation lay in the formation of a reversible intermediate E–S complex in equilibrium with the enzyme and substrate species. It was regarded that reaction consisted of occasional disruptions of the equilibrium, resulting in the transformation of S into products and the release of the intact enzyme molecule. This mechanism is usually represented as

$$\mathrm{E} + \mathrm{S} \underset{k_2}{\overset{k_1}{\rightleftharpoons}} \mathrm{ES} \xrightarrow{k_3} \mathrm{E} + \mathrm{P} \tag{442}$$

which has the same form as the generalized Lindemann–Hinshelwood mechanism for the unimolecular reaction [see Eqs. (166) and (167)] rewritten,

$$\mathrm{A} + \mathrm{M} \underset{k_2}{\overset{k_1}{\rightleftharpoons}} \mathrm{A}^* \xrightarrow{k_3} \mathrm{B} + \mathrm{C}. \tag{443}$$

It has also the same form as the simple bimolecular exchange reaction discussed in Section II, in the treatment of the transition-state theory:

$$X + YZ \rightleftharpoons (X—Y—Z) \longrightarrow XY + Z. \tag{444}$$

In even more general terms we may compare it with the general bimolecular transition-state mechahism,

$$A + B \rightleftharpoons X^* \longrightarrow C. \tag{445}$$

Thus, the E–S complex was placed in a kinetic context which strongly resembles those of the energized or activated unimolecular reactant molecule and the activated complexes of modern absolute rate theory. The bimolecular complexation step with rate constant k_1 corresponds to the activation step of the Lindemann–Hinshelwood mechanism and the association step of the transition-state theory. The unimolecular dissociation of ES with rate constant k_2 corresponds to Lindemann–Hinshelwood deactivation or transition-state-theoretic dissociation. And the unimolecular decomposition with rate constant k_3 corresponds to the decomposition of the activated complex X*. We have already seen, however, the models or theories for deriving the rate expressions in the Lindemann–Hinshelwood and transition-state theory, and we have noted that featured in these are the special natures of the intermediate species. Thus, the activated A* species and the activated complex X* are not exactly normal molecules, though the treatments of normal intermediates are used to some extent for developing associated kinetic models "by analogy." In the case of the intermediate E–S complex, in the immediately preceding section, we have already implied some of its peculiar characteristics and great complexity. The empirical experssion of the kinetics just referred to gave the first indication that, kinetically speaking, the enzyme-catalyzed reaction is, in fact, unique.

The original model devised for explaining the observed kinetic behavior is expressible as the following system of nonlinear differential equations:

$$d[E]/dt = -k_1[E][S] + (k_2 + k_3)[ES], \tag{446}$$

$$d[S]/dt = -k_1[E][S] + k_2[ES], \tag{447}$$

$$d[ES]/dt = k_1[E][S] - (k_2 + k_3)[ES], \tag{448}$$

$$d[P]/dt = k_3[ES], \tag{449}$$

where [P] is the concentration of products; where E_T is total enzyme concentration and initial free-enzyme concentration; S_0 is the initial substrate concentration; and $[P] = 0$ is the initial concentration of products, it is seen that the following conditions hold at all times:

$$E_T = [E] + [ES], \qquad S_0 = [S] + [P] + [ES]. \tag{450}$$

The assumptions (1) that an equilibrium is rapidly established between E, S, and ES, (2) that $k_1, k_2 \gg k_3$, and (3) that $[S] \gg E_T$ constitute the so-called "equilibrium model" which Michaelis and Menten implied in their derivation of the rate expression:

$$v = V[S]/(K' + [S]), \tag{451}$$

where $V \equiv \max_{[ES]} v = k_3 E_T$, v is the rate of formation of P, and K' is equal to k_2/k_1 (reciprocal of the equilibrium constant).

2. *The Steady-State Model of Briggs and Haldane*

Subsequently, Briggs and Haldane [1925] incorporated Eqs. (446)–(449) into a less restrictive model, removing the restriction $k_3 \ll k_1, k_2$; that is, that the rate of decomposition is so slow as to not disturb the continuously maintained equilibrium, and replacing it by the "steady-state" (or, better, "the pseudo-steady-state") assumption, which in turn has drawn the fire of other investigators (see Section IV.B.5).

The essence of the Briggs–Haldane reformulation is contained in the replacement of the third assumption by the condition

$$d[ES]/dt = 0 \tag{452}$$

("pseudo-steady-state assumption"). Substitution of this condition into Eq. (448) combined with Eq. (450) leads to the expression for [ES]

$$[ES] = k_1 E_T[S]/((k_2 + k_3) + k_1[S]) \tag{453}$$

from which the final form of the Michaelis–Menten equation is obtained as

$$v = V[S]/(K_m + [S]), \tag{454}$$

where now

$$K_m = (k_2 + k_3)/k_1 \tag{455}$$

has come to be called "the (generalized) Michaelis constant."

a. Estimation of Parameters. It is because of the impracticability of obtaining kinetic measurements of ES, and because of the lack of closed solutions to the system of differential equation (446)–(449) that the Michaelis–Menten equation is turned to as the basis for the interpretation of enzyme kinetic data. The same pattern seen in this equation persists in the more complicated enzyme-catalyzed reactions referred to earlier. The determinations of the parameters K_m and V are made from kinetic data consisting of the initial pairs $((S_0)_i, v_i)$, where $i = 1, 2, \ldots, R$ corresponding to R reaction runs under identical conditions. Thus, a given v_i is determined as the slope at $t = 0$ of the corresponding graph showing [S] plotted against t in the ith reaction run; and $(S_0)_i$ is the initial concentration of S in this same run. The direct determination of V and K_m from the final plot of v_i versus $(S_0)_i$ ($i =$

1, . . . , *R*) would involve fitting the hyperbola of Eq. (454) to these data. Statistical complications aside, because of the tedious nature of such a non-linear regression theory it has become customary for the experimentalist to turn to the classical procedure of linearizing such expressions. Accordingly, either of the following forms is usually employed:

$$v^{-1} = K_m/V[S]^{-1} + V^{-1} \qquad \text{(Lineweaver–Burk)}, \qquad (456)$$

$$v/[S] = V/K_m - K_m^{-1}v \qquad \text{(Eadie)}. \qquad (457)$$

In the former case K_m and V are obtained as linear regression coefficients of the plot v^{-1} versus $[S]^{-1}$. In the latter case they are determined also by linear regression of the plot of $v/[S]$ versus v. A number of more conscientious statistical estimation procedures have begun to appear in the Biometrics and Technometrics journals.

b. The Integrated Michaelis–Menten Equation (Irreversible Case). The absence of closed solutions of the system of differential equations of the irreversible enzyme–catalyzed reaction has exercised a strong effect not only on the nature of the kinetic data collected in experimental studies and on the modes of statistical analysis of such data, but also on the mode of study of enzyme kinetics. Thus, studies of the kinetics are now fragmented, relative to the total time course of the reaction, into the following four phases: (1) initial, (2) transient or presteady state, (3) steady, or pseudo-steady state, (4) final state. The first and third of these receive the most attention, along the lines indicated in the previous section. It is seen there that these two phases are analyzed in close relationship with each other. Difficulties in experimental techniques have contributed to slower progress made in the second phase. The fourth phase has been delineated because of the peculiarity of the system which gives rise to substrate-saturation effects which can cause inhibition of the reaction. It is not possible here to go into all of these matters, but it is important to call attention to additional points particularly in the third phase which have received a great deal of attention. In any case it is interesting to note that, in a sense, it is the nature of the mathematical difficulties involved in the theoretical treatment of this system, that is, its mathematical model, which has forced enzyme kinetics into a rather unique, perhaps, *ad hoc* operational pattern in chemical kinetics.

An interesting effort to turn away from the pattern of deducing steady-state properties on the basis of initial-phase data has come to be referred to as "the intergrated Michaelis–Menten equation." It should be emphasized that this refers specifically to Eq. (454) and not to the original system of differential equations; it does not imply that a closed solution of the original system has been deduced. The earliest example of this approach is due to Haldane [1930]. It begins by differentiating the second of Eqs. (450) with respect to time t and introducing the pseudo steady-state assumption of Eq. (452) into the result, leading to

$$d[\mathrm{S}]/dt \approx -d[\mathrm{P}]/dt = -v \tag{458}$$

which on substitution into Eq. (454) gives the differential equation

$$d[\mathrm{S}]/dt = V[\mathrm{S}]/(K_\mathrm{m} + [\mathrm{S}]) = k_3 E_\mathrm{T}[\mathrm{S}]/(K_\mathrm{m} + [\mathrm{S}]). \tag{459}$$

Integration then gives

$$k_3 E_\mathrm{T} t = K_\mathrm{m} \log_e (S_0/[\mathrm{S}]) + S_0 - [\mathrm{S}], \tag{460}$$

which is referred to as "the integrated Michaelis–Menten equation." Plots of t versus $\log_e(S_0/[\mathrm{S}])$ may then be used in linear regression to estimate the constants k_3 and K_m. The advantage in obtaining K_m in this way is that it depends on the entire time course of the single reaction, as opposed to initial rates of a sequence of reactions. A number of variants of this procedure have appeared, one of the most recent being by Hommes [1962].

3. *The Reversible Michaelis–Menten Mechanism*

A somewhat larger measure of mathematical success has been obtained in the case of the elementary reversible Michaelis–Menten mechanism,

$$\mathrm{E} + \mathrm{S} \underset{k_2}{\overset{k_1}{\rightleftharpoons}} \mathrm{ES} \underset{k_4}{\overset{k_3}{\rightleftharpoons}} \mathrm{E} + \mathrm{P} \tag{461}$$

obtained from Eq. (442) by "closing the loop," that is, by allowing the reversibility of the final decomposition. If one admists that in theory all reactions are after all reversible, then, Eq. (461) appears much more satisfying than Eq. (442). In addition, as we shall see in Section IV.B.3.a, immediately following, Eq. (461) contains mathematical advantages. But even in science the truism that there are two sides to every argument seems to apply. The fact is that biological reactions (that is, reactions *in vivo*) may be irreversible. As Baldwin [1963, p. 57] states: ". . . often it happens that the products of a given reaction are removed as fast as they are formed so that although the reaction is reversible, in reality it may nevertheless proceed in one direction under biological conditions" (thus pointing to another difference between the *in vivo* and *in vitro* experimental contexts).

a. Alberty's Integrated Michaelis–Menten Equation for the Reversible Reaction. Alberty [1959] obtained a (restricted) reversible counterpart of Haldane's Eq. (460) in the same fashion, employing the corresponding Michaelis–Menten equation and beginning with the differential equation

$$-\frac{d[\mathrm{S}]}{dt} = \frac{d[\mathrm{P}]}{dt} = \frac{(V_\mathrm{S} K_\mathrm{S}^{-1} + V_\mathrm{P} K_\mathrm{P}^{-1})[\mathrm{S}] - V_\mathrm{P} K_\mathrm{P}^{-1} S_0}{1 + [K_\mathrm{S}^{-1} - K_\mathrm{P}^{-1}][\mathrm{S}] + K_\mathrm{P}^{-1} S_0}, \tag{462}$$

where $V_S = k_4 E_T$, $K_S = (k_2 + k_3)/k_1$: $V_P = k_2 E_T$; $K_P = (k_2 + k_3)/k_4$. And provided that $V_S K_P \gg V_P k_3$, he showed

$$(V_S/K_S)t = (K_S^{-1} - K_P^{-1})[P] - [1 + S_0 K_P^{-1}] \log_e\{1 - [P]S_0^{-1}\} \tag{463}$$

which, as he points out, reduces to the result obtained by Walker and Schmidt [1944] for the irreversible case

$$V_S t = [P] - K_S \log_e \{1 - [P]S_0\}. \tag{464}$$

A good discussion of the details of applying this approach to the determination of rate constants in even more complex enzyme-catalyzed reactions may be found in Laidler [1958, Chapters 1, 4].

b. Closed (Exact) Solutions of the Briggs–Haldane Model for the Reversible Michaelis–Menten Mechanism. It is interesting that while exact solutions of the Briggs–Haldane system of differential equations in the irreversible Michaelis–Menten mechanism are lacking, in the reversible Michaelis–Menten mechanism, solutions to the Briggs–Haldane system have been published, though with restrictions on some of the rate constants. Of particular interest here is the solution by Miller and Alberty [1958], for it was in this connection that they showed the corresponding pseudo steady state to be a good approximation if $E_T \ll S_0$, or, if $(E_T + S_0) \ll (k_2 + k_3)/k_1$, a result which is similar to that obtained by Heineken *et al.* [1967] for the irreversible case (to be discussed in Section IV.B.5).

As in the irreversible case, where $E_T = [E] + [ES]$ and $[S] = S_0 - [ES] - [P]$, the interest centers on the following differential equations of the Briggs–Haldane model:

$$d[ES]/dt = k_1[E][S] - (k_2 + k_3)[ES] + k_4[E][P], \tag{465}$$

$$d[P]/dt = k_3[ES] - k_4[E][P]. \tag{466}$$

The use of the preceding material balance conditions on E_T and [S] allow these equations to be written as

$$\begin{aligned} d[ES]/dt = {} & k_1 E_T S_0 + (k_4 - k_1)E_T[P] \\ & - [k_1 E_T + k_1 S_0 + (k_4 - k_1)[P] + k_2 + k_3][ES] + k_1[ES]^2, \end{aligned} \tag{467}$$

$$d[P]/dt = (k_3 + k_4[P])[ES] - k_4 E_T[P]. \tag{468}$$

For the pseudo-steady-state case Haldane had obtained

$$v = \frac{d[P]}{dt} = \frac{(k_3 E_T[S]/K_S) - (k_2 E_T[P]/K_P)}{1 + ([S]/K_S) + ([P]/K_P)}, \tag{469}$$

where

$$K_S = (k_2 + k_3)/k_1 = \text{Michaelis constant for substrate}, \tag{470}$$

$$K_P = (k_2 + k_3)/k_4 = \text{Michaelis constant for product}. \tag{471}$$

In the special case $k_1 = k_4$, Miller and Alberty obtain the exact solution to the system of Eqs. (467)–(468). The implied equality of $K_S = K_P$ makes the result applicable, for example, to the fumarase, enolase, and phosphoglucose isomerase reactions (at a pH which depends on buffer and ionic strengths).

Assuming that at $t = 0$, $[ES] = 0$ (and $k_1 = k_4$), straightforward integration of Eq. (467) gives

$$[ES] = \frac{2k_1 E_T S_0 (1 - e^{-dt})}{d - b + (d + b)e^{-dt}}, \tag{472}$$

where

$$b = -(k_1 E_T + k_1 S_0 + k_2 + k_3),$$
$$c = k_1 E_T S_0,$$
$$d = [(k_1 E_T + k_1 S_0 + k_2 + k_3)^2 - 4k_1{}^2 E_T S_0]^{1/2}.$$

The substitution of Eq. (472) into Eq. (468) then gives

$$[P] = \frac{2k_3 S_0\left(d\left[1 - \exp\left\{-\frac{k_1 E_T(2k_1 S_0 + b - d)t}{b - d}\right\}\right] - \frac{k_1 E_T(2k_1 S_0 + b - d)}{b - d}[1 - e^{-dt}]\right)}{(k_2 + k_3)[d - b + (d + b)e^{-dt}]} \tag{473}$$

Connections of this result with results obtained by others [for example, Laidler, 1955a; Swoboda, 1957; Gutfreund, 1955] for the transient state are also discussed by Miller and Alberty [1958].

4. *Pre-Steady-State (Transient) Kinetics and the Calculation of Individual Rate Constants*

The measurement of individual rate constants in the Michaelis–Menten mechanism pose extremely difficult experimental problems. Some assistance is obtained by expanding further the mathematical implications of the original model, that is, of the original set of differential equations. In this way some progress has been made in the measurement of the bimolecular rate constant k_1. This required the development of two kinds of experimental techniques: (1) rapid-flow techniques (transient, or pre-steady state) and (2) relaxation techniques (post-steady state or equilibrium). The mathematical framework required for implementing the former class of techniques is discussed next.

Differentiating Eq. (449) with respect to time t,

$$d^2[P]/dt^2 = k_3\, d[ES]/dt, \tag{474}$$

followed by substituting Eq. (448) into the right-hand side and making use of the condition $[E] = E_T - [ES]$ gives

$$d^2[P]/dt^2 + (k_1[S] + k_2 + k_3)\, d[P]/dt - k_1 k_3 E_T [S] = 0. \tag{475}$$

Then, taking [S] $\approx S_0$ reduces Eq. (475) to a second-order linear differential equation with constant coefficients, with solution of the form

$$[P] = c_1 + c_2 \exp(-a_1 t) + (a_2/a_1)t, \tag{476}$$

where

$$a_1 = k_1 S_0 + k_2 k_3, \qquad a_2 = k_1 k_3 E_T S_0. \tag{477}$$

Assuming [P] $= P_0$ at $t = 0$ gives $c_2 = P_0 - c_1$. And from Eq. (449) the conditions [ES] $= 0$ at $t = 0$ gives $d[P]/dt = 0$. Using these boundary conditions the general solution obtained is

$$[P] = P_0 - (a_2/a_1^2) + (a_2/a_1^2) \exp(-a_1 t) + (a_2/a_1)t, \tag{478}$$

which amounts to

$$[P] = P_0 + \left(\frac{k_3 E_T S_0}{S_0 + K_m}\right)t + \frac{k_1 k_3 E_T S_0}{(k_1 S_0 + k_2 + k_3)^2} [\exp\{-(k_1 S_0 + k_2 + k_3)t\} - 1]. \tag{479}$$

In the pre-steady-state period, that is, for small t, discarding all terms higher than the second power of t in the expansion of the exponential term gives

$$[P] - P_0 \approx k_1 k_3 E_T S_0 t^2/2. \tag{480}$$

Then substituting $k_3 = V/E_T$ into Eq. (480) yields a convenient quadratic expression for [P] $- P_0$,

$$[P] - P_0 \approx k_1(VS_0/2)t^2. \tag{481}$$

A plot of experimental data([P] $- P_0$) versus t, then allows the estimation of k_1 as a regression coefficient. Substituting k_1 into $K_m = (k_2 + k_3)/k_1$ then yields a corresponding estimate of k_2.

5. *The Pseudo-Steady-State Challenges*

As pointed out earlier, the (pseudo-) steady-state assumption, $d[ES]/dt = 0$, has been subjected to continual criticism in the literature. Most recently Heineken *et al.* [1967a] summarized the argument in extreme terms as follows:

> . . . To the mathematician this hypothesis, known as the pseudo-steady-state hypothesis (p.s.s.h.) is somewhat scandalous. For clearly, $dc/dt = 0$ (εd $\therefore$ $d[ES]/dt = 0$) in the strict sense at only one instant, and it is notorious that to say that c is small does not of itself assert anything about the smallness of dc/dt. To the biochemist the p.s.s.h. is a valued method of simplifying operations, justified by the excellent agreement with experiment that it gives [p. 97].

This has reference to the fact then that the biochemist alledgedly justifies the assumption $d[ES]/dt = 0$ on the basis that, the enzyme being present in relatively small supply: $E_T \ll S_0$, the descending part of the curve [ES] $= f(t)$

must be very gradual, so that for the major part of the reaction the pseudo-steady-stale challenge is justifiable, thus giving rise to the Michaelis–Menten equation.

The point made by Heineken *et al.* (1967a, pp. 95–107), is that the Michaelis–Menten equation may, however, be arrived at asymptotically and in a mathematically more rigorous manner, still on the assumption of the smallness of E_T but by recourse to the singular perturbation theory of differential equations.

Referring to our Eqs. (447) and (448) altered by substitutions from Eq. (450),

$$d[S]/dt = -k_1 E_T[S] + (k_1[S] + k_2)[ES], \qquad (482)$$

$$d[ES]/dt = k_1 E_T[S] - (k_1[S] + k_2 + k_3)[ES], \qquad (483)$$

with an eye toward the perturbation theory format, they introduce the following additional quantities:

$$y = [S]/S_0, \qquad z = [ES]/E_T, \qquad \mu = E_T/S_0, \qquad (484)$$

$$\tau = k_1 E_T t, \qquad \alpha_0 = (k_2 + k_3)/k_1 S_0, \qquad \lambda = k_3/k_1 S_0. \qquad (485)$$

With these additional substitutions, Eqs. (482) and (483) are replaced by

$$dy/d\tau = -y + (y + \alpha_0 - \lambda)z, \qquad (486)$$

$$\mu(dz/d\tau) = y - (y + \alpha_0)z, \qquad (487)$$

with initial conditions

$$y = 1, \ z = 0 \qquad \text{at} \quad \tau = 0. \qquad (488)$$

They note that the limiting condition

$$\mu = E_T/S_0 = 0 \qquad (489)$$

applied to Eq. (487) yields $z = y/(y + \alpha_0)$ which, when substituted into Eq. (486), gives

$$dy/d\tau = \lambda y/(y + \alpha_0), \qquad (490)$$

referred to as "the dimensionless form" of the Michaelis–Menten equation, which is justified by first reintroducing the rate constants, [S], S_0, and K_m from the relations Eq. (485), giving

$$dy/d\tau = -(k_3/k_1 S_0)[S]/(K_m + [S]). \qquad (491)$$

But $dy/d\tau = (dy/dt)(dt/d\tau)$ so that where $\tau = k_1 E_T t$,

$$dy/dt = v = -k_3 E_T[S]/(K_m + [S]). \qquad (492)$$

Then setting $V = k_3 E_T$ and replacing y by $[S]/S_0$ gives

$$d[S]/dt = -V[S]/(K_m + [S]), \qquad (493)$$

the Michaelis–Menten equation.

In the process of identification it is thus established that the Michaelis–Menten equation corresponds to the limiting (zero) value of the central perturbation-theoretic parameter μ. The theory of singular perturbations asserts that under suitable conditions the solution of Eqs. (486) and (487) lies "close" to the solution of Eq. (490), and, in fact, tends to it in the limit as $\mu \longrightarrow 0$. In other words, simply with the classical assumption $E_T \ll S_0$ it is possible to legitimize the pseudo-steady-state hypothesis.

A technical sense of great general importance in dealing with such systems of differential equations in which this convergence may be understood is given in a theorem due to Tikhonov [1952; and quoted also in Vasil'Eva, 1963, p. 21]. To understand the statement of this theorem some preliminary notions are necessary which are stated next with reference to the more general system

$$\mu(dz/dt) = F(z, y, t), \tag{494}$$

$$dy/dt = f(z, y, t). \tag{495}$$

Where μ is a small positive parameter, z is an m-dimensional vector, and y is an n-dimensional vector with initial conditions $z(t_0) = z_0$, $y(t_0) = y_0$, let $z = \phi(y, t)$ be one of the solutions of the system of algebraic equations $F(z, y, t) = 0$ defined on a bounded, closed set D. Associated with the root $z = \phi(y, t)$ is the so-called "degenerate system of equations":

$$dy/dt = f(\phi(y, t), y, t). \tag{496}$$

The solution of this degenerate system satisfying the initial conditions $y(t_0) = y_0$ is indicated by $\bar{y}(t)$, and corresponding to it $\bar{z} = \phi(\bar{y}, t)$.

In general, a root $z = \phi(y, t)$ is called "isolated" on the set D if there exists an $\varepsilon > 0$ such that the system $F(z, y, t) = 0$ has no solution other than $\phi(y, t)$ for $|z - \phi(y, t)| < \varepsilon$.

The equation

$$dz/d\sigma = F(z, y^*, t^*), \tag{497}$$

where y^*, t^* now are regarded as parameters, is called "the adjoined system." The isolated root $z = \phi(y, t)$ will be called "positively stable" in D if, for all points (y^*, t^*) in D, the points corresponding to $z = \phi(y^*, t^*)$ are asymptotically stable stationary points (in the sense of Lyapunov) of the adjoined system, Eq. (497), as $\sigma \longrightarrow \infty$. If the same situation holds as $\sigma \longrightarrow -\infty$, then the root is classified as "negatively stable." The "domain of influence" of an isolated stable root $z = \phi(y, t)$ is the set of points (z^*, y^*, t^*) such that the solution of the adjoined system satisfying the condition $z|_{\sigma=0} = z^*$ tends to the value $\phi(y^*, t^*)$ as $\sigma \longrightarrow \infty$.

Tikhonov Theorem. If some root $z = \phi(y, t)$ of the system of algebraic equations $F(z, y, t) = 0$ is an isolated, positively stable root in some bounded, closed domain D; if the initial point (z_0, y_0, t) belongs to the domain of influence of this root; and if the solution $y = \bar{y}(t)$ of the degenerate system belongs to D for $t_0 \leqslant t \leqslant T$, then the solution $(y(t), z(t))$ of the original system tends to the solution $(\bar{y}(t), \bar{z}(t))$ of the degenerate system as $\mu \longrightarrow 0$, the passage to the limit

$$\lim_{\mu \to 0} z(t, \mu) = \bar{z}(t) = \phi(\bar{y}(t), t), \tag{498}$$

holding for $t_0 < t \leqslant T_0 < T$, and the passage to the limit

$$\lim_{\mu \to 0} y(t, \mu) = \bar{y}(t), \tag{499}$$

holding over the same interval.

In the case of the Michaelis–Menten system of Eqs. (486)–(487), the Tikhonov theorem guarantees that their solution $(y = y(\tau, \mu), z = z(\tau, \mu))$ tends to the solution $(y = \bar{y}(\tau);\ z = \bar{z}(\tau) = \phi(\bar{y}, \tau)$ of the corresponding degenerate equation,

$$dy/d\tau = -y + (y + \alpha_0 - \lambda)\, \phi(y, \tau) \tag{500}$$

(with $y_0 = 1$ at $\tau = 0$) in the sense of the Tikhonov theorem, writing for this simply

$$\lim_{\mu \to 0} y(\tau, \mu) = \bar{y}(\tau), \tag{501}$$

$$\lim_{\mu \leftarrow 0} z(\tau, \mu) = \phi(\bar{y}(\tau), \tau), \tag{502}$$

where $z = \phi(y, \tau)$ is a root of the equation

$$y - (y + \alpha_0)z = 0, \tag{503}$$

that is

$$y - (y + \alpha_0)\, \phi(y, \tau) = 0. \tag{504}$$

The details of constructing the actual asymptotic solution are very laborious. To a first approximation they have been carried out by Heineken *et al.* (1967a, b) who discuss the general procedure involved. Suffice it to remark here that they begin the construction by cleverly introducing the new time parameter σ,

$$\sigma \equiv \tau/\mu, \tag{505}$$

which allows the delineation of two different time scales, σ and τ, with respect to which procedures are given for determining the "inner" and the "outer" solutions. The former corresponds to time close to $\tau = 0$. Thus, if μ is small, then the inner solution will be in the expanded time scale of σ and the outer solution is handled in the scale of τ. For example, since

$$dz/d\sigma = (dz/d\tau)\,(d\tau/d\sigma) = \mu\, dz/d\tau, \tag{506}$$

Eq. (494) becomes

$$dz/d\sigma = F(z, y, \tau) \tag{507}$$

so that if μ is small, the original time parameter becomes vastly accelerated, z being always held to the stable root of $F(z, y, \tau) = 0$. The method of working out the asymptotic inner and outer solutions is due to Vasil'Eva [1963].

C. A Closer Look at Some Active Site Characterizations

The three hypotheses presented earlier in Section IV.A.3.e beg the question of the actual nature of an active-site configuration in an enzyme molecule. Considering the great number of identified and otherwise characterized biological macromolecules, there is a paucity of information on the particulars of an active site. In this section we shall note some outstanding examples of work in this difficult area which after years of persistent investigation is beginning to produce some exciting results. In particular, the lysozyme molecule and the class of enzymes called by Vallee "metalloenzymes" will be referred to at least in sufficient detail to convey the flavor of the investigative methods and approaches which have been tried and some idea of the results and state of progress attained.

1. *Lysozyme*

The many studies of this molecule, like the studies of the structure of DNA by Watson and Crick [1953], have featured the use of hardware models and visual devices because of the overwhelming complexity of detail deduced from x-ray diffraction and binding studies.

Preliminary x-ray diffraction work on low-resolution structure of hen egg-white lysozyme (molecular weight, 17,000) in several different crystal forms was first reported in 1962 [Blake *et al.*, 1962; see also Davies, 1967]. This was followed in 1965 [Blake *et al.*, 1965] by a complete structural determination of the molecule at 2 Å resolution. The molecule was shown to be roughly ellipsoidal with dimensions 45 × 30 × 30 Å and seemed to have a marked cleft. The interpretation of the electron density map was accomplished by making use of the one-dimensional structure, the amino acid sequences of Canfield [1963] and J. Jolles, Jauregui-Adell, and P. Jolles [1963]. Blake *et al.* used essentially the procedure of Kendrew and Watson [see Watson, 1965] for the analysis of myoglobin, including the construction of a scale wire model, peptide by peptide. Only about 25% of the molecule was found to exist in helical form.

a. The Active-Site Cleft. Information on the active-site details was obtained by binding studies and x-ray diffraction work. Phillips [1966] and coworkers determined [see also Dayhoff and Eck, 1968] that known inhibitors were bound in the cleft, the cleft apparently enlarging (in accord with the expectations of induced-fit hypothesis) to permit the binding. From these studies, considering also the action of the enzyme in the hydrolysis of a number of polymers, the position of the active site was placed in the region near aspartic acid 52 and glutamic acid 35. Reasoning from the model the known substrate was considered to fit into the cleft with many weak bonding contacts, though the substrate would have to be distorted in the region in which its bond was known to be broken. The so-called "difference Fourier synthesis substitution method" of x-ray diffraction theory was used to determine the location of the inhibitor-binding sites of the enzyme. Of kinetic interest in these studies is the fact x-ray diffraction evidence of this kind suggested the hypothesis (see below) of the molecular basis of enzyme action for lysozyme.

Inhibitor-binding studies in the crystal were carried out at 6 Å resolution for a number of inhibitors and at 2 Å resolution for polymers such as those containing *N*-acetylglucosaminyl-β-(1-4)–*N*-acetylmuraminyl (NAG–NAM) which showed evidence that the individual sugar residues combine in six positions on the surface of the enzyme molecule and with one exception are collinear with the longitudinal aspect of the cleft. Slight changes in the conformation of the enzyme were observed when either NAG or tri-NAG was bound. Changes in atomic positions were similar in both cases. Residue 62 appeared to move by about 0.75 Å, tending to narrow the cleft; and there were related small shifts particularly to the left of the cleft. It is not believed that the tri-NAG site (*ABC*) is identical with the active site, tri-NAG apparently occupying only one-half of the cleft.

Through a detailed process of elimination it has been suggested by Rupley that the catalytic site occurs in the region between binding sites *D* and *E*. In this region of the cleft there are only two reactive residues, Glu 35 (glutamic acid 35) and Asp 52 (aspartic acid 52). The details of the cleft are shown in remarkable three-dimensional photographs of molecular models constructed from atomic hardware components, obtained by the full color Xograph technique (Harte and Rupley, 1968]. A skeletal model of lysozyme [see Dayhoff and Eck, 1968, pp. 221–223, for the complete composition of chicken lysozyme and bacteriophage T4 lysozyme, and other pertinent references for the structural details] had been built first with Kendrew-type brass rod parts to the scale of 2 cm per Å from coordinates provided by Phillips, North, and Black and their collaborators. This served as the guide to the construction of the model which was made subsequently with space-filling CPK atoms. Use of the new CPK H-bond hydrogen and H-bond acceptor oxygen atoms allowed substantial improvement over an earlier space-filling model. The beautiful color photographs shown in two plates of Harte and Rupley's paper [1968] look directly into the active-site cleft. This feature of the model is formed by the union of the main body of the enzyme with a wing composed of residues 40–85. Also pictured is a saccharide above the cleft in their Plate I.

b. The Mechanism Proposed for the Kinetics of Lysozyme-Catalyzed Hydrolysis.† In follow-up studies Rupley and Gates [1967] proposed the following detailed hypothesis (Fig. 19) of the lysozyme-catalyzed hydrolysis mechanism, which is supported by kinetic data and is consistent with the cleavage by the enzyme of the *N*-acetylglucosamine oligosaccharides. The complexity of this mechanism, compared with the simple Michaelis–Menten mechanism of enzyme kinetics, serves as an example of the more detailed and complex mechanisms of action which will result from the localization of enzyme-substrate interactions. The mechanism is described schematically by the complex reaction shown in Fig. 19, which assumes that lysozyme (E) has for its substrate an *N*-acetylglucosamine oligosaccharide (S) and that the products of the hydrolytic cleavage are, for example, P_1, aglycone moiety of S, and P_2, a glycosyl unit of S. The component reactions, identifiable numerically by the subscripts on the rate constants written over the arrows, are divided into two subsets: (1) cleavage reactions (2, 5, 8, 11) and (2) association–dissociation steps (1, 3, 4, 7, 9, 10, . . .). A is an acceptor other than water, for example, *N*-acetylglucosamine; (ES) may be a whole set of related complexes $\{(ES)_1, \ldots, (ES)_n\}$; $E\text{–}P_2$ indicates an enzyme-bound glycosyl unit;

$$E\begin{matrix} , A \\ \diagdown P_2 \end{matrix}$$

simply a loose association of $E\text{–}P_2$ with A, and so forth.

Designating equilibrium constants by K_i and corresponding forward and reverse rate constants by k_i and k_{-i}, they define an overall association constant

$$K_a \equiv \sum_i K_i, \tag{508}$$

and an overall rate constant

$$k_b \equiv \frac{1}{K_a} \sum_i K_i k_i, \tag{509}$$

that is, as a numerical average of the individual rate constants, each weighted by its normalized equilibrium constant. The mechanism is discussed by analogy with the action of chymotrypsin on low-molecular-weight substrates [see Hein and Niemann, 1962].

As Rupley and Gates emphasize, by no means is their proposed mechanism unique; it is, in fact, one of the simplest possibilities. In its favor is evidence

†From Rupley and Gates [1967, pp. 496–510].

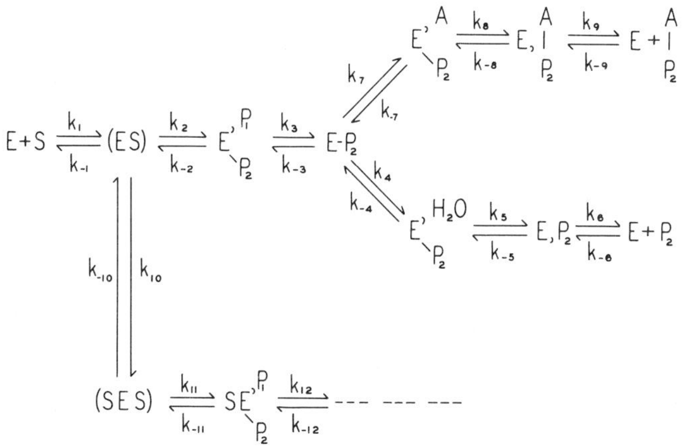

$$E + S \underset{k_{-1}}{\overset{k_1}{\rightleftharpoons}} (ES) \underset{k_{-2}}{\overset{k_2}{\rightleftharpoons}} E\begin{smallmatrix} P_1 \\ P_2 \end{smallmatrix} \underset{k_{-3}}{\overset{k_3}{\rightleftharpoons}} E\text{-}P_2$$

$$E\text{-}P_2 \underset{k_{-7}}{\overset{k_7}{\rightleftharpoons}} E\begin{smallmatrix} A \\ P_2 \end{smallmatrix} \underset{k_{-8}}{\overset{k_8}{\rightleftharpoons}} E,\begin{smallmatrix} A \\ | \\ P_2 \end{smallmatrix} \underset{k_{-9}}{\overset{k_9}{\rightleftharpoons}} E + \begin{smallmatrix} A \\ | \\ P_2 \end{smallmatrix}$$

$$E\text{-}P_2 \underset{k_{-4}}{\overset{k_4}{\rightleftharpoons}} E\begin{smallmatrix} H_2O \\ P_2 \end{smallmatrix} \underset{k_{-5}}{\overset{k_5}{\rightleftharpoons}} E,P_2 \underset{k_{-6}}{\overset{k_6}{\rightleftharpoons}} E + P_2$$

$$(ES) \underset{k_{-10}}{\overset{k_{10}}{\rightleftharpoons}} (SES) \underset{k_{-11}}{\overset{k_{11}}{\rightleftharpoons}} SE\begin{smallmatrix} P_1 \\ P_2 \end{smallmatrix} \underset{k_{-12}}{\overset{k_{12}}{\rightleftharpoons}} \text{--- --- ---}$$

Fig. 19. The Rupley–Gates hypothesized lysozyme-catalyzed reaction mechanism for the hydrolytic cleavage of *N*-acetylglucosamine oligosaccharides. [From Rupley and Gates, 1967, Fig. 4, p. 503.]

produced by them that the cleavage pattern in terms of the saccharide di-, tri, tetra-, penta-, and hexamers support the existence of more than one type of ES complex. Furthermore, the demonstrated dependence of the rate of hydrolysis on chain length of the substrates correlates with the x-ray diffraction characterization of the active site, emphasizing the potential importance of crystallographic evidence in enzyme kinetics.

2. *Metalloenzymes and Metal–Enzyme Complexes*

The importance of the multi- and cross-disciplinary approach to the elucidation of kinetic mechanisms with the active site as a common focal point is further demonstrated by important long-range studies on other enzymes classified as metalloenzymes. Major pioneering and continuous efforts in this direction have been undertaken by Vallee and his co-workers whose fundamental investigations into the role of trace metals such as zinc, magnesium, and copper in these enzymes go back approximately 20 years. Their contributions have produced a large body of results and new concepts calling for integration into the future elaboration of enzyme kinetics.

In a comprehensive review article Vallee [1960] differentiates between "metalloenzymes" and "metal–enzyme complexes" in terms of the degree of association or bond strength between the enzyme and the metal with which it is associated. Thus, the latter refers to a loose association between an enzyme and the metal ion which is necessary for its activity, a high degree of specificity apparently existing. But a metalloenzyme assumes the trace (oligo-) metal to be an integral part of the enzyme molecule, important to both its structure and to its catalytic activity. Some provocative replacement studies have been carried out in which the metal native to the enzyme has been replaced by others, in some cases with increased activity. It is allowed that the two classes of enzymes delineated in this fashion may be two extreme points in a whole spectrum of possible types classifiable in terms of bond strengths, so that present distinctions are not considered to be definitive, being offered more in the spirit of operational definitions. The classes of enzymes considered mainly are the dehydrogenases, carbonic anhydrase, and carboxypeptidase.

Carboxypeptidase. As a case in point consider carboxypeptidase, first crystallized in 1937 and purified from bovine pancreatic juice, with molecular weight 34,300. Spectrographic and chemical studies [Vallee and Neurath, 1954, 1955] have concurred that this is a metalloenzyme containing one atom of zinc, all other metals associated with it being present in insignificant amounts. The symbolism (CPD)Zn has been suggested as a convenient representation in stoichiometric expressions and equations. Among the *ad hoc*, operational criteria satisfied was the concomitant rise in the zinc/protein and activity/protein ratios obtained in following the course of purification of the enzyme. Moreover, other extraneous metals were shown to decrease in concentration as zinc and specific activity continued to increase. (It had been pointed out in an early stage of these studies that the zinc atom is bound sufficiently firmly to the CPD protein that they are isolated together). Removal of zinc was accomplished by orthophenanthroline (OP) a known metal-chelating agent. An experimental accomplishment of note was the reincorporation of one gram-atom of zinc per mole, with full restoration of activity [Vallee *et al.*, 1960; Coleman and Vallee, 1960]. Of additional interest was the discovery that other transition metals could also restore activity in the following order of decreasing magnitudes: Co^{2+}, Ni^{2+}, Zn^{2+}, Mn^{2+}, Fe^{2+}. Isotope exchange experiments showed that Co^{2+} and zinc occupy identical sites on the enzyme surface.

x-Ray crystallographic studies carried out more recently by Ludwig *et al.* [1967] on the tertiary structure and function of carboxypeptidase A (CPA) have elaborated the details of structure. For example, in an electron-density map at 2.8 Å resolution, clearly resolved regions of right-handed helix are discernible. Difference electron-density maps at 6 Å resolution show peaks which locate a substrate-binding site for the substrate glycine-L-tyrosine. Efforts to fit the few published portions of the primary structure; that is of the amino acid sequence, to these x-ray diffraction findings are currently being made. At the same time Vallee and co-workers have produced a number of possible characterizations of the structure in the neighborhood of the zinc-containing active site, based on physicochemical considerations as well as inhibition studies. The combinations of these results with the x-ray diffraction studies centering on the binding between enzyme and substrate, at high resolution, should eventually lead to a better understanding of the elusive kinetic aspects of this physiologically important enzyme.

Laidler [1958] made the following cryptic remark referring to early kinetic studies of carboxypeptidase:

> A detailed investigation of the kinetics of carboxypeptidase-catalyzed reactions has been carried out by Lumry. Smith, and Glantz. The present section will be devoted to this work, since the results of it indicate the previous kinetic studies of carboxypeptidase systems are not entirely reliable [p. 252].

The reasons for the early confusion and anomalies in dealing with the kinetics of this enzyme have become apparent as knowledge of its structure and special features have increased. Current attempts at elucidating its kinetics are based closely on the new findings referred to here. But, even so, suggested mechanisms must still be considered at the level of verifiable hypotheses. In this connection, most recently, Vallee *et al.* [1968] have proposed a mathematical model for the substrate binding and kinetics which correlates well with a number of findings and looks quite promising. The nature of the complexities involved is set forth in portions of their abstract, as follows:

> A schematic model based on multiple modes of substrate binding is proposed . . . . The model proposes that nonidentical but overlapping binding of dipeptides . . . could be one basis for much of the existent data, characterized by substrate and product inhibition and activation. . . . The model has correctly predicted a reciprocal effect of products and certain analogous compounds on the esterase and peptidase activities of the enzyme. Thus, compounds presumably binding to the nonoverlap loci, which constitute the elements of the site, inhibit one activity and activate the other. . . . The model also predicted that certain peptides should be competitive inhibitors of the esterase activity [Vallee *et al.*, 1968, p. 3547].†

Commenting on the x-ray diffraction studies mentioned in the preceding paragraphs, they mention that at least two different but competitive binding sites in the vicinity of the zinc atom can be discerned for the inhibitor β-(p-iodophenyl)propionate. They add that two regions of increased electron density, labeled A and B, are also detected on binding of both glycyl-L-tyrosine and L-lysyl-L-tyrosinamide.

It is certainly clear from such studies that the Michaelis–Menten mechanism and its associated mathematical model in kinetics is by no means a general solution to the problem of characterizing and interpreting the details of enzyme-catalyzed reactions. It would appear that the future of chemical kinetics must develop around the active-site concept and in strong relation to enzyme structure. While Michaelis–Menten kinetics have served to bring

†Reprinted from *Biochemistry*, Volume 7, No. 10, October 1968, p. 3547. Copyright 1968 by the American Chemical Society. Reprinted by permission of the copyright owner.

us to this realization by opening up the field of enzyme kinetics, it seems important to turn to new kinds of formulations.

Examples have been given in this section of the nature of a single active site and its kinetic significance and possible treatment. But this is the simplest of the new situations to consider. Some enzymes are suspected to have more than one active site. These so-called "multivalent" enzymes add at least another order of complexity to enzyme kinetics. In Section IV.D, after some reflections on probabilistic formulations suggested by the active-site concept and considering some further implications of the active-site hypotheses referred to in Section IV.A.3.e, we shall discuss some of the first attempts already on record for characterizing multivalent systems. It will be seen that the nature of the complexities involved in such systems, even at only slightly below the macroscopic molecular level of discussion, require stretching the deterministic approach beyond its natural limitations, the arguments becoming more and more "stochasticized." The first attempts at crossing the border into complete stochastic models are therefore indicated in Sections IV.E and F in the spirit of anticipation.

D. Kinetic Significance of Active Site

1. *The Difficulties Involved in Applying the Idealized Classical Collision Theory*

From our discussion of the active site it is evident that in applying collision theory to enzyme kinetics, one is confronted at once with the fundamental limitations of collision theory itself and with the necessity to discard the idealized conceptions of a collision between the rigid inelastic spheres borrowed from gas kinetics. This refers to the difficulties involved in extending collision theory to reactions beyond second-order molecularity. Theoretical demonstrations of the rarity of a reaction requiring termolecular collisions, even in the simplest case of idealized spherical molecules in which collision is considered to involve contacts or "near contacts" at random points on their surfaces, have induced investigators to explain reactions involving more than two reactants as reactions compounded of unimolecular and bimolecular steps.† If one considers, in addition to enzyme and substrate, cofactors or modifiers, and inhibitors, or if one considers multiactive centers on a single enzyme surface capable of, or requiring the cooperation of, two or more substrate molecules—all in relation to the active-site hypotheses—then it is clear that even greater limitations exist in dealing with enzyme

†Since the completion of this chapter in 1969, a paper on termolecular reactions pertinent to this point has appeared [Bartholomay, 1971].

kinetics than in ordinary chemical kinetics. The matter will be considered further in Section IV.D.1.c.

a. Probabilistic Framework for E–S Collisions. Such specification, of course, causes us to reexamine the whole question of what actually constitututes a collision, a question that can no longer be treated by analogy with gas kinetics. Moving one small step closer to the reality of the situation, suppose that the enzyme molecule is still represented by a spherical idealization of radius r_1, but that a small area σ_1 is specified on its surface, indicating its active site. Correspondingly let a substrate molecule be thought of also as a spherical surface of radius r_2, but again, with a small surface area designated by σ_2. In this context a "collision" might be defined as a one-point contact between the areas σ_1 and σ_2.

Assuming the simplest possible probability space for characterizing this event, we might begin by defining an "effective E–S collision" as a compound event $E_1 \cap E_2$, where E_1 is the event of an ordinary random bimolecular point collision occurring anywhere on the two spheres, and E_2 is the (dependent) event that the point of contact falls within both active areas σ_1 and σ_2 of the E and S molecules. In this case the probability $P(E_1 \cap E_2)$ would be given by

$$P(E_1 \cap E_2) = P(E_1)P(E_2 | E_1), \tag{510}$$

where $P(E_2 | E_1)$ is the conditional probability of event E_2 given that E_1 has occurred.

The expression for $P(E_1)$ would be determined as in ordinary bimolecular collision theory [see Bartholomay, 1962a, Eq. (90), p. 29], namely, as

$$P(E_1) = (8\pi k_B T/m_{12})^{1/2}(r_1 + r_2)^2 \exp\{-mV_0{}^2/2k_B T\}, \tag{511}$$

the probability of a point contact between one of the n_1 molecules E present at time t and one of the n_2 molecules S present at time t, where $V_0 > 0$ is the critical value of the relative velocity between E and S, $m_{12} = m_1 m_2/(m_1 + m_2)$, and m_1 and m_2 are the masses of E and S.

An expression for $P(E_2 | E_1)$ could be obtained by using geometric probability measure as follows: Assume uniform distributions of orientations for the E and S molecules, and a condition of statistical independence between them; then

$$P(E_2 | E_1) = P(\sigma_1)P(\sigma_2), \tag{512}$$

where $P(\sigma_1)$ is the probability that the point of effective contact in the E molecule lies in the active region of surface area σ_1 and $P(\sigma_2)$ is the corresponding probability for the S molecule. In this case,

$$P(\sigma_1) = \sigma_1/4\pi r_1{}^2, \tag{513}$$

$$P(\sigma_2) = \sigma_2/4\pi r_2{}^2, \tag{514}$$

so that

$$P(E_2 \mid E_1) = \sigma_1\sigma_2/16\pi^2 r_1{}^2 r_2{}^2, \tag{515}$$

whence

$$P(E_1 \cap E_2) = (4\sqrt{2}\,\pi^{3/2})^{-1}(k_B T/m_{12})^{1/2}(\sigma_1\sigma_2/r_{12}^2)\exp\{-mV_0{}^2/2k_B T\}, \tag{516}$$

where $r_{12} = r_1 r_2/(r_1 + r_2)$.

Thus, by multiplying the ordinary rate constant by the probability factor $P(E_2 \mid E_1)$ given in Eq. (515), one obtains a slightly more reasonable approximation to the rate constant k_1 corresponding to the association step of the Michaelis–Menten mechanism.

More detailed formulations can be obtained by varying the distributions governing the movements and orientations of the molecules and by varying the shapes of the molecules and active regions and taking into account some of the recent findings on the nature of the active-site–substrate interaction. In any case, the broad outlines of a plan for incorporating such details would be contained in reinterpretations of Eq. (510).

b. Additional Considerations. While a slightly more realistic interpretation of the basic notion of an E–S collision motivated the preceding demonstration, it must also be underscored that purely structural and mechanical interpretations do not suffice. The progression of hypotheses from the lock–key to the induced-fit and the entatic state make clear that an effective E–S collision involves not only inter- and intramolecular topological factors but also the deployment of energy to certain critical bonds. This, of course, is also acknowledged in ordinary collision theory and may be seen most simply by the substitution of E_0, the critical activation energy, for the kinetic energy of relative collision $\frac{1}{2}mV_0{}^2$ appearing in the exponential term of expressions such as Eq. (511). A corresponding replacement in Eq. (576), however, would involve much greater elaboration. Thus, even if the structure in the immediate vicinity of the active site is known, much more information is required to explain the bond-formation mechanism. As Laidler [1958] puts it with reference to the lock–key hypothesis, first the key must fit, but then the key must be turned to open the lock.

Finally, with reference to our $\mathcal{B}$(C, S, E, F, τ) symbolism for a biological system, if one considers the active site as the target "biosystem" $\mathcal{B}$, then, in addition to the composition, structure, and functional aspects, the environmental systemic component E must be elucidated. In this case, the latter systemic component would have to include the coupled E–S system, the total environment, and at least that portion of the remaining overall enzyme molecule system that can in some way influence the interaction. This raises the question of how far into the enzyme molecule the "sphere of influence" of the active site configuration extends. One might include here all side chains and peptide bonds extending to the active site on the surface and possibly the sites of attachment of cofactors and their own "spheres of influence" relative to the enzyme catalytic function F of the system.

2. *Additional Difficulties Presented by the Case of n Active Sites, or Multivalent Kinetics*

In the case of multivalent enzyme with $n \geqslant 2$ active sites, two different types of mathematical models have been suggested in the literature. The first is an extension of deterministic Michaelis–Menten kinetics combined with the usual convention of considering the overall n-molecular mechanisms as composed of a succession of bimolecular associations, with no attention to the special nature of the active-site–substrate interactions or allowances for site–site interactions. The second type is based on analogies with statistical mechanical and adsorption processes, utilizing probabilistic arguments and allowing site–site interactions. The first is demonstrated in Section IV.D.3 and the second type in Sections IV.D.4 and 5.

A large number of alternative models are possible in n-valency kinetics. To assume *a priori* that an algorithm can be found governing the arrangement or placement of the sites in the enzyme would be inconsistent with the high degree of individuality and specificity exhibited by enzymes. Multi-active-site kinetic model constructions may therefore be expected to begin with a statement of the particular arrangement assumed, such as (topologically) linear arrangements, random arrangements or closest-packing arrangements. As we shall see, these are, in fact, the cases that have been studied in the literature.

Furthermore, in considering the mode of action, in even the most macroscopic molecular model, it is necessary to specify the existence or nonexistence of interactions or cooperations between the sites. The simplest cases are: (1) no site–site interactions, (2) nearest-neighbor interactions only. To allow for more than this would be to open the door to a tremendous number of variations, combinatorially speaking.

The collision-theoretic aspects of n-site enzyme kinetics suggest an "n-body problem" which brings to mind the centuries-old n-body problem of analytical or celestical mechanics. It is interesting that n-body problems are of fundamental importance at opposite ends of the spectrum of magnitudes of material objects. The problem of the stability of the solar system as originally formulated by Laplace is one of the earliest forms of the n-body problem in science. Attempts at solving such problems—most particularly that of the collision of three arbitrary celestial bodies in its various forms such as the so-called "restricted problem" have led to a great development in analytical mechanics generally. In this connection Poincaré introduced the topological approach involving considerations of phase space from the point of view of differentiable manifolds and Lie theory. A particularly noteworthy result was obtained by Sundman [1913], who found that if the angular momentum of the bodies is nonzero about every axis through the center of gravity of the three-body system, the least of the three mutual distances will always exceed a specifiable constant, depending on the initial configuration. Thus, even in the case of three bodies in a celestial system, actual triple collision is regarded as a practical impossibility, that is, an event of negligibly small probability. Abraham and Marsden [1967] give a good account of the mathematical machinery and models growing out of these problems, none of which has been solved completely.

The n-molecular collisions of chemical and biochemical kinetics, of course, represent entirely different systems, the principal difference being that the regularity or stability of the celestial system is replaced by molecular chaos, requiring the substitution of stochastic processes in place of (deterministic) mathematical analysis. A brief discussion of termolecular collisions from the stochastic point of view was presented earlier [Bartholomay, 1962a]. Additional results have been presented elsewhere [Bartholomay, 1971].

3. *A Deterministic Michaelis–Menten Type Mathematical Model for n-Valency Kinetics with No Site–Site Interactions*

Given an enzyme molecule with n active sites, without specifying their relative locations or their structural details, assuming the enzyme molecule and its substrate molecules to be idealizations, and assuming that the catalytic activities associated with each site are identical but mutually independent among themselves (that is, with no site–site interactions of any kind), Botts and Morales [1953]† extended the Michaelis–Menten kinetic mechanism accoding to the following system of chemical equations:

$$\left\{\begin{array}{c} E + S \underset{k_2^{(1)}}{\overset{k_1^{(1)}}{\rightleftharpoons}} ES_1 \\ ES_1 + S \underset{k_2^{(2)}}{\overset{k_1^{(2)}}{\rightleftharpoons}} ES_2 \\ \vdots \qquad \vdots \\ ES_{i-1} + S \underset{k_2^{(i)}}{\overset{k_1^{(i)}}{\rightleftharpoons}} ES_i \\ \vdots \qquad \vdots \\ ES_{n-1} + S \underset{k_2^{(n)}}{\overset{k_1^{(n)}}{\rightleftharpoons}} ES_n \end{array}\right\} \left\{\begin{array}{c} ES_1 \overset{k_3^{(1)}}{\longrightarrow} E + P \\ ES_2 \overset{k_3^{(2)}}{\longrightarrow} ES_1 + P \\ \vdots \qquad \vdots \\ ES_i \overset{k_3^{(i)}}{\longrightarrow} ES_{i-1} + P \\ \vdots \qquad \vdots \\ ES_n \overset{k_3^{(n)}}{\longrightarrow} ES_{n-1} + P \end{array}\right\} \tag{517}$$

†From Botts and Morales [1953, pp. 701–702].

In this representation ES_n represents the enzyme with all of its n active sites occupied by substrate molecules; and ES_i ($i = 1, 2, \ldots, n-1$) the enzyme with i sites occupied, so that at any time a mixture of n different species of complexes may be present. Implicitly, it is assumed here that the enzyme adds only one substrate molecule at a time and similarly that dissociations or decompositions occur at only one site at a time; the rarity of termolecular "collisions" make the former condition reasonable. However, if one allows random lifetimes for the complexes these dissociations and decompositions could occur in randomly intermixed groups. For this reason a stochastic model such as the one to be described in Section E would seem to be more realistic. This system would then be discussible in terms of Markovian transition probabilities between the various states. However, a discussion of this approach is postponed.

Considering the complete ith sequence of the extended Michaelis–Menten mechanism

$$ES_{i-1} + S \underset{k_2^{(i)}}{\overset{k_1^{(i)}}{\rightleftharpoons}} ES_i \xrightarrow{k_3^{(i)}} ES_{i-1} + P, \tag{518}$$

the corresponding differential equation is

$$d[ES_i]/dt = k_1^{(i)}[ES_{i-1}][S] - (k_2^{(i)} + k_3^{(i)})[ES_i]. \tag{519}$$

Botts and Morales take as a pseudo-steady-state assumption

$$d[ES_i]/dt = 0, \qquad i = 1, 2, \ldots, n, \tag{520}$$

which leads to

$$[ES_i] = k_1^{(i)}[ES_{i-1}][S]/(k_2^{(i)} + k_3^{(i)}) = K^{(i)}[ES_{i-1}][S], \qquad i = 1, 2, \ldots, n, \tag{521}$$

where

$$K^{(i)} = (K_m^{(i)})^{-1} = k_1^{(i)}/(k_2^{(i)} + k_3^{(i)}). \tag{522}$$

The total rate V of production of product P is the sum of the individual rates $\{V_i\}$ as described in the equation

$$V = \sum_{i=1}^{n} V_i = \sum_{i=1}^{n} k_3^{(i)}[ES_i] \tag{523}$$

$$V = \sum_{i=1}^{n} k_3^{(i)} K^{(i)}[ES_{i-1}][S] \tag{524}$$

But $[ES_{i-1}] = K^{(i-1)}[ES_{i-2}][S]$, so that proceeding recursively,

$$[ES_{i-1}] = \left(\prod_{j=1}^{i-1} K^{(j)}\right)[E][S]^{i-1}. \tag{525}$$

Hence,

$$V = [E] \sum_{i=1}^{n} \left\{k_3^{(i)}[S]^i \prod_{j=1}^{i} K^{(j)}\right\} \tag{526}$$

but

$$[E] = E_T - \sum_{i=1}^{n} [ES_i] \tag{527}$$

$$= E_T - \sum_{i=1}^{n} \left(\prod_{j=1}^{i} K^{(j)}\right)[S]^i[E] \tag{528}$$

so that

$$[E] = E_T\left\{1 + \sum_{i=1}^{n} \left(\prod_{j=1}^{i} K^{(j)}\right)[S]^i\right\}^{-1}. \tag{529}$$

Introducing for convenience

$$\phi_i \equiv [S]^i \left(\prod_{j=1}^{i} K^{(j)}\right), \tag{530}$$

from Eq. (526) is obtained

$$V = \left[E_T \middle/ \left(1 + \sum_{i=1}^{n} \phi_i\right)\right] \sum_{i=1}^{n} k_3^{(i)} \phi_i. \tag{531}$$

a. An Alternative Form. An interesting reduction to the Michaelis-Menten pattern is obtained by introducing three "intrinsic rate constants" k_1, k_2, k_3 in terms of which the three sets $\{k_1^{(i)}\}$, $\{k_2^{(i)}\}$, $\{k_3^{(i)}\}$ may be expressed.

(1) THE INTRINSIC BIMOLECULAR COMPLEXATION RATE CONSTANT k_1. In the ith complexation step,

$$ES_{i-1} + S \xrightarrow{k_1^{(i)}} ES_i,$$

there are available to a waiting substrate "customer" (see Section IV.F.2), $(n - i + 1)$ open sites. If each of these possible sites has associated with it an "intrinsic" rate constant k_1, then $k_1^{(i)}$ is computable in terms of the number of microscopically distinct ways in which the complexation at this stage can occur as a simple summation,

$$k_1^{(i)} = (n - i + 1)k_1, \qquad i = 1, 2, \ldots, n. \tag{532}$$

This may also be arrived at by reasoning stochastically, considering, as in the E–S stochastic model of Section IV.E.1, the $k_1^{(i)}$'s as probability parameters. Then considering that there are $(n - i + 1)$ mutually exclusive (alternative) ways of accomplishing the next complexation, each with probability k_1, the probability of a single complexation is therefore obtained by summing over all the individual probabilities of complexation.

(2) THE INTRINSIC UNIMOLECULAR DISSOCIATION CONSTANT k_2. For the reverse reaction,

$$ES_i \xrightarrow{k_2^{(i)}} ES_{i-1} + S.$$

there are i alternative complexed sites at which the single dissociation may occur. Again, associating with each of these the same intrinsic rate constant k_2, it follows that

$$k_2^{(i)} = ik_2, \qquad i = 1, 2, \ldots, n. \tag{533}$$

(3) THE INTRINSIC UNIMOLECULAR DECOMPOSITION CONSTANT k_3. For the decomposition

$$ES_i \xrightarrow{k_3^{(i)}} ES_{i-1} + P,$$

associating with each of the i occupied sites, the intrinsic rate constant k_3, it follows that

$$k_3^{(i)} = ik_3, \qquad i = 1, 2, \ldots, n. \tag{534}$$

(4) THE INTRINSIC RECIPROCAL MICHAELIS CONSTANT K. This is defined by

$$K \equiv k_1/(k_2 + k_3). \tag{535}$$

It may be shown that K is directly related to $K^{(1)}$; specifically,

$$K = n^{-1}K^{(1)}. \tag{536}$$

Furthermore,

$$K^{(j)} = (n - j + 1)K/j, \tag{537}$$

so that Eq. (530) becomes

$$\phi_i = K^i[S]^i \prod_{j=1}^{i} [(n - j + 1)/j]. \tag{538}$$

But

$$\prod_{j=1}^{i} \left(\frac{n - j + 1}{j}\right) \equiv \frac{n!}{i!(n - i)!} \equiv \binom{n}{i}, \tag{539}$$

whence

$$\phi_i = K^i[S]^i \binom{n}{i}. \tag{540}$$

Then substituting Eqs. (540). and (534) into Eq, (531), one obtains

$$V = \frac{k_3 E_T}{1 + \sum_{i=1}^{n} K^i[S]^i \binom{n}{i}} \sum_{i=1}^{n} iK^i[S]^i \binom{n}{i}. \tag{541}$$

A simplification of the denominator is deducible from the binomial expansion

$$(1 + K[S])^n = 1 + \sum_{i=1}^{n} \binom{n}{i} 1^{n-i}K^i[S]^i, \tag{542}$$

which gives

$$\sum_{i=1}^{n} \binom{n}{i} K^i[\mathrm{S}]^i = (1 + K[\mathrm{S}])^n - 1. \tag{543}$$

And, with reference to the summation in the numerator, differentation of both sides of Eq. (542) with respect to $K[\mathrm{S}]$ gives

$$n(1 + K[\mathrm{S}])^{n-1} = \sum_{i=1}^{n} \binom{n}{i} iK^{i-1}[\mathrm{S}]^{i-1}. \tag{544}$$

Then multiplying both sides of Eq. (544) by $K[\mathrm{S}]$,

$$nK[\mathrm{S}](1 + K[\mathrm{S}])^{n-1} = \sum_{i=1}^{n} \binom{n}{i} iK^i[\mathrm{S}]^i. \tag{545}$$

Substitution of Eqs. (543) and (545) into Eq. (541) gives

$$V = \frac{k_3 E_\mathrm{T}}{(1 + K[\mathrm{S}])^n} nK[\mathrm{S}](1 + K[\mathrm{S}])^{n-1} \tag{546}$$

or

$$V = \frac{K[\mathrm{S}]}{1 + K[\mathrm{S}]} k_3(nE_\mathrm{T}). \tag{547}$$

Note that when $n = 1$, this reduces to the Michaelis–Menten expression for the univalent case. And even in the general n-valent case, the use of intrinsic constants brings out the fact the Michaelis–Menten pattern is completely reproduced if E_T is replaced by nE_T equal to the total number of available active sites in the system. It is thus the result one would arrive at intuitively by considering the n-valent system as n copies of a univalent E–S system.

4. *A Mathematical Model for the Kinetics of an n-Valent Enzyme with Linear Arrangement of the n-Sites and Nearest-Neighbor Interactions Permitted*

Botts and Morales [1953]† extended their analysis of n-valency kinetics to the case of a set of n linearly arranged sites with nearest-neighbor interactions between sites. This physical model is analogous to a class of problems in physics in which Ising's [1925] well-known model in the theory of ferromagnetism is generally applied. For example, it has been applied to the problem of approximating the thermodynamic properties of so-called regular binary solution mixtures, though a more satisfactory treatment of the problem has been obtained using the method of quasichemical equilibrium (see following Section IV.D.5). This approach, in combination with results obtained earlier by Morales *et al.* [1948] in the course of deriving equilibrium equations for

†From Botts and Morales [1953, pp. 702–705].

a model of antibody–antigen combination (antibody corresponding to univalent substrate; and antigen to n-valent enzyme), led Botts and Morales to formulate and solve the n-valency system just described as follows.

In addition to assuming nearest-neighbor interactions among the complexed substrate molecules in linear array, it is also assumed that the individual E–S dissociations occur much more rapidly than the decompositions:

$$k_3^{(j)} \ll k_2^{(j)}, \tag{548}$$

so that

$$K^{(j)} \approx k_1^{(j)}/k_2^{(j)} \equiv K_e^{(j)}, \qquad j = 1, 2, \ldots, n, \tag{549}$$

which implies that $[ES_j]$ will be maintained at near equilibrium values $[ES_j]_e$.

Drawing on the antibody–antigen analogy they deduced that at equilibrium the expression ϕ_i defined in Eq. (530) becomes in this case

$$\phi_i = (K^{(1)}[S]/n)^i \sum_{p=1}^{i} W_i^{(p)} \exp\{-pF_{SS}/k_B T\}. \tag{550}$$

The term $K^{(1)}/n$ in parentheses results from the fact that $k_1 = n^{-1}k_1^{(1)}$, $k_2 = k_2^{(1)}$, $k_3 = k_3^{(1)}$ so that $K = n^{-1}K^{(1)}$, F_{SS} is the free energy of the S–S interactions and, the term $W_i^{(p)}$ is the number of microscopically distinct ways in which i sites (of a total of n) may be filled in such a way that there are p nearest-neighbor pairs among the filled sites, the following expression for it having been deduced by Ising:

$$W_i^{(p)} = \frac{(i-1)!(n-i+1)!}{(i-p-1)!p!(i-p)!(n-2i+p+1)!}. \tag{551}$$

Substitution of this into Eq. (550) and, in turn, use of these adjusted ϕ_i's in Eq. (531) in effect adjusts that part of the rate expression having to do with the $[ES_i]_e$ equilibrium values to the nearest-neighbor S–S interactions. Further ajdustments must be made in the $k_3^{(i)}$'s to allow for the presence of interaction. They accomplished this as follows:

The ES_i complexes are classified according to the number p of nearest-neighbor pairs among the i complexed substrate molecules, so that where $ES_i^{(p)}$ is an ith-order E–S complex, with p nearest-neighbor pairs,

$$\sum_p [ES_i^{(p)}] = [ES_i]. \tag{552}$$

It is then considered that the decomposition of a single substrate molecule of an $ES_i^{(p)}$ species degrades it to one of the classes, $ES_{i-1}^{(p)}$, $ES_{i-1}^{(p-1)}$, $ES_{i-1}^{(p-2)}$, depending on whether the substrate molecule was

Class 0: isolated (having no nearest neighbor),

Class I: singly neighbored,

Class II: doubly neighbored.

Then, independently of i and p, Botts and Morales introduce the rate constants ${}^{0}k_3$, ${}^{\mathrm{I}}k_3$, ${}^{\mathrm{II}}k_3$ corresponding to these decomposition states.

If ${}^{0}X_i^{(p)}$, ${}^{\mathrm{I}}X_i^{(p)}$, ${}^{\mathrm{II}}X_i^{(p)}$ are the fractions of substrate molecules complexed on $\mathrm{ES}_i^{(p)}$ corresponding to states 0, I, II, where

$$ {}^{0}X_i^{(p)} + {}^{\mathrm{I}}X_i^{(p)} + {}^{\mathrm{II}}X_i^{(p)} = 1, \tag{553} $$

the concentration of these different types of molecules are, respectively,

$$ \begin{aligned} i\,{}^{0}X_i^{(p)}[\mathrm{ES}_i^{(p)}] &= \text{concentration of 0-type S molecules,} \\ i\,{}^{\mathrm{I}}X_i^{(p)}[\mathrm{ES}_i^{(p)}] &= \text{concentration of I-type S molecules,} \\ i\,{}^{\mathrm{II}}X_i^{(p)}[\mathrm{ES}_i^{(p)}] &= \text{concentration of II-type S molecules.} \end{aligned} \tag{554} $$

So, the contribution of the $\mathrm{ES}_i^{(p)}$ complexes to the rate of product formation is

$$ i({}^{0}k_3\,{}^{0}X_i^{(p)} + {}^{\mathrm{I}}k_3\,{}^{\mathrm{I}}X_i^{(p)} + {}^{\mathrm{II}}k_3\,{}^{\mathrm{II}}X_i^{(p)})[\mathrm{ES}_i^{(p)}]. \tag{555} $$

After tedious combinatorial computation, they obtain the results,

$$ \begin{aligned} {}^{0}X_i^{(p)} &= (i-p)(i-p-1)/i(i-1), \\ {}^{\mathrm{I}}X_i^{(p)} &= 2(i-p)p/i(i-1), \\ {}^{\mathrm{II}}X_i^{(p)} &= p(p-1)/i(i-1). \end{aligned} \tag{556} $$

Eventually, they obtain

$$ k_3^{(i)} = i \sum_p [({}^{0}k_3\,{}^{0}X_i^{(p)} + {}^{\mathrm{I}}k_3\,{}^{\mathrm{I}}X_i^{(p)} + {}^{\mathrm{II}}k_3\,{}^{\mathrm{II}}X_i^{(p)})Z_i^{(p)}/Z_i], \tag{557} $$

where $Z_i^{(p)}$ is the configurational partition function for all enzyme molecules on which i substrate molecules are complexed in such a way that any of them are p nearest-neighbor pairs; that is,

$$ \frac{Z_i^{(p)}}{Z_i} = \frac{W_i^{(p)} \exp\{-pF_{\mathrm{SS}}/k_{\mathrm{B}}T\}}{\sum_p W_i^{(p)} \exp\{-pF_{\mathrm{SS}}/k_{\mathrm{B}}T\}} = \frac{[\mathrm{ES}_i^{(p)}]_{\mathrm{e}}}{[\mathrm{ES}_i]} \tag{558} $$

(considering the quasiequilibrium state here which allows the Maxwell–Boltzmann statistics to be used). Then, combining Eqs. (531), (550), and (557) yields

$$ \frac{V}{E_{\mathrm{T}}} = \frac{\sum_{i=1}^{n} \{i(K^{(1)}[\mathrm{S}]/n)^i \sum_p [{}^{0}k_3\,{}^{0}X_i^{(p)} + {}^{\mathrm{I}}k_3\,{}^{\mathrm{I}}X_i^{(p)} + {}^{\mathrm{II}}k_3\,{}^{\mathrm{II}}X_i^{(p)}]Z_i^{(p)}\}}{1 + \sum_{i=1}^{n} \{(K^{(1)}[\mathrm{S}]/n)^i Z_i\}}. \tag{559} $$

The case in which $k_3^{(i)}$ is not negligible relative to $k_2^{(i)}$ has also been worked out and leads to an expression for $k_3^{(i)}$ identical with Eq. (557). Also in this case, an analogous expression for $k_2^{(i)}$ in terms of intrinsic constants ${}^{0}k_2$, ${}^{\mathrm{I}}k_2$, ${}^{\mathrm{II}}k_2$ has been found, which amounts to substituting these quantities for the k_3 quantities in Eq. (557). The case of $k_1^{(i)}$ required very extensive further combinatorial analysis.

Some numerical studies of the expression for v have been made by Botts and Morales for special cases of n and simplifying conditions in the rate constants, on the basis of which it was demonstrated that nearest-neighbor interactions can result in reduced velocities of reaction, substrate inhibition at high substrate concentration being particularly apparent. It was also noted that in an experimental study of urease-catalyzed hydrolysis of urea (communicated to the authors by Kistiakowsky and Rosenberg) hydrolysis inhibition at high substrate concentration could be interpreted analytically by their results if one assumed such interactions between sites.

5. *Hill's Monolayer Adsorption Model for Closest Packing of Active Sites and Nearest-Neighbor Interactions*

a. The Langmuir (Idealized) Adsorption Isotherm Analogy. In the earlier work [Bartholomay, 1962a, pp. 73–75] the derivation due to Laidler [1950, pp. 145, 148, 305] of the Michaelis–Menten expression using the Langmuir adsorption isotherm as the basis of a physical model of the E–S complexation process was presented. In this derivation advantage was taken of the important work of Langmuir which did much to erase a superficial kind of distinction that had grown up between "physical" adsorption and "chemical adsorption," or, "chemisorption." This derivation carried forward this same spirit into enzyme kinetics by considering the E–S complexation process as a case of the adsorption of a substrate molecule onto the surface of the enzyme molecule. The derivation cited, however, was at the macroscopic level, making use of a number of analogies. Thus, corresponding to the total adsorbing surface in adsorption theory was the total enzyme concentration E_T; corresponding to the set of adsorbable gas molecules was the set of substrate molecules; corresponding to gas pressure p was the substrate concentration [S]; corresponding to the fraction θ of surface covered by adsorbed molecules was the fraction $\theta = [ES]/E_T$ of enzyme molecules complexed by substrate molecules; corresponding to the rate of condensation v_c on the surface was the rate of complexation (rate constant k_1); corresponding to v_e, the rate of evaporation, was the rate of dissociation of ES complexes (rate constant k_2); and, corresponding to the equilibrium state defined by $v_c = v_e$ was the equilibrium between E, S, and ES. Reference to the Langmuir isotherm,

$$\theta = k_1 p/(k_2 + k_1 p), \tag{560}$$

allowed the expression of [ES] in the form

$$[ES] = E_T K_e[S]/(1 + K_e[S]), \tag{561}$$

where $K_e = k_1/k_2 = (K')^{-1}$, in which case, as before,

$$v = k_3[\mathrm{ES}] = \frac{k_3 E_T[\mathrm{S}]}{K' + [\mathrm{S}]} = \frac{V[\mathrm{S}]}{K' + [\mathrm{S}]}. \tag{562}$$

b. The More General Localized Monolayer Adsorption Theory. Thus, a precedent and rationale already existed for looking to monolayer gas adsorption theory as a source for mathematical models of the enzyme catalytic process. Just as this simplest adsorption isotherm was followed by statistico-thermodynamical models of the adsorption process in physics and physical chemistry, the enzyme catalytic model constructed by analogy was followed by a statistical thermodynamical model by Hill [1952]. We have already seen in the preceding Botts–Morales model that interactions between adsorbed molecules are characterizable thermodynamically in terms of free energy and partition functions. The statistical thermodynamical counterparts to these considerations are obtained from the statistical thermodynamical theory of localized (as opposed to mobile) adsorbed monolayers, namely, adsorbed layers of molecules of at most one molecule in thickness, attached to the surface of a solid or liquid in which the molecules are practically insoluble [see Fowler and Guggenheim, 1939, Chapter 10]. The localized monolayer specification is taken as the case analogous to n-valent enzyme complexation because the localized monolayer model specifies definite points of attachment on the solid surface, each capable of accomodating just one adsorbed molecule. These points of attachment are in fact classified in the general theory as "sites" further underscoring the analogy with enzyme kinetics.

With reference to general adsorption theory, the result described in Section a, preceding, is really the simple case of an idealized, local monolayer in the sense that an assumption of negligible interactions between adsorbed molecules has been made.

(1) The Case of No Site–Site Interactions, the Idealized Langmuir Isotherm for Localized Monolayers. In localized monolayer theory, considering first the case in which no interactions are assumed between adsorbed molecules, let the partition function for the internal degrees of freedom of a molecule of type A (including vibrations) relative to its average position on a site be denoted by $f_A(T)$. And, let N_S be the number of sites on the surface; $N_A, N_B, N_C, \ldots$, the number of these occupied by molecules of types A, B, C, . . . ; $(N_S - \sum_A N_A)$ the symbol for the number of unoccupied sites; and, $g(N_A, N_B, N_C, \ldots)$ the number of different ways of distributing the adsorbed molecules over the N_S sites (referring to the occupancy problem of dividing N_S sites into subgroups of $N_A, N_B, N_C, \ldots, N_S - \sum_A N_A$) Then, from combinatorial analysis,

$$g(N_A, N_B, N_C, \ldots) = \frac{N_S!}{N_A! N_B! \cdots (N_S - \sum_A N_A)!}, \tag{563}$$

in which case the partition function for the monolayer becomes

$$f(T) = \frac{N_S!}{N_A!N_B! \cdots (N_S - \sum_A N_A)!} [f_A(T)]^{N_A}[f_B(T)]^{N_B} \cdots . \tag{564}$$

For the free energy E of the adsorbed monolayer one then obtains

$$E/k_B T = -\log_e f(T) = -\log_e (N_S!) + \sum_A \log_e (N_A!) + \log_e \{(N_S - \sum_A N_A)!\} - \sum_A N_A \log_e f_A(T). \tag{565}$$

For large N assuming the approximation $\log_e N! = N \log_e N - N$,

$$E/k_B T = -N_S \log_e N_S + \sum_A N_A \log_e N_A + (N_S - \sum_A N_A) \log_e (N_S - \sum_A N_A) - \sum_A N_A \log_e f_A(T). \tag{566}$$

Then, calculating in the usual thermodynamic fashion the partial potential μ_A of molecules A, where $\mu_A = (\partial E/\partial N_A)_{T, N_S, N_B}$

$$\mu_A = k_B T \left[\log_e \frac{N_A}{N_S - \sum_A N_A} - \log_e f_A(T) \right]. \tag{567}$$

Where $\theta_A, \theta_B, \ldots$ denote the fractions of sites occupied by molecules of type A, B, . . . , that is, $\theta_A = N_A/N_S$, Eq. (567) is written

$$\mu_A = k_B T \left[\log_e \left(\frac{\theta_A}{1 - \sum_A \theta_A} \right) - \log_e f_A(T) \right]. \tag{567'}$$

Where $\lambda_A^{(m)}$, the "absolute activity" of the molecules of type A is defined as

$$\lambda_A^{(m)} \equiv \exp(\mu_A/k_B T), \tag{568}$$

from Eq. (568) it follows that

$$\lambda_A^{(m)} = [f_A(T)]^{-1} \frac{\theta_A}{1 - \sum_A \theta_A}. \tag{568'}$$

It can be shown that in the vapor phase [Fowler and Guggenheim, 1939, Chapter 5] the absolute activity $\lambda_A^{(v)}$ is given by

$$\lambda_A^{(v)} = \frac{p_A}{k_B T} \frac{h^3}{(2\pi m_A k_B T)^{3/2} g_A(T)}, \tag{569}$$

where p_A represents partial pressure for type A gaseous molecule and $g_A(T)$ represents the partition function of internal degree of freedom of type A molecule. The condition $\lambda_A^{(m)} = \lambda_A^{(v)}$ for equilibrium between the monolayer and vapor phase yields

$$\frac{\theta_A}{1 - \sum_A \theta_A} = \frac{p_A}{p_0}, \tag{570}$$

where

$$p_0 = \frac{k_B T}{f_A(T)} \frac{(2\pi m_A k_B T)^{3/2} g_A(T)}{h^3}. \tag{571}$$

(Note that for a single type of adsorbed molecule this becomes

$$\theta = p/(p_0 + p), \tag{572}$$

another form of Eq. (560), the Langmuir isotherm.) This derivation makes it clear that the isotherm conditions must hold, whatever the kinetics of the processes, provided only that the molecules are adsorbed onto definite sites which do not interact with each other, no interaction terms having been incorporated into the derivation.

(2) THE CASE OF SITE–SITE INTERACTIONS. In the case of interactions allowed between adsorbed molecules in the monolayer the problem is of considerably greater mathematical magnitude, requiring extensive approximation. One of the more reliable approximation methods is the so-called "quasi-chemical method" [Guggenheim, 1948] which evolved from the thermodynamic treatment of so-called "regular binary solutions." It was also applied to the study of order–disorder phenomena in crystals, the latter taking into account the arrangement of atoms or molecules on proper lattice points in crystalline solids. It is interesting that whereas the quasi-chemical method was worked out in relation to chemical solutions [see Fowler and Guggenheim, 1939, Chapter 8], then applied to monolayer adsorption studies [Fowler and Guggenheim, 1939, Chapter 10] and then to crystalline order–disorder studies, the idea for characterizing a system of two or more chemical molecular species in solution was suggested by the closest packing arrangements of atoms and molecules associated with crystalline lattices. In the idealized case of spherical molecules this means that such problems in turn are formulable on the basis of the packing of spheres—a point of view which has been quite profitable in crystal-structure characterization.

(*a*) *The Closest Packing of Spheres as a Geometrical Idealization of the Arrangement of Adsorption Sites and Crystalline Molecular Arrangements.* The problem of the packing of spheres in ways which leave the minimum of interstitial space has interested many investigators during the past century [see for example, Barlow, 1883a, b, 1894, 1898; Sohncke, 1884; Coxeter, 1962; Fejes Toth, 1959; Vayo, 1965].

The closest packing of spheres in a single layer is achievable by the familiar, so-called hexagonal arrangement in which each sphere is in contact with six other equivalent spheres. Going to three-dimensional layers, to achieve closest packing, a second layer can be superposed on a plane parallel to the hexagonally arranged monolayer so that each sphere is in contact with three spheres in the adjacent monolayer. But, here, as discovered by Barlow, one differentiates between two different arrangements relative to the monolayer: (1) the so-called "cubic symmetry" and (2) the "hexagonal" symmetry. The latter case is symbolized by the string of letters ABABAB . . . , indicating that the structure repeats itself every two layers; the former case by ABCABC . . . , indicating repetition of the arrangement every three layers. In each of these cases it can be seen that each internal sphere is in contact with twelve others: has "twelve nearest neighbors," namely, a hexagon of six in the same layer and two triangles of three each in the adjacent layers (above and below). In the hexagonal spatial case the upper triangle has the same orientation as the lower triangle; but in the cubic case the upper triangle is rotated 60°. Many other arrangements can be found which differ slightly from these two optimum cases [see Pauling, 1939, p. 407].

(*b*) *Strictly Regular Binary Solutions*. Given a molecular species A forming in solution a cubic-packing arrangement and a species B with the same packing characteristic, so that both are characterized individually as having the same "coordination number" $z = 12$, that is, the same number of nearest neighbors to a given molecule, a "strictly regular binary solution" is said to be formed by combining them, subject to a number of additional assumptions [see Fowler and Guggenheim, 1939, p. 351]. For example, when the two liquids are mixed at a given temperature and pressure, the "molecular volumes" V_A and V_B and the "free volumes" v_A, v_B of both remain unaltered. The following energy considerations for such a solution apply also to mixed crystals quite directly.

Where an A molecule of average potential energy $(-\chi_A)$ has z nearest neighbors, $(-2\chi_A/z)$ is taken as the average energy of interaction between two A molecules. Similarly the average energy of interaction between two B molecules is $(-2\chi_B/z)$. A "mixing energy" w_{AB} is defined by the condition that, starting with two pure liquids and interchanging an interior A molecule with an interior B molecule produces a total increase in potential enegry equal to $2w_{AB}$. In such an interchange z pairs AA and z pairs BB are destroyed and $2z$ pairs AB are created. Hence, according to the definition of w_{AB},

Average potential energy of AB pair

$$= (-2\chi_A - 2\chi_B + 2w_{AB})/2z \tag{573}$$

$$= (-\chi_A - \chi_B + w_{AB})/z. \tag{574}$$

Next, consider a particular configuration of the mixture of N_A molecules A and N_B molecules B in which the number of AB pairs of nearest neighbors is (zX). Then the number of neighbors of A molecules which are also A type is $z(N_A - X)$. The number of AA pairs is $\frac{1}{2}z(N_A - X)$; and the number of BB pairs is $\frac{1}{2}z(N_B - X)$. In this case the total potential energy W is therefore given by

$$W = \tfrac{1}{2}z(N_A - X)(-2\chi_A/z) + \tfrac{1}{2}z(N_B - X)(-2\chi_B/z) + zX(-\chi_A - \chi_B + w_{AB})/z, \tag{575}$$

which reduces to

$$W = -N_A\chi_A - N_B\chi_B + Xw_{AB}. \tag{575'}$$

Fowler and Guggenheim [1939, Chapter VIII] had shown that the free energy F of an assembly of N molecules of mass m confined to volume V is given by

$$F = -Nk_BT\log_e(2\pi mk_BT)^{3/2}/h^3 - Nk_BT\log_e j(T) - k_BT\log_e\Omega(T), \tag{576}$$

where $j(T)$ is the partition function for all internal (including rotational, vibrational, electronic, and nuclear spin) degrees of freedom and

$$\Omega(T) = (N!)^{-1}\iint\cdots\int\exp(-W/k_BT)(d\omega)^N, \tag{577}$$

where now the symbolism $(d\omega)^N$ refers to N elements of volume to which the centers of the N molecules belong. The multiple integration extends over the whole positional phase space accessible to the molecules. For the case of binary mixtures of liquids, they later deduce the free energy F to be

$$F = -N_Ak_BT\log_e\phi_A(T) - Nk_BT\log_e\phi_B(T) - k_BT\log_e\Omega(T), \tag{578}$$

where

$$\phi(T) = [(2\pi mk_BT)^{3/2}/h^3]\,j(T) \tag{579}$$

and

$$\Omega(T) = (N_A!\,N_B!)^{-1}\iint\cdots\int\exp(-W/k_BT)(d\omega_A)^{N_A}(d\omega_B)^{N_B} \tag{580}$$

which upon substitution of Eq. (575) for W becomes

$$\Omega(T) = \frac{\exp\{(N_A\chi_A + N_B\chi_B)/k_BT\}}{N_A!\,N_B!} \times\iint\cdots\int\exp(-Xw_{AB}/k_BT)(d\omega_A)^{N_A}(d\omega_B)^{N_B}. \tag{581}$$

This expression holds for so-called strictly regular, nonperfect ($w_{AB} \neq 0$) solutions.

A quantity $\bar{\bar{X}}$ is then defined by the relation,

$$\exp(-\bar{\bar{X}}w_{AB}/k_BT)\iint\cdots\int(d\omega_A)^{N_A}(d\omega_B)^{N_B} = \iint\cdots\int\exp(-Xw_{AB}/k_BT)(d\omega_A)^{N_A}(d\omega_B)^{N_B} \tag{582}$$

from which, after evaluation of the integral on the left-hand side and substitution into Eq. (581) the result obtained is

$$\Omega(T) = [eV_{\mathrm{A}} \exp(\chi_{\mathrm{A}}/k_{\mathrm{B}}T)]^{N_{\mathrm{A}}}[eV_{\mathrm{B}} \exp(\chi_{\mathrm{B}}/k_{\mathrm{B}}T)]^{N_{\mathrm{B}}} \times [(N_A + N_B)!/N_A!\, N_B!] \exp(-\bar{\bar{X}} w_{\mathrm{AB}}/k_{\mathrm{B}}T). \tag{583}$$

Substitution of Eq. (583) into Eq. (578) gives

$$F = N_{\mathrm{A}}\{-\chi_{\mathrm{A}} - k_{\mathrm{B}}T \log_e(\phi_{\mathrm{A}}V_{\mathrm{A}}) - k_{\mathrm{B}}T + k_{\mathrm{B}}T \log_e[(N_{\mathrm{A}}/(N_{\mathrm{A}} + N_{\mathrm{B}})]\} + N_{\mathrm{B}}\{-\chi_{\mathrm{B}} - k_{\mathrm{B}}T \log_e(\phi_{\mathrm{B}}V_{\mathrm{B}}) - k_{\mathrm{B}}T + k_{\mathrm{B}}T \log_e[(N_{\mathrm{B}}/(N_{\mathrm{A}} + N_{\mathrm{B}}]\} + \bar{\bar{X}} w_{\mathrm{AB}}. \tag{584}$$

Then, introducing an "equilibrium value" $\bar{X}$ for X, the number of AB pairs of nearest neighbors:

$$\bar{X} = \frac{\iint \cdots \int X \exp(-Xw_{\mathrm{AB}}/k_{\mathrm{B}}T)(d\omega_{\mathrm{A}})^{N_{\mathrm{A}}}(d\omega_{\mathrm{B}})^{N_{\mathrm{B}}}}{\iint \cdots \int \exp(-Xw_{\mathrm{AB}}/k_{\mathrm{B}}T)(d\omega_{\mathrm{A}})^{N_{\mathrm{A}}}(d\omega_{\mathrm{B}})^{N_{\mathrm{B}}}} \tag{585}$$

and treating $\bar{X}$ as the mean of a function X of the system of A and B molecules with respect to a probability density function f_N which corresponds to the location of the A and B molecules treated as random variables yields

$$f_N \equiv \frac{\exp(-Xw_{\mathrm{AB}}/k_{\mathrm{B}}T)}{\iint \cdots \int \exp(-Xw_{\mathrm{AB}}/k_{\mathrm{B}}T)(d\omega_{\mathrm{A}})^{N_{\mathrm{A}}}(d\omega_{\mathrm{B}})^{N_{\mathrm{B}}}}. \tag{586}$$

It is seen that $\bar{X}$ and $\bar{\bar{X}}$ are related by

$$\bar{X} = \bar{\bar{X}} - T\frac{\partial \bar{\bar{X}}}{\partial T} = \frac{\partial(\bar{\bar{X}}/T)}{\partial(1/T)}. \tag{587}$$

THE QUASI-CHEMICAL EQUILIBRIUM ASSUMPTION: It can be seen from this discussion that in this way the problem of computing the free energy, and so all equilibrium quantities is then reduced to that of determining $\bar{\bar{X}}$, or alternatively $\bar{X}$, in a given case. The difficulty in obtaining a convenient expression for $\bar{X}$ as defined by Eq. (585) is avoided by turning to Guggenheim's quasi-chemical method of defining the equilibrium state, which in this case amounts to approximating $\bar{X}$ by redefining it as the solution of the following equation:

$$\bar{X}^2 = (N_{\mathrm{A}} - \bar{X})(N_{\mathrm{B}} - \bar{X}) \exp(-2w_{\mathrm{AB}}/zk_{\mathrm{B}}T) \tag{588}$$

so that

$$\bar{X} = (N_{\mathrm{A}} + N_{\mathrm{B}})\tfrac{1}{2}(\beta - 1)(\exp(2w_{\mathrm{AB}}/zk_{\mathrm{B}}T) - 1)^{-1} \tag{589}$$

in which

$$\beta = \left\{1 + \frac{4N_{\mathrm{A}}N_{\mathrm{B}}[\exp(2w_{\mathrm{AB}}/zk_{\mathrm{B}}T) - 1]}{(N_{\mathrm{A}} + N_{\mathrm{B}})^2}\right\}^{1/2} \tag{590}$$

As Rushbrooke [1949, p. 310] points out, the quasi-chemical approximation takes its name from a formal analogy between the partition function expression of the equilibrium condition for a symbolic bimolecular chemical reaction,

$$\mathrm{AA} + \mathrm{BB} \rightleftharpoons 2\mathrm{AB}, \tag{591}$$

namely,

$$n_{\mathrm{AA}} n_{\mathrm{BB}} / n_{\mathrm{AB}}^2 = f_{\mathrm{AA}}(T) f_{\mathrm{BB}}(T) / f_{\mathrm{AB}}^2(T), \tag{592}$$

and the evaluation of the partition function of the whole assembly (system) of N_{A} and N_{B} molecules. In the latter case an explicit form for the occupancy function $g(N_{\mathrm{A}}, N_{\mathrm{B}}, N_{\mathrm{AB}})$ equal to the number of different arrangements of the N_{A} molecules A and the N_{B} molecules B, with specified value N_{AB} is required. And the quasi-chemical equilibrium assumption amounts to taking

$$g(N_{\mathrm{A}}, N_{\mathrm{B}}, N_{\mathrm{AB}}) = \left[\frac{N_{\mathrm{A}}!\, N_{\mathrm{B}}!}{\{N_{\mathrm{A}} - (N_{\mathrm{AB}}/z)\}!\{N_{\mathrm{B}} - (N_{\mathrm{AB}}/z)!(N_{\mathrm{AB}}/z!)^2}\right]^{z/2} \times \left[\frac{(N_{\mathrm{A}} + N_{\mathrm{B}})!}{N_{\mathrm{A}}!\, N_{\mathrm{B}}!}\right]^{1-z/2} \tag{593}$$

in which

$$(N_{\mathrm{A}} + N_{\mathrm{B}})!/N_{\mathrm{A}}!\, N_{\mathrm{B}}! = \sum_{N_{\mathrm{AB}}} g(N_{\mathrm{A}}, N_{\mathrm{B}}, N_{\mathrm{AB}}). \tag{594}$$

(*c*) *Application of the Results of* (*b*) *to the Localized Monolayer Adsorption System.* The proceeds of these theoretical results for the binary solution case have been applied to the monolayer adsorption problem to obtain improvements in the accuracy of Langmuir-type adsorption isotherms by allowing interactions between adsorbed molecules. In this case only one type of molecule, call it A, is considered. However, as in the binary-solution case the treatment due to Fowler and Guggenheim [1939, Chapter 10, Section 1006], which is followed here, assumes that each adsorbed site has z nearest neighbors in the general case (six in the monolayer hexagonal closest packing) and that the total energy of interaction case be expressed as the sum of contributions from pairs of nearest neighbors.

Suppose the layer has a total of N_{S} sites of which N_{A} are occupied and $N_{\mathrm{S}} - N_{\mathrm{A}}$ are empty. Again, letting N_{AA} be the number of pairs of neighboring molecules in a particular configuration of the monolayer, the contribution to the interaction energy by these pairs is written in the form $(2N_{\mathrm{AA}}w/z)$ (the factor $2/z$ being introduced simply for mathematical convenience in some of the derivations). The complete partition function of the monolayer is

$$\sum_{N_{\mathrm{AA}}} g(N_{\mathrm{A}}, N_{\mathrm{AA}}) \exp(-N_{\mathrm{AA}} 2w/zk_{\mathrm{B}}T)[f(T)]^{N_{\mathrm{A}}}, \tag{595}$$

where $g(N_{\mathrm{A}}, N_{\mathrm{AA}})$ is the number of different configurations with N_{AA} pairs of nearest neighbors among the N_{A} adsorbed molecules, and

$$\sum_{N_{\mathrm{AA}}} g(N_{\mathrm{A}}, N_{\mathrm{AA}}) = \binom{N_{\mathrm{S}}}{N_{\mathrm{A}}} = N_{\mathrm{S}}!/N_{\mathrm{A}}!(N_{\mathrm{S}} - N_{\mathrm{A}})!, \tag{596}$$

the total number of distinguishable configurations in a monolayer of N_A adsorbed molecules.

Then in direct analogy with the binary solution case a quantity $\bar{\bar{N}}_{AA}$ is defined by the expression,

$$\sum_{N_{AA}} g(N_A, N_{AA}) \exp(-N_{AA}2w/zk_BT) \equiv \exp(-\bar{\bar{N}}_{AA}2w/zk_BT) \sum_{N_{AA}} g(N_A, N_{AA}), \tag{597}$$

$$\sum_{N_{AA}} g(N_A, \bar{\bar{N}}_{AA}) \exp(-N_{AA}2w/zk_BT) \doteqdot \exp(-\bar{\bar{N}}_{AA}2w/zk_BT) N_S!/N_A!(N_S - N_A)!. \tag{598}$$

Then, the free energy F of the monolayer is given by

$$F/k_BT = -N_S \log_e N_S + N_A \log_e N_A + (N_S - N_A) \log_e(N_S - N_A) - N_A \log_e f(T) + \bar{\bar{N}}_{AA}2w/k_BT. \tag{599}$$

Again, the equilibrium value $\bar{N}_{AA}$ of N_{AA} is defined via

$$\sum_{N_{AA}} N_{AA}\, g(N_A, N_{AA}) \exp(-N_{AA}2w/zk_BT) \equiv \bar{N}_{AA} \sum_{N_{AA}} g(N_A, N_{AA}) \exp(-N_{AA}2w/zk_BT) \tag{600}$$

from which it may be seen that

$$\bar{N}_{AA} = -T^2 \frac{\partial(\bar{\bar{N}}_{AA}/T)}{\partial T}. \tag{601}$$

So, once again the construction of all thermodynamic quantities is referred to $\bar{\bar{N}}_{AA}$, or alternatively, because of Eq. (601), to $\bar{N}_{AA}$ which, in turn, is approximated according to the quasi-chemical equilibrium method by the equality

$$2\bar{N}_{AA}(zN_S - 2zN_A + 2\bar{N}_{AA}) = (zN_A - 2\bar{N}_{AA})^2 \exp(-2w/zk_BT), \tag{602}$$

the derivation of which is given in Fowler and Guggenheim [1939, pp. 439–441].

Introducing the new dimensionless variable

$$X \equiv N_{AA}/\tfrac{1}{2}zN_S \tag{603}$$

and taking

$$\theta = N_A/N_S, \tag{604}$$

transform Eq. (602) to

$$\bar{X}(1 - 2\theta + \bar{X}) = (\theta - \bar{X})^2 \exp(-2w/zk_BT), \tag{605}$$

the meaningful solution of which is the root,

$$\bar{X} = \theta - 2\theta(1 - \theta)/(\beta + 1) \qquad (= \bar{N}_{AA}/\tfrac{1}{2}zN_S), \tag{606}$$

where now

$$\beta = \{1 - 4\theta(1 - \theta)[1 - \exp(-2w/zk_BT)]\}^{1/2}. \tag{607}$$

It is these results which have been applied to the n-valent enzyme catalysis problem by Hill.

c. Hill's Model for n-Valency Enzyme Kinetics. As a prologue to the Hill model, the preceding results will be examined along the lines of Hill's adaptation of them to the enzyme-kinetics model.

(1) PROBABILITIES OF SITE–SITE INTERACTIONS DEDUCIBLE FROM THE LOCALIZED MONOLAYER ADSORPTION THEORY. Where $\bar{N}_{AA}$ is the quasichemical equilibrium approximation to the number of pairs of molecules A and $\frac{1}{2}zN_A$ is the number of pairs of molecules that can be formed with the N_A adsorbed molecules A (division by 2, since zN_A counts potential pairs twice, once for each molecule of the pair),

$$\frac{\bar{N}_{AA}}{\frac{1}{2}zN_A} = \begin{cases}\text{the fraction of possibile pairs that actually} \\ \text{are formed at quasi-chemical equilibrium.}\end{cases} \tag{608}$$

The frequency ratio in Eq. (608) can be considered as an approximation to the probability q of finding a neighboring site occupied. And where $N_A = \theta N_S$, combining this with Eq. (609) leads to the expression for q,

$$q = \bar{N}_{AA}/\tfrac{1}{2}z\theta N_S = \theta^{-1}\bar{X} = 1 - 2(1 - \theta)/(\beta + 1). \tag{609}$$

Thus, taking $p \equiv 1 - q$ as the probability that a given adsorbed molecule A has an unoccupied neighboring site,

$$p \equiv 2(1 - \theta)/(1 + \beta), \tag{610}$$

is interpretable as the probability that a nearest-neighbor site of an adsorbed molecule is empty.

Similarly, one can deduce an expression for r, the probability that a site neighboring an empty site will be filled, as follows. Let

(i) $N_S - N_A$ be the number of empty sites;

(ii) $\frac{1}{2}z(N_S - N_A)$ be the number of nearest-neighbor pairs of sites in which each pair has at most one adsorbed molecule;

(iii) $\frac{1}{2}zN_A$ be the number of nearest-neighbor pairs each of which has at least one filled site;

(iv) $\bar{N}_{AA}$ be the number of nearest-neighbor pairs (at quasichemical equilibrium).

Then, $(\frac{1}{2}zN_A - \bar{N}_{AA})$ is equal to the number of nearest-neighbor pairs of sites each of which has exactly one adsorbed molecule and $[\frac{1}{2}z(N_S - N_A) - (\frac{1}{2}zN_A - N_{AA})]$ is equal to the number of nearest-neighbor pairs of empty sites. Thus,

$$r \equiv \frac{\frac{1}{2}z(N_S - N_A) - (\frac{1}{2}zN_A - \bar{N}_{AA})}{\frac{1}{2}z(N_S - N_A)} \tag{611}$$

may be interpreted as an approximation to the probability that a site that is nearest neighbor to an empty site is also empty. Eq. (611) becomes

$$r = 1 - \frac{N_A}{N_S - N_A} + \frac{\bar{N}_{AA}}{\frac{1}{2}z(N_S - N_A)}. \tag{612}$$

Substitution of $\theta = N_A/N_S$ into Eq. (612) gives

$$r = 1 - \frac{\theta}{1-\theta} + \frac{\bar{N}_{AA}/N_S}{\frac{1}{2}z(1-\theta)}. \tag{613}$$

Then substituting for $(\bar{N}_{AA}/\frac{1}{2}zN_S)$ Eq. (606) reduces this to

$$r = 1 - \frac{2\theta}{\beta+1} = \frac{\beta+1-2\theta}{\beta+1}. \tag{614}$$

(2) HILL'S MODEL.† Hill likens the simple, irreversible enzyme-catalyzed reaction for an n-valent enzyme to a heterogeneous reaction of the surface-catalyzed type, both enzyme and analogous surface obeying the closest-packing principle with respect to the substrate molecules. By analogy, then, it is considered that the mechanism of conversion of substrate S to product P is describable schematically as

$$S \underset{k_2}{\overset{k_1}{\rightleftharpoons}} S_a \overset{k_3}{\longrightarrow} P, \tag{615}$$

consisting, as in the Michaelis–Menten mechanism, of three reaction steps:

(*a*) *Adsorption.* $S \longrightarrow S_a$, in which an S molecule on being adsorbed (complexed) to an adsorption site (active site) becomes an activated molecule S_a, the rate process having a (bimolecular) rate constant k_1.

(*b*) *Desorption.* $S_a \longrightarrow S$, in which an adsorbed (complexed) molecule escapes from its site back to its normal form, that is, without transformation to product P, the rate process being characterized by a unimolecular rate constant k_2.

(*c*) *Decomposition.* $S_a \longrightarrow P$, in which the activated, adsorbed molecule is transformed into the product species P in the form of a unimolecular reaction with rate constant k_3.

It is further assumed that the "evaporation" of product molecules from this surface is fast enough that the number of product molecules on the surface at any time is negligible. Two cases are distinguished.

CASE 1 NO SITE–SITE INTERACTION: Where $\theta = (N_A/N_S)$ and c = concentration of substrate molecules, the case is treated by recourse to the Langmuir isotherm of Section IV.D.5.a, in which Eq. (560) follows from the "noninteraction" rates $v_1{}^0$, $v_2{}^0$, $v_3{}^0$, corresponding to steps 1, 2, 3, respectively, of the mechanism in Eq. (615),

$$v_1{}^0 = k_1{}^0\, c(1-\theta), \tag{616}$$

$$v_2{}^0 = k_2{}^0\, \theta, \tag{617}$$

$$v_3{}^0 = k_3{}^0\, \theta, \tag{618}$$

†From Hill [1952, pp. 4710–4711].

the corresponding rate constants $k_1{}^0$, $k_2{}^0$, $k_3{}^0$ being referred to as the "noninteraction rate constants."

CASE 2 NEAREST-NEIGHBOR SITE INTERACTIONS: In this case deduced directly from the probabilistic interpretations given in Section IV.D.5.c(1), Hill's model for the kinetics was constructed as follows:

Writing w_a (instead of $2w/z$) for the interaction free energy between nearest-neighbor S molecules (the type A molecules) and w_a' for that between nearest-neighbor S and P molecules, and assuming a condition of Maxwell–Boltzmann "internal equilibrium" for the adsorbed phase, which he approximated by identifying it with the condition of quasi-chemical equilibrium, as in Eq. (610), he defined

$$p \equiv 2(1 - \theta)/(\beta + 1) \tag{619}$$

as the probability that a given one of the z sites nearest neighbor to a filled site would be empty, where

$$\beta = \{1 - 4\theta(1 - \theta)(1 - \exp(-w_a/k_B T)\}^{1/2}. \tag{620}$$

The complement $(1 - p)$ is then the probability that a given one of the z sites nearest neighbor to a filled site will be empty. Then, using the binomial probability function, he introduced

$$P_j^{(f)} \equiv \binom{z}{i} p^{z-j}(1 - p)^j \tag{621}$$

as the probability that exactly j of the z sites nearest neighbor to a filled site would be filled.

In accord with Eq. (614),

$$r = (\beta + 1 - 2\theta)/(\beta + 1) \tag{622}$$

is taken as the probability that a given one of the z sites nearest neighbor to an empty site will also be empty, so that $(1 - r)$ gives the probability that that such a site will be filled. And again, from the binomial probability function, it is seen that

$$P_j^{(e)} = \binom{z}{j} r^{z-j}(1 - r)^j \tag{623}$$

is the probability that next to an empty site exactly j of the nearest-neighbor sites will be filled.

Then, Hill assumed that the rate constant $k_1(j)$ for the complexation of an S molecule at an empty site with j nearest–heignbor filled sites is related to the corresponding noninteractive rate constant $k_1{}^0$ by

$$k_1(j) \equiv k_1{}^0 \exp(-jw_b/k_B T), \tag{624}$$

where w_b is the free energy of activation per nearest neighbor. Similarly, the rate constant for the dissociation (desorption) of an adsorbed S_a molecule from a site with j nearest-neighbor filled sites is related to the corresponding noninteractive $k_2{}^0$ constant according to

$$k_2(j) \equiv k_2{}^0 \exp(-j(w_a - w_b)/k_B T). \tag{625}$$

Finally, the rate constant for the decomposition, that is for a P molecule leaving a site with j nearest-neighbor filled (with S) sites, relative to the non-interactive constant k_3 is taken as

$$k_3(j) \equiv k_3{}^0 \exp(j(w_a' - w_b')/k_B T), \tag{626}$$

where the primed energy terms refer to the interaction energies between S and P molecules. Equations (616)–(618) for the noninteractive system are replaced in this case by

$$v_1 = c(1 - \theta) \sum_{j=0}^{z} k_1(j) P_j^{(e)}, \tag{627}$$

$$v_2 = \theta \sum_{j=0}^{z} k_2(j) P_j^{(f)}, \tag{628}$$

$$v_3 = \theta \sum_{j=0}^{z} k_3(j) P_j^{(f)}, \tag{629}$$

so that the rate constants k_1, k_2, k_3 for the process are

$$k_1 = \sum_{j=0}^{z} k_1(j)\, P_j^{(e)}, \tag{630}$$

$$k_2 = \sum_{j=0}^{z} k_2(j)\, P_j^{(f)}, \tag{631}$$

$$k_3 = \sum_{j=0}^{z} k_3(j)\, P_j^{(f)}, \tag{632}$$

Equations (627)–(629) have been simplified by Hill to the following forms:

$$v_1(\theta) = k_1{}^0 c(1 - \theta)\, f(\theta), \tag{633}$$

$$v_2(\theta) = k_2{}^0 \theta \left[\frac{2(1 - \theta)}{\beta + 1 - 2\theta}\right]^z f(\theta), \tag{634}$$

$$v_3(\theta) = k_3{}^0 \theta \left[\frac{2(1 - \theta)}{\beta + 1 - 2\theta}\right]^z g(\theta), \tag{635}$$

where

$$f(\theta) = \left[1 + \frac{2\theta(\exp\{-w_b/k_B T\} - 1)}{1 + \beta}\right]^z \tag{636}$$

and

$$g(\theta) = \left[1 + \frac{2\theta(\exp\{(w_a' - w_b' - w_a)/k_B T\} - 1)}{1 + \beta}\right]^z. \tag{637}$$

Equation (635) then gives the desired rate v of the appearance of products, expressed as a function of θ as the parameter. To obtain v as a function of c, this latter may be determined also as a function of θ, in which case the pair

$$v = v(\theta), \qquad c = c(\theta), \tag{638}$$

is the parametric representation of v as a function of c. For this purpose Hill introduces as a "steady-state" condition

$$v_1 = v_2 + v_3, \tag{639}$$

that is,

$$k_1{}^0 c(1 - \theta) f(\theta) = \theta \left[\frac{2(1 - \theta)}{\beta + 1 - 2\theta} \right]^z [k_2{}^0 f(\theta) + k_3{}^0 g(\theta)], \tag{640}$$

from which the "steady-state value" of c is

$$c = \frac{\theta}{1 - \theta} \left[\frac{2(1 - \theta)}{\beta + 1 - 2\theta} \right]^z \left[\frac{k_2{}^0}{k_1{}^0} + \frac{k_3{}^0}{k_1{}^0} \frac{g(\theta)}{f(\theta)} \right]. \tag{641}$$

E. Stochastic Models of the Michaelis–Menten Mechanism

In Section IV.D, we brought together some of the first steps that have been taken toward the development of microscopic models of enzyme-catalyzed reactions, the point of penetration being the active site, where the complexation of enzyme and substrate takes place. Even in the case of a single active site, approached from a highly idealized structural viewpoint and at the most "macroscopic" microscopic level it is seen that binary E–S collisions contain an additional probabilistic component. And in the case of Hill's model it is seen that approximations and analogies have been employed in order to estimate interaction energies for n-valency systems, even treated from the simplest, idealized point of view of closest-packing arrangements restricted to nearest-neighbor interactions. Again, in this case, we are confronted with uncertainties requiring the introduction of probabilistic arguments. It seems clear from these beginnings then that the descent into microscopic enzyme kinetics will give rise to stochastic models just as it did in the case of microscopic models of the basic unimolecular reaction.

As in the unimolecular case, at the macroscopic level of enzyme kinetics, stochastic models have begun to appear which should then form a bridge between the microscopic theory and the phenomenological approaches associated with macroscopic kinetics. Some attention has altready been paid [see Bartholomay, 1962b, c] to such interconnections, which can be further expanded with attention to the active-site considerations presented in Section D, preceding. But mainly, these stochastic models have taken as their point of departure the Michaelis–Menten mechanism, being in fact, stochastic reformulations of that mechanism. The first of these to appear is due to Bartholomay [1957, 1962b, c] and is discussed next.

1. *A Stochastic Model of the Irreversible Michaelis–Menten Mechanism*†

The principles of construction of the stochastic models of the macroscopic kinetics of the elementary unimolecular and bimolecular reactions may be generalized and extended, as seen in Section III.F.5–7 [see also Bartholomay, 1957, Introduction, and 1962b]. It will be seen next that an application of this general method to the irreversible Michaelis–Menten mechanism therefore leads to a stochastic model which is at once consistent with the models representing the individual bimolecular and unimolecular steps of the mechanism and also with the deterministic model ("in the mean").

The construction of the model in this case begins with consideration as random (stochastic) variables those which have for their values

n_1 = the number of free enzyme molecules E,
n_2 = the number of substrate molecules S,
n_3 = the number of E–S molecules,
n_4 = the number of product molecules P,

each of these being the number present per constant volume at time t after initiation of reaction. All information about the stochastic model will be contained in the multivariate probability function $p(n_1, n_2, n_3, n_4; t)$, specifying the joint probabilities of n_1 free E molecules, n_2 S molecules, n_3 (E–S) molecules, and n_4 P molecules at time t, the underlying probability space for which will be in effect the Cartesian product of the elementary probability spaces corresponding to each of the these component reaction steps described below.

R1. *The Probability of a Complexation* T_1: $E + S \longrightarrow ES$. Given n_1 molecules of E and n_2 molecules of S present at time t, there are $(n_1 n_2)$ possible ES complexes. Associated with each such possibility is the same probability $[\mu_1\, \Delta t + o(\Delta t)]$ of "collision" (which, as seen in Section D.1. can be adjusted to a site–substrate collision). Consider then that associated with this process is the simplest possible (most macroscopic) probability space $\mathcal{S}_1$ consisting of two points or outcomes: Q_1, actual collision, and Q_2, failure to collide in time $(t, t + \Delta t)$ independently of t. Assuming, then, Bernoulli trials, and correspondingly, the Cartesian product probability space made up of $(n_1 n_2)$ copies of $\mathcal{S}_1$, the probability of exactly one collision in time $(t, t + \Delta t)$ is given by

$$b(n_1 n_2; 1, [\mu_1\, \Delta t + o(\Delta t)]) = \mu_1 n_1 n_2\, \Delta t + o(\Delta t), \tag{642}$$

†From Bartholomay [1962c, pp. 223–230].

where $o(\Delta t)$ is a higher order infinitesinal. And it may be shown that $b(n_1 n_2; k, [\mu_1 \Delta t + o(\Delta t)] = o(\Delta t)$, so that the probability of more than one such collision in this time is negligible. Note that according to ordinary bimolecular collision theory [Bartholomay, 1962a, Eq. (90); or Eq. (511) in the present chapter] μ_1 would correspond to

$$(8\pi k_B T/m_{12})^{1/2}(r_1 + r_2)^2 \exp(-E_0/k_B T) \tag{643}$$

or to the active-site corrected expression corresponding to Eq. (516). In any case it is intended here that μ_1 in the stochastic model correspond to k_1 in the deterministic model.

R2. *The Probability of a Dissociation* T_2: ES $\longrightarrow$ E + S. This is treated as in the elementary unimolecular stochastic model of Section III. In other words, the underlying probability space $\mathcal{S}_2$ associated with T_2 again consists of two outcomes corresponding to a dissociation or no dissociation of the ES complex in time $(t, t + \Delta t)$, the probability assigned to the former outcome being taken as $[\mu_2 \Delta t + o(\Delta t)]$, the same for each of the n_3 of the ES molecules present at time t, independently of time t. Similarly, it may be seen that the probability of exactly one dissociation in this time is $[\mu_2 n_3 \Delta t + o(\Delta t)]$ and that of more than one is negligible that is, $o(\Delta t)$. Again, μ_2 would correspond to the unimolecular deterministic rate constant k_2, though it is more important in this context to stress its relation to the stochastic models of microscopic kinetics discussed in Section III.

R3. *The Probability of a Decomposition* T_3: ES $\longrightarrow$ E + S. This is treated as the preceding case for T_2, the underlying probability space $\mathcal{S}_3$, having one point corresponding to a successful decomposition in time $(t, t + \Delta t)$ for each of the n_3 ES molecules present at time t and one to "no decomposition," the probability of the former being taken as $[\mu_3 \Delta t + o(\Delta t)]$ and that of the latter as its complement. And again, it can be seen that the probability of exactly one such decomposition of the n_3 ES molecules in time $(t, t + \Delta t)$ equals $[\mu_3 n_3 \Delta t + o(\Delta t)]$; that of more than one equals $o(\Delta t)$.

R4. *The Probability Space* $\mathcal{S}$ *for the Total Complex Reaction.* This, then may be taken as the Cartesian product: $\mathcal{S} = \mathcal{S}_1 \times \mathcal{S}_2 \times \mathcal{S}_3$ of the three component spaces and the probabilities assigned to $\mathcal{S}$ are assigned by assuming independence of the three processes, so that the probability of a component event or outcome $Q = (Q^{(1)}, Q^{(2)}, Q^{(3)})$, where $Q^{(i)}$ $(i = 1, 2, 3,)$ is an event associated with T_i, is obtained as the product

$$P(Q) = \prod_{i=1}^{3} P(Q^{(i)}). \tag{644}$$

The differential equation $(\partial/\partial t)\, p(n_1, n_2, n_3, n_4; t)$ is set up in the usual fashion, beginning with deducing an expression for the difference,

$$p(n_1, n_2, n_3, n_4; t + \Delta t) - p(n_1, n_2, n_3, n_4; t). \tag{645}$$

The probability rules R1–R4 are used to enumerate and assign probabilities to all possible events leading to $p(n_1, n_2, n_3, n_4; t + \Delta t)$. The work is simplified [see Bartholomay, 1962b, for details] by taking advantage of the interrelationships between the random variables

$$n_1 = n_0 - n_3 \tag{646}$$

$$n_4 = n_{20} - (n_2 + n_3) \tag{647}$$

(zero subscripts for initial values), where the ranges of the variables are

$$0 \leqslant n_1 \leqslant n_{10} \leqslant n_{20}, \qquad 0 \leqslant n_2 \leqslant n_{20},$$
$$0 \leqslant n_3 \leqslant n_{10}, \qquad 0 \leqslant n_4 \leqslant n_{20}. \tag{648}$$

Then with the random variables corresponding to n_2 and n_3 taken as the ultimate variables, the result is sought in the form $p(n_2, n_3; t)$. The corresponding time-differential equation is

$$\begin{aligned} \partial p(n_2, n_3; t)/\partial t = {} & \mu_3(n_3 + 1)p(n_2, n_3 + 1; t) + \mu_2(n_3 + 1) \\ & \times p(n_2 - 1, n_3 + 1; t) - \mu_2 n_1(n_{10} - n_3) \\ & \times p(n_2, n_3; t) - (\mu_2 + \mu_3)n_3 p(n_2, n_3; t) \\ & + \mu_1(n_2 + 1)(n_{10} - n_3 + 1) \\ & \times p(n_2 + 1, n_3 - 1; t). \end{aligned} \tag{649}$$

This, in turn, may be shown to be equivalent to the following second-order partial differential equation of the process:

$$\begin{aligned} \partial\phi/\partial t = {} & (\mu_3 + \mu_2 s_2 - \mu_2 s_3)(\partial\phi/\partial s_3) \\ & + \mu_1 n_{10}(s_3 - s_2)(\partial\phi/\partial s_2) + \mu_1 s_3(s_2 - s_3)(\partial^2\phi/\partial s_2 \partial s_3), \end{aligned} \tag{650}$$

where $\phi(s_2, s_3; t) = \sum_{n_2=0}^{n_{20}} \sum_{n_3=0}^{n_{10}} p(n_2, n_3; t)s_2^{n_2}s_3^{n_3}$ is the "probability-generating function" of the process and s_2, s_3 arbitrary variables, with $|s_2|, |s_3| \leqslant 1$.

No solution of this partial differential equation has been obtained. However, the mathematical difficulty here cannot be taken as an argument against the use of stochastic models and in favor of deterministic models, since the corresponding deterministic model, as we have noted, consists of nonlinear ordinary differential equations which, likewise, have not been solved. In fact, because of the interconnections between the two models, a general solution of one might suggest a solution of the other, for in a certain sense the stochastic model, as in the case of the unimolecular process, may be said to be "consistent in the mean" with the deterministic model. Thus, consider the mean values of n_2 and n_3 with respect to the probability function $p(n_2, n_3; t)$:

$$\bar{n}_2 = \sum_{n_2=0}^{n_{20}} \sum_{n_3=0}^{n_{10}} n_2\, p(n_2, n_3; t), \tag{651}$$

$$\bar{n}_3 = \sum_{n_2=0}^{n_{20}} \sum_{n_3=0}^{n_{10}} n_3\, p(n_2, n_3; t). \tag{652}$$

Differentiating $\bar{n}_2$ and $\bar{n}_3$ with respect to time gives

$$d\bar{n}_2/dt = \sum_{n_2=0}^{n_{20}} \sum_{n_3=0}^{n_{10}} n_2\, \partial p(n_2, n_3; t)/\partial t, \tag{653}$$

$$d\bar{n}_3/dt = \sum_{n_2=0}^{n_{20}} \sum_{n_3=0}^{n_{10}} n_3\, \partial p(n_2, n_3; t)/\partial t. \tag{654}$$

Substitution of expression (649), (653), and (654) after algebraic rearrangements, leads eventually to

$$\begin{aligned} d\bar{n}_2/dt = {} & \mu_2 \sum_{n_2=0}^{n_{20}} \sum_{n_3=0}^{n_{10}} n_3 p(n_2, n_3; t) \\ & - \mu_1 \sum_{n_2=0}^{n_{20}} \sum_{n_3=0}^{n_{10}} (n_{10} - n_3) n_2\, p(n_2, n_3; t), \end{aligned} \tag{655}$$

$$\begin{aligned} d\bar{n}_3/dt = {} & \mu_1 \sum_{n_2=0}^{n_{20}} \sum_{n_3=0}^{n_{10}} (n_{10} - \mu_3) n_2\, p(n_2, n_3; t) \\ & - (\mu_2 + \mu_3) \sum_{n_2=0}^{n_{20}} \sum_{n_3=0}^{n_{10}} n_3\, p(n_3, n_3; t), \end{aligned} \tag{656}$$

which amount to

$$d\bar{n}_2/dt = \mu_2 \bar{n}_3 - \mu_1 \overline{n_1 n_2}, \tag{657}$$

$$d\bar{n}_3/dt = \mu_1 \overline{n_1 n_2} - (\mu_2 + \mu_3)\bar{n}_3, \tag{658}$$

where $\overline{n_1 n_2} = \sum_{n_2=0}^{n_{20}} \sum_{n_3=0}^{n_{10}} n_1 n_2\, p(n_2, n_3; t)$. Comparison with the canonical differential equations of the deterministic model,

$$d[\mathrm{S}]/dt = k_2[\mathrm{ES}] - k_1[\mathrm{E}][\mathrm{S}], \tag{659}$$

$$d[\mathrm{ES}]/dt = k_1[\mathrm{E}][\mathrm{S}] - (k_2 + k_3)[\mathrm{ES}], \tag{660}$$

shows that the identifications [S] with $\bar{n}_2$, [ES] with $\bar{n}_3$, [E][S] with $\overline{n_1 n_2}$, and (k_1, k_2, k_3) with (μ_1, μ_2, μ_3) allow Eqs (657) and (658) to be identified with Eqs. (659) and (660), thus validating the mean-consistency criterion.

2. *Extension to the Reversible Michaelis–Menten Mechanism*

Darvey and Staff [1967] extended the preceding model to the reversible case; that is, adding the fourth step,

$$T_4\colon \mathrm{E} + \mathrm{P} \longrightarrow \mathrm{ES}. \tag{661}$$

As in the deterministic treatment of this same case by Miller and Alberty (discussed in Section IV.B.3) for the special case of $\mu_1 = \mu_4$ and at equilibrium, (corresponding to large t) Darvey and Staff obtained a solution to the differential equation of the corresponding joint probability function. The ultimate variables selected were n_2 and n_4, corresponding to which the time derivative of the joint probability function is

$$\begin{aligned}
&\partial p(n_2, n_4; t)/\partial t \\
&\quad = \mu_1(n_{10} - n_{20} + n_2 + n_4 + 1)(n_2 + 1)\, p(n_2 + 1, n + 4; t) \\
&\qquad + \mu_2(n_{20} - n_2 - n_4 + 1)\, p(n_2 - 1, n_4; t) \\
&\qquad + \mu_3(n_{20} - n_2 - n_4 + 1)\, p(n_2, n_4 - 1; t) \\
&\qquad + \mu_4(n_{10} - n_{20} + n_2 + n_4 + 1)(n_4 + 1)\, p(n_2, n_4 + 1, t) \\
&\qquad - [\mu_1(n_{10} - n_{20} + n_2 + n_4)n_2 + (\mu_2 + \mu_3)(n_{20} - n_2 - n_4) \\
&\qquad + \mu_4(n_{10} - n_{20} + n_2 + n_4)n_4\, p(n_2, n_4; t)]
\end{aligned} \tag{662}$$

with the following initial conditions: $p(n_2, 0; 0) = 1$ for $n_2 = n_{20}$; $p(n_2, 0; 0) = 0$ otherwise. The probability generating function in this case,

$$G(s_2, s_4; t) = \sum_{n_2=0}^{n_{20}} \sum_{n_4=0}^{n_{40}} p(n_2, n_4; t) s_2^{n_2} s_4^{n_4}, \tag{663}$$

has for its partial differential equation

$$\begin{aligned}
\partial G/\partial t &= \mu_1 s_2(1 - s_2)\, \partial^2 G/\partial s_2{}^2 + \mu_4 s_4(1 - s_4)\, \partial^2 G/\partial s_4{}^2 \\
&\quad + [\mu_1 s_4(1 - s_2) + \mu_4 s_2(1 - s_4)]\, \partial^2 G/\partial s_4 \partial s_2 \\
&\quad + [\mu_1(n_{10} - n_{20} + 1)(1 - s_2) + \mu_2 s_2(1 - s_2)]\, \partial G/\partial s_2 \\
&\quad + \mu_3 s_2(1 - s_4)\, \partial G/\partial s_2 \\
&\quad + [\mu_4(n_{10} - n_{20} + 1)(1 - s_4) + \mu_2 s_4(1 - s_2)]\, \partial G/\partial s_4 \\
&\quad + \mu_3 s_4(1 - s_4)\, \partial G/\partial s_4 \\
&\quad - s_0[\mu_2(1 - s_2) + \mu_3(1 - s_4)] G(s_2, s_4; t),
\end{aligned} \tag{664}$$

the initial condition being $G(s_2, s_4; 0) = s_2^{n_{20}}$.

The pair of variables (s_2, s_4) and t being separable, G may be expressed in the form

$$G(s_2, s_4; t) = H(s_2, s_4)\, T(t), \tag{665}$$

in which case the substitution of Eq. (665) into Eq. (664) leads to a pair of equations; namely, an ordinary differential equation in t:

$$T^{-1}\, dT/dt = -\lambda_n \tag{666}$$

(where $-\lambda_n$ is the "separation constant") the solution of which is

$$T = k = \exp(-\lambda_n t) \tag{667}$$

and a second-order hyperbolic, partial differential equation in H, not solved.

An interesting property of this model emerges from the expressions for $d\bar{n}_2/dt$ and $d\bar{n}_4/dt$ which show that the Darvey-staff model does not satisfy everywhere (that is, at all times t) the mean-consistency criterion relative to the deterministic model. Specifically, $d\bar{n}_2/dt$ and $d\bar{n}_4/dt$ differ from $d[\mathrm{S}]/dt$

and $d[\mathrm{P}]/dt$ of the deterministic model by the factors

$$\mu_1[\sigma_2^2(t) + \sigma_{2,4}(t)], \qquad \mu_4[\sigma_4^2(t) + \sigma_{2,4}(t)] \tag{668}$$

where $\sigma_2^2(t)$ is the variance of n_2, $\sigma_4^2(t)$ is that of n_4, and $\sigma_{2,4}(t)$ is the covariance of n_2, n_4 with respect to $p(n_2, n_4; t)$. However, it is deduced that for most experimental studies *in vitro*, where the initial substrate concentration is much greater than initial enzyme concentration the mean consistency criterion is approximately satisfied.

While a general solution for all t of Eq. (664) has not been obtained, an equilibrium solution: $\lim t \longrightarrow \infty\ G(s_2, s_4; t)$ has been obtained for the special case of $\mu_1 = \mu_4$, just as Miller and Alberty [1958] were able to obtain a solution of the reversible deterministic model for the case, $k_1 = k_4$. Expressions for the distributions of all species at equilibrium under these conditions are available.

F. Some Further Examples of the Use of Stochastic Models in Dealing with Complex Biochemical Kinetics

1. *DNA-Replication Kinetics*

One of the most complex enzyme-catalyzed biochemical kinetic systems of particular interest at the present time, to which our comments on the feasibility of stochastic formulation seem to apply is that of the DNA-replication system. Here the usual uncertainties associated with the enzyme involved, which have already been discussed in general terms, are compounded with additional uncertainties associated with other complex macromolecules within the system, such as the nucleic acid molecules, our total knowledge of the composition, structure, and function of the system being very limited in terms of verifiable details.

The replication system as studied *in vitro*, for example, by Kornberg *et al.* [1956; see also Kornberg, 1960] is considered to consist at least of the following component species and conditions: (1) all four of the substrate deoxynucleotide species which form the adenine–thymine and guanine–cytosine pairs must be present; (2) moreover, the substrates must be the tri-, and not the diphosphates; (3) even in the presence of all four triphosphates, if DNA is absent, then no reaction will take place; (4) in fact, the DNA seems to function as a primer and/or as a template: (5) an enzyme such as the *Escherichia coli* polymerase enzyme must be present, as well as ATP and, say, $MgCl_2$.

It is presumed that the enzyme performs its catalytic functions with respect to the triphosphate nucleotides as substrates in a complex fashion, being involved in some way with the breaking of hydrogen bonds between the adenosine–tyrosine and guanine–cytosine doublets in the original template

DNA parent strands in preparation for the reformations of the bonds with the available substrate nucleotides, and with the maintenance of an association between the two newly forming daughter DNA helixes and the newly dissociating parent helixes. Thus, the complex replication process includes such activities as the diffusion of nucleotide bases to appropriately freed sites on the disjoined parent helixes coupled with the unwinding of the complementary chains of the parents by the splitting of bonds, followed by the formation of pairs of separate new bonds with appropriate new bases of both strands of the original parent. Schematically, the process may be represented as in Fig. 20, which shows the original DNA template, partially unwound, the pair of newly forming daughter helixes, each in association with one member of the template pair of complementary parent helixes, and the enzyme surface simply as an ellipsoidal surface moving in one direction over the template.

The Zimmerman–Simha Stochastic Model. Zimmerman and Simha [1965, 1966] have proposed the following hypothesis for the replicating system with respect to which they have constructed a stochastic mathematical model. They assume that the enzyme molecule moves along the template in a fixed

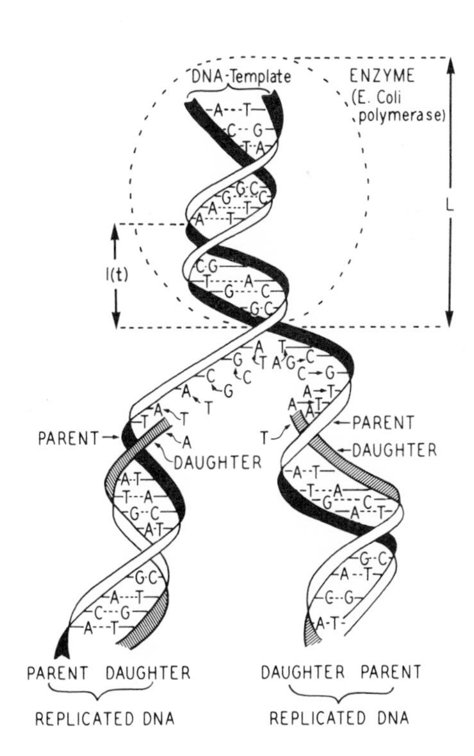

Fig. 20. Enzyme-catalyzed DNA-replication system, showing uncoupling of original complementary strands, diffusing nucleotide bases, and recoupling of separated parent strands with newly forming complementary strands.

direction, the active sites of the enzyme molecule being so arranged as to cover a fixed number of template sites. It is presumed that the molecule begins its crawling motion along the template at the initial point of separation of the two strands. As it progresses "upwards" along the template, the complementary helixes begin separating via the breaking of bonds, followed immediately by the reformation of the bonds with appropriate nucleotide bases which are forming the new daughter helixes, each replacing the complementary member of the strand with which it becomes associated. They further assume that each template (or primer) DNA molecule has a fixed length N (that is, number of paired bases) and at any time t a given base pair at site position j ($j = 1, 2, \ldots, N$) along the template either (1) still possesses an intact hydrogen bond or (2) this bond has been broken, providing a free template site or (3) the open site has incorporated the monomeric base unit into a newly replicated daughter helix.

The preceding conception of the replication system leads to a complex stochastic model; that is one consisting of a pair of interdependent stochastic processes, the equations of which will not be recounted here. Some of the basic probabilistic considerations which went into the derivation are (1) the arrival of the monomeric substrate units at the jth position on the template is describable as a first-order diffusion process, (2) the probability of the binding of monomers of daughter helixes amounts to the probability of transition from length j to length $(j + 1)$ of a daughter molecule, independently of the jth position, (3) associated with the availability of base pairs is a time-dependent probability function. In the process of constructing the stochastic model the probability $\phi_k(t)$ that there are k successive pairs of unbound template sites at time t is utilized to obtain an important expression,

$$l(t) = \sum_{k=1}^{L} k\, \phi_k(t), \tag{669}$$

for the average length of the region of the unbound base pairs. And the random fluctuations in length of the region are discussed in terms of the second moment

$$\mu_2 = \sum_{k=1}^{L} k^2\, \phi_k(t). \tag{670}$$

In this connection the observation was made that the problem of the determination of $l(t)$ is formally equivalent to a single-server queueing problem, where $l(t)$ is identified with the length of the queue of the polymerization process, and the "service" consists of the unwinding of the parent template helix.

The model obtained has been utilized in detailed and informative *in numero* studies, and also in the interpretation of additional *in vitro* and *in vivo* experimental studies of the underlying replication system. Two resultants of their studies have been (1) the quantification of the degree to which the unwinding of the template inhibits replication and (2) the demonstration of the dependency of the latter on specific parameters of the model.

2. *The General Catalytic Queue Process (GCQP)*†

A number of basic points of similarty between enzyme catalysis, the epidemological aspects of communicable desease, and the great variety of processes encompassed in queueing theory led to the formulation of the so-called general catalytic queue process by Bartholomay [1964], originally conceived as a generalization of the concept of catalysis in biokinetics. It may be seen that the DNA-replication system just discussed is another realization of such a process, a possible point of contact between it and queueing theory itself, having also been noted by Zimmerman and Simha [1965, 1966] (see above).

The GCQP may, in fact, be thought of in the spirit of an algorithm or *ad hoc* operational procedure to assist in promoting the translation of the stochastic–catalytic aspects of a biological system into a comprehensive stochastic model or interdependent union of stochastic processes capable of incorporating the uncertain characteristics of such complex underlying systems into probabilistically formulated predictions of the kinetics of the system. The stochastic model of the Michaelis–Menten mechanism discussed in Section IV.E is easily formulated in these terms, as well as more complex enzyme kinetic systems of many components. The treatment of epidemiological processes along these same lines raises the possibility of a general mathematical biological kinetics and has been discussed elsewhere [Bartholomay, 1964b]. Here, the main features of the GCQP will be introduced and related to enzyme catalysis. To elaborate the analogy with queueing theory, the terminology of that theory is used to describe the GCQP. However, certain differences will be apparent, the most basic of which, methodologically speaking, is that, whereas queueing theory refers generally to the congestion of the service mechanism manifested by a line of waiting customers at a given fixed service counter, the GCQP allows more complex topology, the "customers" being discussed more in the sense of "potential" customers, randomly dispersed, and in random movement, in a "waiting area," and thus ultimately making a random selection of any one of a number of equivalent service counters. The GCQP thus is not limited to a congestion process in the usual sense; it constitutes, in effect, an extension of the queueing-theory concept.

a. General Description of the GCQP. Specifically, it is considered that initially there are n_0 "customers" each of which will ultimately have a "service" performed by one of a number m_0 of "servers," each of which is

†From Bartholomay [1964b, pp. 101–145].

capable of serving at least one customer at a time. The number of "service counters" ν associated with a given server in some cases is a fixed integer; in other cases, it may be an intergral-valued random variable. In addition to the assumption of a systematic driving force of affinity between server S and customer C, it is also assumed that superimposed on this force is a random force manifested by relative random motions of customers and random arrivals and departures at serving counters until effective contact resulting in completion of the service mechanism is accomplished. The latter has reference to the fact that associated with the system composed of customers and servers is a major activity or service requiring a "service time" which may be a random variable characterized by a probability density function $f(t)$. However, even after appropriate contact is made between S and C, several outcomes are possible: (1) if C is "dissatisfied" the service period ends with C's reentrance into the waiting area, once again as a potential customer; (2) if satisfied, the service period ends with the reidentification of C; and (3) in some kinds of processes (for example, autocatalytic or epidemiologic) C may transformed into a server. In the simplest cases, associated with the initiation of the service phase are certain prior probabilities p of "success" and $q = 1 - p$ of "failure."

As in the case of queueing theory, given the initial numbers n_0 and m_0, the central mathematical problem is the determination of the time-dependent (distribution) probability function of the "queue size"; this is the probability function of the number of waiting customers in the service area as a function of time t. Another important parameter to be determined is the time T required for all customers to be serviced.

b. Some Variants of the GCQP. Certain variants and extensions of the basic catalytic process are also allowed in GCQP, for example:

(1) THE SERVER–CUSTOMER–COFACTOR PROCESS. The service mechanism requires the co-operation of an additional species, referred to as the "cofactor" or "attendant" or modifier.

(2) THE COMPETITIVE OR SERVER–CUSTOMER VERSUS CUSTOMER PROCESS. Each server is capable of serving two different classes of customers. In this case, a "mixed waiting area" is involved (similar to the binary solution concept treated in Section IV.D.5) and the calculation of the "queue size" requires a bivariate or joint probability function $p(n_1, n_2, t)$, where n_1 gives the cardinality of the first class C; and n_2, that of the second class.

(3) THE SERVER–CUSTOMER–INHIBITOR PROCESS. A third species I of "inhibitors" is present which can block the S–C service mechanism, say, either by preoccupying the active service counter or indirectly, that is, removed from the service counter.

(4) THE COMPETITIVE SERVER VERSUS SERVER–CUSTOMER PROCESS. Two different classes S_1 and S_2 of servers compete for the same customer C, who may be indifferent to S_1 and S_2 or else exercise selectivity (unequal probabilities case) between them.

(5) THE MULTIPLE-SERVICE COUNTER PROCESS. Each server can serve simultaneously more than one customer for example, (the multiactive site or multivalent enzyme case).

Mixtures and obvious extensions of these basic types are also intended to be included in the GCQP concept.

c. The Michaelis–Menten Mechanism as GCQP. The particularization of the general process to the case of an enzyme-catalyzed reaction begins with the identification of: customer with substrate molecule, server with enzyme molecule, cofactor or attendant with coenzyme or other enzymic modifiers or activator chemical species, inhibitors with enzyme inhibitors, serving counter with active site, and so forth. Some of the particulars follows.

(1) SERVING CAPABILITY OR TURNOVER NUMBER. A given enzyme molecule, after catalyzing the transformation of one substrate molecule into products P of reaction, is ready to serve another that happens to collide with it effectively. That a given enzyme molecule is extremely "busy" is clear from the so-called turnover number of an enzyme referred to earlier, indicating in some cases service times of the order of milliseconds. Because of this, an *in vitro* enzyme catalysis can be carried out effectively by mixing a relatively small number m_0 of server E molecules with a number $n_0 \gg m_0$ of substrate customer molecules.

(2) ARRIVALS. A mathematical analysis or characterization of the arrival of substrate molecules at active sites (service counters) on enzyme molecules is suggested by considering each (idealized) molecule (E or S) as a "free particle" in the same Brownian-motion-theoretic sense referred to in the Kramers model of Section III. In that case, the driving force of reaction and the random aspects would be combined in a Langevin, mixed stochastic-differential equation, such as Eq. (244) of Section II.E.3.a. Here $(x(t), y(t), z(t))$ gives the position of the molecule at time t and $\mathbf{u}$ is the velocity vector $(\dot{x}, \dot{y}, \dot{z})$; $-\beta\mathbf{u}$, the systematic component; and $\mathbf{A}(t)$, the random component characteristic of Brownian motion. In accord with Stokes Law, the coefficient β would be defined by

$$\beta \equiv 6\pi\eta/m \tag{671}$$

as a function of the mass m of the idealized spherical molecule and η, the coefficient of viscosity of the surrounding fluid.

As indicated in Section II, the solution of Eq. (244) is a probability density function $f(\mathbf{u}, t, \mathbf{u}_0)$. It is specified generally by imposing constraints such as

$$\lim_{t\to\infty} f(\mathbf{u}, t; \mathbf{u}_0) = (m/2\pi k_B T)^{3/2} \exp(-mu^2/2k_B T), \tag{672}$$

that is, a Maxewllian distribution is to be satisfied in the limit. This particular constraint leads to the solution,

$$f(\mathbf{u}, t; \mathbf{u}_0) = \{m/2\pi k_B T[1 - \exp(-\beta t)]\}^{3/2} \times \exp\{-m\,|\mathbf{u} - \mathbf{u}_0 \exp(-\beta t)|^2/2k_B T[1 - \exp(-2\beta t)]\}. \tag{673}$$

If the usuall convention is adopted, namely, the equilibrium hypothesis that the Maxwellian distribution holds throughout the process, then it is found that the probability that one enzyme molecule of mass m_1 has a velocity in the range $(\mathbf{u}_1, \mathbf{u}_1 + d\mathbf{u}_1) = (\dot{x}_1, \dot{x}_1 + d\dot{x}_1; \dot{y}_1, \dot{y}_1 + d\dot{y}_1; \dot{z}_1, \dot{z}_1 + d\dot{z}_1)$ is given by

$$p_{m_1} = \left(\frac{m_1}{2\pi k_B T}\right)^{3/2} \exp\left\{-\frac{m_1}{2} \frac{\dot{x}_1{}^2 + \dot{y}_1{}^2 + \dot{z}_1{}^2}{k_B T}\right\} d\dot{x}_1\, d\dot{y}_1\, d\dot{z}_1. \tag{674}$$

The corresponding probability for the substrate molecule of mass m_2 becomes

$$p_{m_2} = \left(\frac{m_2}{2\pi k_B T}\right)^{3/2} \exp\left\{-\frac{m_2}{2} \frac{\dot{x}_2{}^2 + \dot{y}_2{}^2 + \dot{z}_2{}^2}{k_B T}\right\} d\dot{x}_2\, d\dot{y}_2\, d\dot{z}_2. \tag{675}$$

From of this point on the present discussion merges with that given in Section IV.E.1 (R1) leading to the determination of the stochastic parameter μ_1 in terms of Eq. (643) [Bartholomay, 1962a, b, c].

(3) Service Mechanism. Satisfaction of the substrate (customer) results in the transformation ES $\longrightarrow$ E + P, associated with which, as in R3 of Section IV.E.1 is the probability $p_3 = \mu_3\, \Delta t + o(\Delta t)$. Dissatisfaction results in the dissociation ES $\longrightarrow$ E + S; and as in R2 of Section IV.E.1, the probability associated with this eventuality is $p_2 = \mu_2\, \Delta t + o(\Delta t)$. The probability that the service mechanism is not terminated may therefore be taken as

$$p_1 = 1 - (\mu_2 + \mu_3)\, \Delta t + o(\Delta t). \tag{676}$$

Then, considering that each of the n_3 ES complexes present at time t is undergoing independent trinomial trials, the probability that there will be r_2 dissatisfactions, r_3 satisfactions, and r_1 continuations in time $(t, t + \Delta t)$ is given by the trinomial probability function,

$$p(r_2, r_3, r_1) = (n_3!/r_1!\, r_2!\, r_3!) p_1^{r_1} p_2^{r_2} p_3^{r_3}, \tag{677}$$

where $n_3 = r_1 + r_2 + r_3$. Thus, in agreement with the determinations of Section IV.E, the probability of one dissatisfaction, no satisfaction, and $(n_3 - 1)$ continuations is $[\mu_2 n_3\, \Delta t + o(\Delta t)]$; that of no dissatisfaction, one satisfaction, and $(n_3 - 1)$ continuations is $[\mu_3 n_3\, \Delta t + o(\Delta t)]$. The probability that all n_3 service mechanisms continue throughout the period is $[1 - n_3(\mu_2 + \mu_3)\, \Delta t + o(\Delta t)]$. And the probability of more than one dissatisfaction or satisfaction, or of one or more pairs of satisfactions or dissatisfactions is an infinitesional of higher order, that is, negligible in the limit.

In this way, then, paraphrasing the model of Section IV.E one arrives at the differential equation of the probability function $p(n_2, n_3; t)$ of the joint "queue sizes."

References

Abraham, R., and Marsden, J. E. [1967]. "Foundations of Mechanics." Benjamin, New York.

Alberty, R. A. [1959]. The rate equation for an enzymic reaction, *in* "The Enzymes" (P. D., Boyer, H. Lardy, and K. Myrback, eds.), Vol. 1, Chapter 3, 2nd ed. Academic Press, New York.

Arrhenius, S. [1889]. Ueber die Reaktionsgeschwindigkeit bei der Inversion von Rohrzucker durch Säuren, *Z. Phys. Chem.* **4**, 226.

Baldwin, E. [1963]. "Dynamic Aspects of Biochemistry." Cambridge Univ. Press, London and New York.

Barlow, W. [1883a]. Probable nature of the internal symmetry of crystals (1), *Nature* **29**, 186.

Barlow, W. [1883b]. Probable nature of the internal symmetry of crystals (2), *Nature* **29**, 205.

Barlow, W. [1884]. Reply to Sohncke's critique, *Nature* **29**,404.

Barlow, W. [1894]. Über die geometrischen eigenschaften homogener staner Structuren und ihre anmendung auf Krystalle, *Z. Krist.* **23**, 1.

Barlow, W. [1898]. Geometrischen Untersuchung uber eine mechanische Ursache der Homogenitaet der Structuren und der Symmetrie mit besonderer Anwendung auf Kristallisation und chemische Verbindung, *Z. Krist.* **29**, 433.

Bartholomay, A. F. [1957]. A stochastic approach to chemical reaction kinetics. Doctoral Thesis, Harvard Univ., Cambridge.

Bartholomay, A. F. [1958]. Stochastic models for chemical reactions: I. Theory of the unimolecular reaction process, *Bull. Math. Biophys.* **20**, 175.

Bartholomay, A. F. [1959]. Stochastic models for chemical reactions: II. The unimolecular rate constant, *Bull. Math. Biophys.* **21**, 363.

Bartholomay, A. F. [1960]. Molecular set theory: A mathematical representation for chemical reaction mechanisms, *Bull. Math. Biophys.* **22**, 285.

Bartholomay, A. F. [1962a]. Physico-mathematical foundations of reaction rate theory, *in* "Physicomathematical Aspects of Biology" (N. Rashevsky, ed.), pp. 5–86. Academic Press, New York.

Bartholomay, A. F. [1962b]. Enzymatic reaction rate theory: A stochastic approach, *Ann. N.Y. Acad. Sci.* **96**, 897.

Bartholomay, A. F. [1962c]. A stochastic approach to statistical kinetics with application to enzyme kinetics, *Biochemistry* **1**, 223.

Bartholomay, A. F. [1964a]. The mathematical approach to the study of discrete biological events, *6th IBM Med. Symp.* pp. 325–350. IBM, Poughkeepsie, New York.

Bartholomay, A. F. [1964b]. The general catalytic queue process, *in* "Stochastic Models in Medicine and Biology" (J. Gurland, ed.), pp. 102–145. Univ. of Wisconsin Press, Madison, Wisconsin (© 1964 by the Regents of The University of Wisconsin).

Bartholomay, A. F. [1965]. Molecular set theory: II. An aspect of the biomathematical theory of sets, *Bull. Math. Biophys.* **27**, 235.

Bartholomay, A. F. [1968a]. Some general ideas on deterministic and stochastic models of biological systems, *in* "Quanitative Biology of Metabolism" (*3rd Int. Symp. Helgoland, Sept. 1967*) (A. Locker, ed.), pp. 45–66. Springer-Verlag, Berlin and New York.

Bartholomay, A. F. [1968b]. The case for mathematical biology, or the mathematical destiny of biology. *Bioscience* **18**, 717.

Bartholomay, A. [1971]. A stochastic solution of the 3-body problem of chemical kinetics: The termolecular stochastic process I, *Bull. Math. Biophys.* **33**, 67.

Bharucha-Reid. A. T. [1960]. *In* "Elements of the Theory of Markov Processes and Their Applications," Chapter 8. McGraw-Hill, New York.

Blake, C. C. F.. Fenn. R. H., North. A. C. T., Phillips. D. C., and Poljak. R. T. [1962]. Structure of lysozyme, a fourier map of the electron density at 6 Å resolution obtained by X-ray diffraction, *Nature* **196**, 1173.

Blake, C. C. F., Koenig. C. F., Mair. G. A., North. A. C. T., Phillips. D. C., and Sarma, V. R. [1965]. Structure of hen egg-white lysozyme, A three-dimensional Fourier synthesis at 2 Å resolution, *Nature* **206**, 757.

Bohr, N., and Wheeler, J. A. [1939]. The fission of protoactinium, *Phys. Rev.* **56**, 426.

Born, M., and Oppenheimer, J. R. [1927]. Quantum theory of the molecule, *Ann. Phys.* **84**, 457.

Botts, J., and Morales, M. F. [1953]. Analytical description of the effects of modifiers and of enzyme multivalency upon the steady state catalyzed reaction rate, *Trans. Faraday Soc.* **49**, 696.

Boyer, P. D., Lardy. H., and Myrback, K. (eds.) [1959]. "The Enzymes," Vol. 1. Academic Press, New York.

Bray, H. G., and White, K. [1966]. "Kinetics and Thermodynamics in Biochemistry." Churchill, London.

Briggs, G. E., and Haldane, J. B. S. [1925]. Note on the kinetics of enzyme action, *Biochem. J.* **19**, 338.

Bunker, D. L. [1962]. Monte Carlo calculation of triatomic dissociation rates. I. N_2O and O_3, *J. Chem. Phys.* **37**, 393.

Bunker, D. L. [1964a]. Monte Carlo calculations. 4. Further studies of unimolecular dissociation, *J. Chem. Phys.* **40**, 1946.

Bunker, D. L. [1964b]. Computer experiments in chemistry, *Sci. Amer.* **211**, 100.

Canfield, R. E. [1963]. The amino acid sequence of egg white lysozyme, *J. Biol. Chem.* **238**, 2698.

Chandrasekhar, S. [1943]. Stochastic problems in physics and astronomy, *Rev. Mod. Phys.* **15**, 1.

Christiansen, J. A. [1936]. Uber eine Erweiterung der Arrheniusschen Auffassung der chemischen Reaktion, *Z. Phys. Chem. B* **33**, 145.

Coleman, J. E., and Vallee, B. L. [1960]. Metallocarboxypepitidases, *J. Biol. Chem.* **235**, 390.

Condon, E. U. [1927]. Wave mechanics and the normal state of the hydrogen molecule, *Proc. Nat. Acad. Sci.* **18**, 466.

Coxeter, H. M. S. [1962]. The problem of packing a number of equal nonoverlapping circles on a sphere, *Trans. N.Y. Acad. Sci.* **24**, 320.

Crick, F. H. C., and Kendrew, J. C. [1957] X-ray analysis and protein structure, *Advan. Prot. Chem.* **12**, 133.

Darvey, I. G., and Staff, P. J. [1966]. Stochastic approach to first-order chemical reaction kinetics, *J. Chem. Phys.* **44**, 990.

Darvey, I. G., and Staff, P. J. [1967]. The application of the theory of markov processes to the reversible one subtrate–one intermediate–one product enzymic mechanism, *J. Theoret. Biol.* **14**, 157.

Davies, D. R. [1967]. X-ray diffraction studies of macromolecules, *Ann. Rev. Biochem.* **36**, 321.

Davis, J. C. [1965]. "Advanced Physical Chemistry," pp. 219–224. Ronald Press, New York.

Dayhoff, M. O., and Eck, R. V. [1968]. "Atlas of Protein Sequence and Structure." Nat. Biomed. Res. Foundation, Silver Springs, Maryland.

Delbruck, M. [1940]. Statistical fluctuations in autocatalytic reactions. *J. Chem. Phys.* **8**, 120.

Doob, J. L. [1953]. "Stochastic Processes." Wiley, New York.

Eliason, M. A., and Hirschfelder, J. O. [1959]. General collision theory treatment for the rate of bimolecular, gas phase reactions, *J. Chem. Phys.* **30**, 1426.

Eyring, H. [1931]. The energy of activation for bimolecular reactions involving hydrogen and halogens, according to the Quantum Mechanics, *J. Amer. Chem. Soc.* **53**, 2537.

Eyring, H. [1932a]. Steric hinderance and collision diameters, *J. Amer. Chem. Soc.* **54**, 3191.

Eyring, H. [1932b]. Quantum mechanics and chemical reactions, *Chem. Rev.* **10**, 103.

Eyring, H. [1935a]. The activated complex in chemical reactions, *J. Chem. Phys.* **3**, 107.

Eyring, H. [1935b]. The activated complex and the absolute rate of chemical reactions, *Chem. Rev.* **17**, 65.

Eyring, H., and Polanyi, M. [1931]. Simple gas reactions, *Z. Phys. Chem. B* **12**, 279.

Fejes Toth, T. L. [1959]. Kugelunterdeckunger und Kugeluberdeckunger in Raumen konstanter Krummung, *Arch. Math.* **10**, 310.

Fowler, R. H., and Guggenheim, E. A. [1939]. "Statistical Thermodynamics." Cambridge Univ. Press, London and New York.

Frost, A. A., and Pearson, R. G. [1961]. "Kinetics and Mechanism." Wiley, New York.

Gibbs, J. W. [1948]. "The Collected Works of J. Willard Gibbs." Yale Univ. Press, New Haven, Connecticut.

Glasstone, S., Laidler, K. J., and Eyring, H. [1941]. "The Theory of Rate Processes." McGraw-Hill, New York.

Guggenheim, E. A. [1948]. Statistical thermodynamics of co-operative systems. A generalization of the quasi-chemical method, *Trans. Faraday Soc.* **44**, 1007.

Gutfreund, H. [1955]. Steps in the formation and decomposition of Some enzyme-substrate complexes, *Discuss. Faraday Soc.* **20**, 167.

Haldane, J. B. S. [1930]. "Enzymes," Chapter 5. Longmans, Green, London.

Harte, R. A., and Rupley, J. A. [1968]. Three-dimensional pictures of molecular models—Lysozyme, *J. Biol. Chem.* **243**, 1663.

Hein, G. E., and Niemann, C. [1962]. Steric course and specificity of chymotrypsin-catalyzed reactions, *J. Amer. Chem. Soc.* **84**, 4495.

Heineken, F. G., Tsuchiya, H. M., and Aris, R. B. [1967a]. On the mathematical status of the pseudo-steady state hypothesis of biochemical kinetics, *Math. Biosci.* **1**, 95.

Heineken, F. G., Tsuchiya, H. M., and Aris, R. B. [1967b]. On the accuracy of determining rate constants in enzymatic reactions, *Math. Biosci.* **1**, 115.

Heitler, W., and London, F. [1927]. Interaction of neutral atoms and homopolar binding according to the quantum mechanics, *Z. Physik* **44**, 455.

Hill, T. L. [1952]. Effect of nearest neighbor substrate interactions on the rate of enzyme and catalytic reaction, *J. Amer. Chem. Soc.* **74**, 4710.

Hinshelwood, C. N. [1927]. Quasi-unimolecular reactions. The decomposition of di-ethyl ether in the gaseous state, *Proc. Roy. Soc.* (*London*) **A114**, 84.

Hinshelwood, C. N. [1940]. "The Kinetics of Chemical Change." Oxford Univ. Press (Clarendon), London and New York.

Hommes, F. A. [1962]. Analog computer studies of a simple enzyme-catalyzed reaction, *Arch. Biochem. Biophys.* **96**, 28.

Ising, E. [1925]. The theory of ferromagnetism, *Z. Physik* **31**, 253.

Jacobs, L. E. [1963]. Kinetics of Small Systems. A Master's Thesis submitted to Michigan State Univ., Dept. of Chem., Lansing, Michigan.

Jolles, J., Jauregui-Adell, J., and Jolles, P. [1963]. The chemical structure of egg-white lysozyme: the detailed study, *Biochim. Biophys. Acta.* **78**, 668.

Kassel, L. S. [1928]. Studies in homogeneous gas reactions, *J. Phys. Chem.* **32**, 225.

Katchalsky, A., and Curran, P. F. [1965]. "Nonequilibrium Thermodynamics in Biophysics." Harvard Univ. Press, Cambridge.

Kendall, D. G. [1950]. An artificial realization of a simple birth and death process, *J. Roy. Stat. Soc. B* **12**, 116.

Kornberg, A. [1960]. Biologic synthesis of deoxyribonucleic acid, *Science* **131**, 1503.

Kornberg, A., Lehman, I. R., Bessman, M. J., and Simms, E. S. [1956]. Enzymic synthesis of deoxyribonucleic acid, *Biochem. Biophys. Acta* **21**, 197.

Kramers, H. A. [1940]. Brownian motion in a field of force and the diffusion model of chemical reactions, *Physica* **8**, 284.

Krieger, I. M., and Gans, P. J. [1960]. First-order stochastic processes, *J. Chem. Phys.* **32**, 247.

Laidler, K. J. [1950.] "Chemical Kinetics." McGraw-Hill, New York.

Laidler, K. J. [1955a.] Theory of the transient phase in kinetics, with special reference to enzyme systems, *Canad. J. Chem.* **33**, 1614.

Laidler, K. J. [1955b]. "The Chemical Kinetics of Excited States." Oxford Univ. Press, London and New York.

Laidler, K. J. [1958]. "The Chemical Kinetics of Enzyme Action." Oxford Univ. Press, London and New York.

Lindemann, F. A. [1922]. The radiation theory of chemical action, *Trans. Faraday Soc.* **17**, 598.

London, F. [1928]. "Probleme der modernen Physik Sommerfeld Festschrift," p. 104. Hirzel, Leipzig.

London, W. P. [1968] Steady state kinetics of an enzyme reaction with one substrate and one modifier, *Bull. Math. Biophys.* **30**, 253.

Low, B.W. [1961]. Peptide synthesis and protein structure, *J. Polym. Sci.* **49**, 153.

Ludwig, M. L., Hartsuck, J. A., Steitz, T. A., Muirhead, H., Coppola, J. C., Reeke, G. N., and Lipscomb, W. N. [1967]. The structure of carboxypeptidase A. IV. Preliminary results at 2.8-Å resolution, and a substrate complex at 6-Å resolution, *Proc. Nat. Acad. Sci. U.S.* **57**, 511.

Mahler, H. R., and Cordes, E. H. [1966]. "Biological Chemistry." Harper, New York.

Marcus, R. A. [1952]. Unimolecular dissociations and free radical recombination reactions, *J. Chem. Phys.* **20**, 359.

Marcus, R. A. [1965]. Additivity of heats of combustion, L.C.A.O. resonance energies, and bond orders of conformal sets of conjugated compounds, *J. Chem. Phys.* **43**, 2658.

Marcus, R. A. [1968]. Remarks on the generalization of activated complex theory, *in* "Chemische Elementarprozesse" (H. Hartman, ed.), pp. 23–25. Springer-Verlag, Berlin and New York.

Michaelis, L., and Menten, M. [1913]. Kinetics of invertase action, *Biochem. Z.* **49**, 333.

Miller, W. G., and Alberty, R. A. [1958]. Kinetics of the reversible Michaelis–Menten mechanism and the applicability of the steady-state approximation, *J. Amer. Chem. Soc.* **80**, 5146.

Montroll, E. W., and Shuler, K. E. [1957]. "Studies in nonequilibrium rate processes. I. The relaxation of a system of harmonic oscillators, *J. Chem. Phys.* **26**, 454.
Montroll, E. W., and Shuler, K. E. [1958]. The application of the theory of stochastic processes to chemical kinetics, *Advan. Chem. Phys.* **1**, 361.
Morales, M. F., Botts, J., and Hill, T. L. [1948]. Equilibrium equations for a model of antibody–antigen combination, *J. Amer. Chem. Soc.* **70**, 2339.
Morse, P. M. [1929]. Diatomic molecules according to the wave mechanics. II. Vibrational levels, *Phys. Rev.* **34**, 57.
McQuarrie, D. A. [1963] Kinetics of small systems, I, *J. Chem. Phys.* **38**, 433.
McQuarrie, D. A. [1967]. Stochastic approach to chemical kinetics, *J. Appl. Probability* **4**, 413.
Pauling, L. [1939]. "The Nature of the Chemical Bond." Cornell Univ. Press, Ithaca, New York.
Pauling, L., and Corey, R. B. [1953]. Two rippled-sheet configurations of polypeptide chains and a note about the pleated sheets, *Proc. Nat. Acad. Sci. U.S.* **39**, 253.
Pauling, L., Corey, R. B., and Branson, H. R. [1951]. The structure of proteins: Two hydrogen-bonded helical configurations of the polypeptide chain, *Proc. Nat. Acad. Sci. U.S.* **37**, 205.
Pennycuick, S. W. [1926]. The unimolecularity of the inversion process, *J. Amer. Chem. Soc.* **48**, 6.
Phillips, D. C. [1966]. The three-dimensional structure of an enzyme molecule, *Sci. Amer.* **215**, 78.
Prigogine, I., and Mahieu, M. [1950]. Sur la perturbation de la distribution de Maxwell par des reactions chimiques en phase gazeuse, *Physica* **16**, 51.
Prigogine, I., and Xhrouet, E. [1949]. On the perturbation of Maxwell distribution function by chemical reactions in gases, *Physica* **15**, 913.
Raley, J. H., Rust, F. F., and Vaughan, W. E. [1948]. Decomposition of di-*t*-alkyl peroxides. I. Kinetics, *J. Amer. Chem. Soc.* **70**, 88.
Rice, O. K. [1962]. *In* "Energy and Transfer in Gases, 12th Solvay Conference." Wiley (Interscience), New York.
Rice, O. K., and Ramsperger, H. C. [1927]. Theories of unimolecular gas reactions at low pressures, *J. Amer. Chem. Soc.* **49**, 1617.
Rice, O. K., and Ramsperger, H. C. [1928]. Theories of unimolecular gas reactions at low pressures, II, *J. Amer. Chem. Soc.* **50**, 617.
Rosen, R. [1968]. Some comments on the physico-chemical description of biological activity." *J. Theoret. Biol.* **18**, 380.
Rupley, J. A., and Gates, V. [1967]. Enzymic activity of lysozyme II. The hydrolysis and transfer reactions of *N*-acetylglucosamine oligosaccharides, *Proc. Nat. Acad. Sci. U.S* **57**, 496.
Rushbrooke, G. S. [1949]. "Introduction to Statistical Mechanics." Oxford Univ. Press, London and New York.
Schroedinger, E. [1945]. "What is Life? The Physical Aspect of the Living Cell." Macmillan, New York.
Shea, S. M., and Bartholomay, A. F. [1965]. In numero studies in a cell renewal system: Periodically adjusted cell renewal process, *J. Theoret. Biol.* **9**, 389.
Shuler, K. E. [1958]. On the perturbation of the vibrational equilibrium distribution of reactant molecules by chemical reactions, *7th Symp. Combustion*, pp. 87–92.
Shuler, K. E. [1959]. Vibrational distribution functions in bimolecular dissociation reactions, *J. Chem. Phys.* **31**, 1375.

Singer, K. [1953]. Application of the theory of stochastic processes to the study of irreproducible chemical reactions and nucleation processes, *J. Roy. Stat. Soc.* **B 15**, 92.

Slater, N. B. [1959]. "Theory of Unimolecular Reactions." Methuen, London, and Cornell Univ. Press, Ithaca, New York.

Sohncke, L. [1884]. Commentary on Barlow's work, *Nature* **29**, 383.

Sugiura, Y. [1927]. The properties of the hydrogen molecule in the fundamental state, *Z. Phys.* **45**, 484.

Sundman, K. [1913]. Mémoire sur le probleme des trois corps, *Acta Math.* **36**, 105.

Swoboda, P. A. T. [1957]. Kinetics of enzyme action, *Biochem. Biophys. Acta* **23**, 70.

Thiele, E. [1961]. Comparison of the classical theories of unimolecular reactions, *J. Chem. Phys.* **36**, 1466.

Tikhonov, A. N. [1952]. Systems of differential equations containing small parameters multiplying some of the derivatives, *Mat. Sb.* **31**, 575.

Vallee, B. L., and Neurath, H. J. [1954]. Carboxypeptidase, a zinc metalloprotein, *J. Amer. Chem. Soc.* **76**, 5006.

Vallee, B. L., and Neurath, H. J. [1955]. Carboxypeptidase, a zinc metalloprotein, *J. Amer. Chem. Soc.* **217**, 253.

Vallee, B. L. [1960]. Metal and enzyme interactions: correlation of composition, function, and structure, *in* "The Enzymes" (P. D. Boyer, H. Lardy, K. Myrbäck, eds.), Vol. 3, Chapter 15. Academic Press, New York.

Vallee, B. L., and Williams, R. J. P. [1968]. Metalloenzymes: The entatic nature of their active sites, *Proc. Nat. Acad. Sci. U.S.* **59**, 498.

Vallee, B. L., Rupley, J. A., Coombs, T. L., and Neurath, H. J. [1960]. Role of zinc in carboxypeptidase, *J. Biol. Chem.* **235**, 64.

Vallee, B. L., Riordan, J. E., Bethune, J. L., Coombs, T. L., Auld, D. S., and Sokolovsky, M. [1968]. A model for substrate binding and kinetics of carboxypeptidase, *Biochemistry* **7**, 3547.

van Hove, L. [1957]. The approach to equilibrium in quantum statistics, *Physica* **23**, 441.

van't Hoff, J. H. [1884]. "Études de dynamique chimique." Amsterdam.

Vasil'Eva, A. B. [1963]. Asymptotic behavior of solutions to certain problems involving non-linear differential equations containing a small parameter multiplying the highest derivatives, *Russ. Math. Surv.* **18**, 13.

Vayo, H. W. [1965]. The size of spherical virus capsomeres, *Bull. Math. Biophys.* **27**, 161.

Walker, A. C., and Schmidt, C. L. A. [1944]. Histidase, *Arch. Biochem.* **5**, 445.

Wall, F. T., and Porter, R. N. [1962]. General potential-energy function for exchange reactions, *J. Chem. Phys.* **36**, 3256.

Wall, F. T., Hiller, L. A., and Mazur, J. [1958]. Statistical computation of reaction probabilities, *J. Chem. Phys.* **29**, 255.

Wang, S. D. [1928]. The problem of the normal hydrogen molecule in the new quantum mechanics, *Phys. Rev.* **31**, 579.

Watson, J. D. [1965]. "Molecular Biology of the Gene." Benjamin, New York.

Watson, J. D., and Crick, F. H. C. [1953]. Genetical implications of the structure of deoxyribosenucleic acid, *Nature* **171**, 964.

Wieder, G. M., and Marcus, R. A. [1962]. Dissociation and isomerization of vibrationally excited species. II. Unimolecular reaction rate theory and its application, *J. Chem. Phys.* **37**, 1835.

Zimmerman, J. H., and Simha, R. [1965]. The kinetics of multicenter macromolecule growth along a template, *J. Theoret. Biol.* **9**, 156.

Zimmerman, J. H., and Simha, R. [1966]. The kinetics of cooperative unwinding and template replication of biological macromolecules, *J. Theoret. Biol.* **13**, 106.
Zimmerman, J. H., Simha, R., and Moacanin, J., Jr. [1963]. Polymerization kinetics of biological macromolecules on tmplates, *J. Chem. Phys.* **39**, 1239.
Zwolinski, B. J., and Eyring, H. [1947] The non-equilibrium theory of absolute rates of reaction, *J. Amer. Chem. Soc.* **69**, 2702.

Chapter 3
QUANTUM GENETICS

Robert Rosen

Center for Theoretical Biology
State University of New York at Buffalo
Amherst, New York

I. The Basic Genetic Questions

The concept of the *gene* has been one of the most fecund ideas in biology, and has been preserved essentially intact through the recent explosive development of molecular biology and biochemical genetics. The essentials of classical genetics can be distilled into a set of five propositions, which express the basic facts with which we shall be dealing.

1. The gene is a stable hereditary entity which can be replicated and transmitted intact (barring mutation) to all subsequent generations.
2. The gene contains coded information characterizing a phenotypic quality; this information must be correctly "read out" at an appropriate time in order for that phenotypic quality to be expressed.
3. In higher organisms, a gene occupies a specific discrete site or locus located on a chromosome in the nucleus of each cell of the organism; genes

are distributed linearly along chromosomes. In somatic cells chromosomes are typically present in homologous pairs; in germ cells they occur singly.

4. A gene can mutate, giving rise to one of a number of possible alleles. These alleles occupy the same locus as did the original gene.

5. Recombination or crossing over can occur between homologous chromosomes. Such recombination can only occur between gene loci; hence, in particular, a pair of alleles can never be separated by recombination.

For the moment, we will notice only one characteristic of these five propositions. Namely, the fourth and fifth of them provide us in effect with two different operational definitions of the gene. On the one hand, the gene is a unit of mutation. Indeed, it is impossible to discover an unmutated gene in classical genetics; genetics cannot even begin until two or more alleles are available. On the other hand, the gene is a unit of recombination. In classical genetics, we make an assumption relating the probability of recombination between different genes on a chromosome to their physical separation on the chromosome; on this basis it is possible, using the linear distribution of genes, to construct genetic maps of chromosomes. These maps, based entirely on recombination, indicate the relative position of the genes on the chromosomes.

Thus, any gene must contain information pertaining to the three essential genetic mechanisms set forth in our basic propositions. The gene must *replicate*; the gene must *code phenotypic information*, and the gene must *tell the cell where it is*, in some sense, relative to the other genes on its chromosome. But replication, the expression of phenotypic information, and recombination can only take place when the gene is surrounded by the cell itself, and all the bewildering machinery contained therein. Therefore the three different kinds of information carried by the gene must be *transmitted* to the cell (or appropriate parts thereof), and in order for this to happen at all, the gene must be *observed* by the cell.

In other words, the first step which must occur in each of the three information-processing chains in which the gene is involved is the observation of the gene itself. This much seems too trivial to bear special mention, but as we shall see, it has significant consequences indeed. However, by itself, it is not enough to begin an analysis of the processing of primary genetic information. We need a further assumption characterizing some aspect of the physical properties of the gene itself and the machinery with which it interacts. As is well known, conventional molecular biology makes an immediate identification of the gene with nucleic acids (particularly DNA) and of the surrounding cellular machinery with appropriate enzymes (replicases, transferases, and so forth). We shall not do this. Instead we shall make only one assumption:

Postulate I. The transmission of genetic information, regarding replication, phenotypic characteristics, and recombination, involves microphysical events in an essential way.

The reader may very well ask at this point why one should deliberately ignore the patiently gathered structural information gathered over the past decades and replace this information by an apparently far more primitive and less specific proposition like Postulate I. Part of the answer will lie in the subsequent exposition. Another part of the answer is this: We are seeking to learn what we can capture about the storage and transmission of primary genetic information (that is, how much we can retain of the Propositions 1–5 above) *without* making any specific *structural* assumption about the genetic material, but retaining the basic *functional* characteristic of the gene and its interaction with its environment. By making such a strategic retreat, it should be possible to gain much insight into the specific role of the nucleic acids in primary genetic processes, into what the special structure of the nucleic acids makes possible that would not in general be possible without them, and into exactly how far the specific structure of the nucleic acids is necessary for a viable genetic mechanism. Such questions can hardly be answered by making an immediate structural identification of the gene with nucleic acids, but can only be explored by taking a functional (instead of a structural) approach of the kind to be developed below. Moreover, it should be recalled that numerous authors, concerned with problems of the origin of life (both terrestrial and otherwise) have pointed out that the intricate DNA-mediated coding mechanism revealed by molecular biology to be universal in existing organisms, could not have been employed by the most archaic life forms.

The reader should have one further question about Postulate I: namely, what is a microphysical event? In the next few sections, we proceed to give a (partial) answer to this question. We thus begin a prolonged detour into a development of the concepts necessary for the study of microphysical events necessary to the study of processing of primary genetic information according to Postulate I; we shall return to the specific genetic questions in Section IV.

II. Systems, States, and Observables

Crucial to our discussion are three closely related concepts which must be carefully distinguished and clearly understood: the concepts of *system*, of *state* of a system, and of *observable*. To simply define these logically would not be of much help; therefore we shall proceed intuitively and base our further work on these intuitions (which are, after all, the essential thing).

By a *system* we roughly mean a portion of the physical universe which we have isolated for separate study. A harmonic oscillator; the three bodies represented by the earth, sun, and moon; a hydrogen atom, are all examples of what we consider systems.

We study such systems by allowing them to interact with physical devices (which are also systems themselves) which we call measuring instruments or meters. The outcomes of such interactions are usually expressed numerically and, because we know the relevant properties of our meters (having built them ourselves), these outcomes give us information concerning the system under study. The interactions themselves are called *measurements.*

We usually give special names to the quantities our meters actually measure. Thus, intuitively, we use one kind of meter to determine the displacement of a particle from a reference position at a particular instant, another kind of meter to determine what we call the total energy of the system, still another kind to measure what call angular momentum, still another kind to measure what we call spin, and so forth. Quantities of this kind, determined by specific kinds of meters or measuring procedures, are called observables of the system and we shall conventionally denote a particular meter, and the observable it measures, by the same symbol.

Each such measurement, however, must be performed at a specific time, and thus only gives information about what the system is doing at the time of the measurement. That is, our observing procedures do not give us information about the extended behavior of our system in time, but only about its properties at a specific instant. This is interesting because we intuitively recognize that physical systems do have a temporal extension, and indeed it is the study of this temporal extension that is the main business of physics. Part of this recognition lies in the fact that we know that we are dealing with the same system even though the same meters, interacted with the system at different times, may give us different answers. For instance, the earth–sun–moon system will exhibit different displacements at different times, yet we know that these different displacements all refer to the same system. The kinetic energy of a harmonic oscillator will likewise be different at different times as will the total energy of a hydrogen atom before and after it interacts with radiation. From this we can recognize that the same system may present different aspects to us (or to our measuring instruments) as time changes. These different aspects are what we call the *states* of our system.

Thus, we never "really" observe a system; all that we really observe are states of the system at different instants of time. And all that we can observe at a specific instant of time are the numbers representing the outcomes of interactions of the system with out various meters at that instant. Let us call such a number, obtained by interacting our system with a meter A at an instant of time t, *the value of the observable* A at time t. This number obviously serves to partially specify the *state* of our system at time t. It may not completely specify this state, because we could imagine two copies of our system, both of which would give the same A value at time t, but which would give different values when simultaneously observed with a B meter at the same time t. Thus, assuming that the A meter and the B meter can in fact observe the system simultaneously at time t (which in quantum theory is a large assumption, as we shall see), it is clear that the A value and the B value at time t give more information about the state at time t than either value alone. Likewise, if we have a C meter which again can observe the system simultaneously with our other two meters, we get still more information about the state of our system at time t by considering the A, B, and C values at time t, and so on. Thus, there is a sense in which we can say that *the state of a system at an instant of time can be regarded as the totality of values obtained by simultaneously interacting our system at time t with all conceivable meters.*

Our recognition of the temporal extension of physical systems means that the successive states of a system in time are not independent of one another, but that the state of a system at a particular instant determines, to some degree at least, what the state of the system will be at each later instant. It is the task of dynamics to specify in detail the nature of this relationship. These relationships generally take the form of *equations of motion*, which specify the manner in which the states of the system, expressed in terms of values of specific observables, change in time; indeed, it is correct to say that the equations of motion of a system *represent* its temporal extension, just as much as the values of specific observables at an instant *represent* its state at that instant.

From the foregoing discussion, then, it is correct to pose the following:

1. Any physical system determines, and is determined by, a family of observables; these observables are each defined by a meter or measuring apparatus with which the system can interact, and every conceivable meter with which the system can interact determines an observable of the system.
2. The states of the system are represented in some sense by the values which can be assumed at a particular instant of time by the observables of the system.
3. The temporal extension of the system is determined by equations of motion which specify the manner in which the numerical values of particular observables change in time. Knowing an initial state of the system, and its equations of motion, determines the state of the system at any later time.

These propositions will be basic to our subsequent discussion.

A. Macrophysical and Microphysical Systems

The generalities expressed in the preceding section are valid for all kinds of physical systems. When we look more closely into the process of observation of a system, that is, the interactions between a physical system of interest and the meters used to study it (and by implication, between the system of interest and any other system), we find that we may make an important distinction. On the one hand, we have a class of systems with the property that it is possible to observe them in such a way that the system itself is changed only negligibly by the observation process. For instance, if we wish to measure the position of a billiard ball on a table at a specific instant, it is sufficient to *see* the billiard ball (for example, by taking a photograph of the ball at the instant in question) and then to perform the appropriate measurements. The illumination of the billiard ball with light obviously does change the energy of the ball and many other of its observable properties, but the effect of these changes is utterly negligible. Systems of this kind, which are obviously *large* in some sense, are called *macrophysical* and they are defined precisely by the property that all of their physical properties can be determined in such a fashion that the system itself is essentially undisturbed by the observation process.

On the other hand, if we keep subdividing the billiard ball into smaller and smaller pieces, and keep attempting to measure the positions of these smaller pieces, we ultimately reach a point at which the very act of seeing a fragment of our billiard ball, that is, illuminating it with radiant energy, *significantly* changes the properties of the fragment. It may indeed be possible for us to observe the position of such a fragment without affecting the position itself, but the fragment is so small that some essential physical characteristic of the system is significantly disturbed by the observation. Systems of this kind, which are obviously *small*, and are characterized by the fact that *any* observation of the system produces significant changes in some characteristic of the system, are called *microphysical*.

B. A More Accurate Formation of the Notion of State

In any kind of physical system, if we know the equations of motion of the system and an initial state, then we can in principle *solve* the equations of motion to obtain full information about every future (and past) state of the system. Even for macrophysical systems, however, it is quite impossible to obtain full information about the state of a system at a particular instant of time. For one thing, no meter can be read with infinite accuracy; for another thing, no measurement process can be carried out instantaneously; the interaction of a system with a measuring apparatus always lasts a finite time. Therefore, even in classical physics (that is, the physics of macroscopic systems) we cannot ever really make a prediction of the form, "The value of the observable A at time t will be so-and-so," or equivalently, "The state of our system at time t will be α." All that we can really predict from observational measurements, and all that we can verify observationally, is a proposition of the form, "The probability that the observable A will have such and such a value at time t is so-and-so," or equivalently, "The probability that the system will be in state α at time t is so-and-so."

Thus, when we enter the world of real measurements and real observations, the concept of *state* that we have developed becomes more and more obviously an idealization. We can never know what the value of an observable is in a particular state, partly because we can never *identify* a state exactly and partly because, even if we could identify an initial state exactly, we could never exactly determine by observation the value of any observable of that state. The suggestion then is that we acknowledge our human fallibility, and replace the idealized notion of a state involving precise numerical values by a more accurate notion, involving the probabilistic aspects introduced by our imperfect meters in an essential way.

Accordingly, we shall now say, "A system is in the state α if the *probability* that the observable A, applied to the system in that state, leads to the value such and such is so-and-so." Thus, the states will henceforth be characterized, not by absolutely sharp numerical values assumed by observables, but rather by probability distributions. If these distributions are very sharp, so that the probability that an observation in a state leads to a definite value with probability nearly one, the new concept of state is very close to the old concept. However, of course, this need not be the case, and is decidedly not the case when we deal with microphysical systems. In any case, the propositions of all physics, whether macrophysical or microphysical, must now be expressed in the mathematical form, $p(A, \alpha, E)$ which, in words, may be stated as "the probability that the observable A, measured on the state α, has a value lying in the subset E of set of real numbers."

C. Some Properties of $p(A, \alpha, E)$

From the results of the preceding sections, we see that *any* physical system, macroscopic or microscopic, can be regarded as a set of three things:

a. A set of *observables*, representing all the possible meters, or measuring instruments, with which the system can interact.

b. A set of *states* (in the sense of the preceding section), on which the observables assume numerical values with definite probability distributions, and which represent what the system is doing at an instant of time;

c. *Equations of motion*, which represent the dynamic aspects of system behavior and tell us how the states (or, what is the same thing, the values assumed by observables on the states) change in time.

For the moment, we neglect the dynamic aspects, represented by the equations of motion, and concentrate on system statics, that is, on system description at an instant of time, on states and observables. And as we have seen, the system statics are all summed up in the numbers $p(A, \alpha, E)$, the probability that the observable A measured when the system is in state α gives a value lying in the set E of real numbers. Therefore, we shall concentrate on the properties of this set of numbers and we shall see that all aspects of system statics are expressible in terms of them.

First, some generalities. Note that the expression $p(A, \alpha, E)$ may be regarded as a function of three variables. The first variable represents an observable of the system, the second a state of the system, and the third is a (for the moment) arbitrary set of real numbers. We shall explore the properites of the numbers $p(A, \alpha, E)$ by the usual device of fixing two of the variables and allowing only the third to vary at any one time.

1. *A and α Fixed, E Variable*

By the very definition of $p(A, \alpha, E)$, we know that if E is the empty set $\varnothing$, then

$$p(A, \alpha, \varnothing) = 0 \tag{1}$$

for every state α and observable A. Likewise, if $E = R$, the full set of real numbers, then

$$p(A, \alpha, R) = 1 \tag{2}$$

for every α and A. It is likewise obvious that if E_1 and E_2 are disjoint sets (that is, if $E_1 \cap E_2 = \varnothing$) and if we write $E = E_1 \cup E_2$, then

$$p(A, \alpha, E) = p(A, \alpha, E_1) + p(A, \alpha, E_2)$$

or, more generally, if $E_1, E_2, \ldots$ is any sequence of pairwise disjoint sets, with union E, then

$$p(A, \alpha, E) = \sum_{i=1}^{\infty} p(A, \alpha, E_i). \tag{3}$$

We must of course allow the possibility that the expression $p(A, \alpha, E)$ is underfined for some subsets E of the real line. Henceforth we shall always tacitly suppose that the sets E which arise in our discussion are those for which the expressions $p(A, \alpha, E)$ in which they are involved are defined. We shall also assume that our expressions $p(A, \alpha, E)$ are always defined when E is an open or closed interval, or built out of open and closed intervals by a countable number of union and intersection operations (that is, when E is a Borel set). With that understanding, we may note in passing for the mathematically more sophisticated reader that the expression $p(A, \alpha, E)$ with A and α fixed can be regarded as a mapping, which we can denote by $A_\alpha(E)$, which assigns to each subset E of the real line for which it is defined a real number between zero and one; the properties of Eqs. (1)–(3) turn this function into a *probability measure*.

2. *A and E Fixed, α Variable*

We are now going to use the properties of $p(A, \alpha, E)$ to define operations on the states of our system, which will allow us to construct new states from given ones. Let α_1, α_2 be states, and let t_1, t_2 be positive real numbers such that $t_1 + t_2 = 1$. Then we can *define* a new state α by saying that, for any observable A and any set E of real numbers, we always have

$$p(A, \alpha, E) = t_1 p(A, \alpha_1, E) + t_2 p(A, \alpha_2, E)$$

(the reader may readily verify that, for this α, Eqs. (1)–(3) are verified). Under the circumstances, it is natural to say that the system is in state α when it is in state α_1 with probability t_1 and in state α_2 with probability t_2. We shall some times say that α is a *mixed state* or *mixture* of the states α_1 and α_2 with probability or proportions t_1 and t_2. We shall write

$$\alpha = t_1\alpha_1 + t_2\alpha_2.$$

More generally, if $\alpha_1, \alpha_2, \ldots$ are states, and if $t_1, t_2, \ldots$ is any set of positive numbers which add up to unity, we can define the mixed state

$$\alpha = \sum_{i=1}^{\infty} t_i \alpha_i$$

by means of the property that

$$p(A, \alpha, E) = \sum_{i=1}^{\infty} t_i \, p(A, \alpha_i, E).$$

A state which cannot be expressed as a mixture of other states in this fashion will be called a *pure state*.

3. *α and E Fixed, A Variable*

The next property of the expressions $p(A, \alpha, E)$, and the system which they define, is the following. We have stated in the beginning of Section II. that any kind of meter which could be applied to our system defines an observable of our system. Let us suppose that we have a meter measuring the observable A on the states of our system. Let us suppose that we attach this meter to a computer which, given any numerical value x, computes the value $f(x)$, where f is some given function (for example, $f(x) = x^2$, $f(x) = x^m + x^n$,

$f(x) = e^x$). The result will be a meter measuring, by hypothesis, an observable of our system with the property that, if the A meter reads a value x when applied to a state of our system, then the new meter will read the value $f(x)$. It is natural to denote this new observable by the symbol $f(A)$. Moreover, the defining probabilities are the same in this case, whether we are measuring the observable A or the observable $f(A)$. From this it follows that, for every set E, we must have

$$p(f(A), \alpha, f(E)) = p(A, \alpha, E) \tag{4}$$

for every set E.

Thus, if our system has any observable A at all, it must have a great number of observables, of the form $f(A)$. Note in particular that, if A is an observable, expressions of the form A^2, $A^m + A^n$, e^A, and many others, are all meaningful. In particular, if $f(x)$ is the constant function, which always assumes the value C, the $f(A)$ is independent of A; we shall call $f(A)$ in this case the constant observable, with value C. We draw special attention to the observable I (defined for $C = 1$) and the observable 0 (defined for $C = 0$).

Next we shall introduce a bit of notation which will be most useful. We know that $p(A, \alpha, E)$ designates the probability that an observation of the observable A in the state α gives a value lying in E. But what happens to the state of the system after the observation (that is, the interaction of the state α with the A meter) takes place? Intuitively, the system is left in some particular state immediately after the interaction, which should depend only on A (that is, the meter involved) and on α, the state being observed. For a variety of heuristic reasons, we shall call this state $A(\alpha)$ or sometimes just $A\alpha$; more formally, we can write that $A(\alpha)$ is that state for which

$$p(A, A(\alpha), E) = p(A^2, \alpha, E)$$

for every state α and set E.

Before we leave this section, we draw explicit attention to two facts: (a) everything we have done so far applies equally well to macroscopic and microscopic systems; (b) we have not so far attached any meaning to expressions of the form $f(A, B)$, where $f(x, y)$ is a function of *two* variables and A and B are arbitrary observables. Indeed, as we shall see, the main distinction between the physics of microscopic and macroscopic systems arises precisely at this point, and specifically with regard to the function $f(x, y) = xy$.

D. "Questions"

Let A be an observable, E a subset of the reals. Let us consider how we would go about answering the question, "Did the measurement of A on the state α yield a value lying in E?" At the moment we have only probabilistic statements of the form $p(A, \alpha, E)$, but these do not help us answer our question. On the other hand, the question itself is a perfectly natural one, and we would expect to be able to answer it in any theory claiming to be a comprehensive theory of physical observation.

Observe that the only information we can obtain about a system, and the measurements performed on it, lies in the states of the system and the values attained by observables on those states. Therefore, in order to answer our question, we need to invent an auxiliary observable of the system, which takes on the value 1, on the state α say, if the answer to our question is "yes," and the value 0 if the answer to our question is "no."

We can show that an observable with these properties is already available within the system, and that this observable is unique. If E is any subset of the real line (our given one, for instance), then we can define a function $\chi_E(x)$ (the characteristic function of E) as follows:

$$\chi_E(x) = 0 \qquad \text{if } x \text{ is not in } E;$$

$$\chi_E(x) = 1 \qquad \text{if } x \text{ is in } E.$$

Clearly the expression $\chi_E(A)$ is defined, and is an observable of our system. But by definition, this observable takes on the value one on a state α if and only if the value of A on α lies in E, and the value zero otherwise.

Now let us note some important properites of the observable $\chi_E(A)$, which we shall denote by $Q_E{}^A$. First, since the only values which $Q_E{}^A$ can assume on any state are 0 or 1, it follows that

$$p(Q_E{}^A, \alpha, \{0, 1\}) = 1$$

for every state α. Another way of saying this is that the probability measure α_A, defined in Section II.A, is concentrated in the set $\{0, 1\}$; that is, $\alpha_A(\{0, 1\}) = 1$ for every state α. Next, it is easy to see that the observable $Q_E{}^A$ has the property

$$(Q_E{}^A)^2 = Q_E{}^A.$$

We may interpret this, in a physical sense, by saying that the application of the $Q_E{}^A$ meter twice in immediate succession always gives the same answer as applying it only once, since the square of the characteristic function $x_E(x)$ of the set E is again the characteristic function of E, by a straightforward application of Eq. (4). Mathematically, a quantity like $Q_E{}^A$, which is equal to its own square, is called *idempotent*.

On the other hand, *any* idempotent observable A has the property

$$p(A, \alpha, \{0, 1\}) = 1$$

for every state α. (Why?) Therefore, the idempotent observables are precisely those whose probability measures are concentrated in the set $\{0,1\}$. Let us denote by Q the totality of all idempotent observables; Q is certainly not empty, since it contains all observables of the form $Q_E{}^A$, where A is an arbitrary observable and E is an arbitrary subset of the real numbers.

By virtue of the motivation for this entire discussion, observables in Q that is, observables like $Q_E{}^A$ will be called *questions*. It is out of these particularly simple observables that we will construct arbitrary observables. Indeed, we can already see that there is a close relationship between the observable A and the totality of observables of the form $Q_E{}^A$, where E is an arbitrary subset of the real line [subject of course, as always, to the proviso that $p(A, \alpha, E)$ is defined].

E. Some Properties of the Set of Questions

Let A be an observable, and let E_1, E_2 be subsets of the real line, such that $E_1 \cap E_2 = \varnothing$. Suppose we ask the question, "Did the measurement of A in state α give rise to a value lying in $E = E_1 \cup E_2$?" Clearly this question must be answered by an idempotent observable of the type we have been considering and which, according to our conventions. we must denote by $Q_E{}^A$. It is natural to call $Q_E{}^A$ in this case the *sum* of the observables $Q^A_{E_1}$, $Q^A_{E_2}$.

Note that the disjointness of E_1, E_2 in the preceding paragraph means that the questions $Q^A_{E_1}$, $Q^A_{E_2}$, corresponding to the questions, "Did the measurement of A lead to a value lying in E_1," and "Did the measurement of A lead to a value lying in E_2," respectively, cannot simultaneously be answered "yes." For this reason the corresponding questions are themselves called disjoint. What we have said, then, is that the pair of disjoint questions of the form $Q^A_{E_1}$, $Q^A_{E_2}$ always have a sum, which is itself a question; in fact, it is the question which has the answer "yes" on a state α, if and only if either $Q^A_{E_1}$ or $Q^A_{E_2}$ has the answer "yes" on α.

Generalizing, we may say that every set of disjoint questions of the form $Q_E{}^A$, for a fixed observable A, has a sum. Since all these questions are themselves observables, we may multiply them by real numbers [using Eq. (4)], and hence if $[E_i]$ i $= 1, \ldots$ is any family of pairwise disjoint sets of real numbers, the expression

$$\sum_{i=1}^{\infty} C_i Q^A_{E_i}$$

has a definite meaning.

F. Questions and Observables

We have seen in the preceding sections that to each observable A, we may associate a family of questions of the form $Q_E{}^A$, one such question for every subset E of real numbers for which $p(A, \alpha, E)$ is defined for every state α.

This correspondence between sets E and questions $Q_E{}^A$ may be denoted as a function whose domain is the set of subsets of the real numbers and whose range is the set of questions. This correspondence has the following properties:

a. $E \cap F = \varnothing$ implies $Q_E{}^A$, $Q_F{}^A$ are disjoint;
b. If $[E_i]_{i=1}^{\infty}$ is any family of pairwise disjoint sets, then $Q^A_{E_1 \cup E_2 \cup \cdots} = Q^A_{E_2} + Q^A_{E_2} + \ldots$;
c. $Q_\varnothing{}^A = 0$, $Q_R{}^A = 1$.

Such a correspondence is sometimes called a *question-valued measure* on the set of real numbers.

Conversely, if we are given a correspondence satisfying Properties a, b, and c, that is, if we are given a question-valued measure, then we can use it to *define* an observable as follows: If $Q(E)$ represents the question corresponding to the set E, identify it physically with the question, "Did the measurement of the observable yield a value in the set E?" Hence, there is a one-to-one correspondence between the observables of our system, and the question-valued measures, that is, between the observables A and the corresponding families of questions $[Q_E{}^A]$.

Now let us introduce some terminology which should sound familiar to those familiar with quantum theory. Let us suppose that A is an observable and that $Q_E{}^A = 0$, that is, that $p(Q_E{}^A, \alpha, E) = 0$ for every set E. This means that there is no state α of the system such that the observable A measured on α gives a value lying in E with nonzero probability. We shall say that such a set is of A-measure zero. Let us consider the union of all sets of A-measure zero. This is again, obviously, a set of A-measure zero. The complement of this set is called the *spectrum* of the observable A, and will be denoted by $\sigma(A)$. Clearly, the answer to the question $Q_E{}^A$ is "yes" if and only if E intersects the spectrum of A.

We shall say that a number x belongs to the point spectrum of A if and only if

$$Q_x{}^A = 1;$$

that is, there is a state α such that $p(A, \alpha, \{x\}) = 1$. We shall denote the point spectrum of A by $\sigma_p(A)$. The difference $\sigma_c(A) = \sigma(A) - \sigma_p(A)$ will be called the *continuous spectrum* of A.

EXERCISE: Any state α such that $p(A, \alpha, \{x\}) = 1$ is a pure state.

Finally, we shall say that the observable A is *bounded* if and only if its spectrum $\sigma(A)$is contained in a finite interval; if A is a bounded observable, then the least number N such that $|x| < N$ for every x in $\sigma(A)$ will be called the *norm* of A, denoted by $||A||$.

If Q is a question, then $\sigma(Q) = \sigma_p(Q) = \{0, 1\}$.

We can now make the correspondence between observables A and question-valued measures more precise. For each number λ, we may have either

a. a question $Q_\lambda{}^A$ such that $Q_\lambda{}^A = 1$;
b. $Q_E{}^A = 0$ for any sufficiently small set containing λ;
c. $Q_E{}^A \neq 0$ for every set containing λ.

If $\lambda_1, \ldots, \lambda_n$ is set of numbers, the expression

$$\sum_{i=1}^{n} \lambda_i Q_{\lambda_i}^A$$

makes sense. As we let $n \longrightarrow \infty$ in such a way that the set $\{\lambda_i\}$ becomes dense in R, we have

$$\sum_{i=1}^{n} \lambda_i Q_{\lambda_i}^A \longrightarrow \int_R \lambda \, dQ^A(\lambda).$$

This expression clearly *represents* the observable A, and the identification of observables A with such expressions forms the content of the *spectral theorem*.

G. Expected Values of Observables on States

Suppose first that Q is a question. Then we have seen that Q has pure point spectrum $\sigma_p(Q) = \{0, 1\}$. There are two kinds of pure states: states $\bar{\alpha}$ such that

$$p(Q, \bar{\alpha}, 1) = 1, \qquad p(Q, \bar{\alpha}, 0) = 0,$$

and states $\bar{\bar{\alpha}}$ such that

$$p(Q, \bar{\bar{\alpha}}, 1) = 0, \qquad p(Q, \bar{\bar{\alpha}}, 0) = 1.$$

Let t_1, t_2, be positive real numbers such that $t_1 + t_2 = 1$, and let us see what Q does on a state

$$\alpha = t_1\bar{\alpha} + t_2\bar{\bar{\alpha}}$$

which is a mixture of the two kinds of states. We have

$$p(Q, \alpha, 1) = t_1, \qquad p(Q, \alpha, 0) = t_2 = 1 - t_1.$$

In general, if

$$\alpha = \sum_{i=1}^{\infty} t_i \alpha_i$$

is any mixture formed from pure states of types $\bar{\alpha}$ and $\bar{\bar{\alpha}}$, then

$$p(Q, \alpha, 1) = S, \qquad p(Q, \alpha, 0) = 1 - S,$$

where S is the sum of those t_i appearing as coefficients of pure states of type $\bar{\alpha}$.

More generally, if E is any set such that $E \cap \sigma_1 = 1$, then

$$p(Q, \alpha, E) = S;$$

if $E \cap 0, 1 = 0$, then

$$p(Q, \alpha, E) = 1 - S;$$

if $E \cap 0, 1 = 0, 1$, then

$$p(Q, \lambda, E) = 1;$$

if $E \cap 0, 1 = \varnothing$, then

$$p(Q, \alpha, E) = 0.$$

Thus, the value S completely determines the behavior of Q on the state α, and physically represents *the value we expect to observe* when we apply the meter Q to the state α. For this reason the number

$$m_\alpha(Q) = S$$

is called the *expectation* or *expected value* of the question Q in the state α.

These expected values have some pleasant algebraic properties, which follow directly from the definitions:

1. If A is an observable, and $E_1, E_2, \ldots$ are pairwise disjoint sets with union E, then

$$m_\alpha(Q_E{}^A) = \sum_{i=1}^{\infty} m_\alpha(Q_{Ei}^A)$$

for any state α.

2. If E_1, E_2 are disjoint, then

$$m_\alpha(Q_{E_1}^A) + m_\alpha(Q_{E_2}^A) \leq 1$$

for any state α, and conversely.

3. $m_\alpha(0) = 0$, $m_\alpha(1) = 1$; $0 \leq m_\alpha(Q) \leq 1$.

We have seen in the preceding section how an arbitrary observable A could be "approximated" by algebraic sums of questions. Hence, it is natural to try to define

$$m_\alpha(A)$$

the expected value of an arbitrary observable, in terms of the expected values of the associated family of questions $[Q_E{}^A]$. Write (if A is bounded)

$$m_\alpha(A) = \int_{-\infty}^{\infty} \lambda\, dp(A, \alpha, \{\lambda\}).$$

In particular, if A is a question Q, then

$$\begin{aligned} \int_{-\infty}^{\infty} \lambda\, dp(Q, \alpha, [\lambda]) &= 0 \cdot p(Q, \alpha, \{0\}) + 1 \cdot p(Q, \alpha, 1) \\ &= p(Q, \alpha, \{1\}) \end{aligned}$$

as before.

H. Simultaneous Observations, Commutation and the Uncertainty Relations

We repeat once more that everything which has been done up to this point is perfectly general, and applies equally well to macroscopic and to microscopic systems. It is now time to come to the properties which mark the specific distinction between macrophysics and microphysics, and at the same time, to return to the question we left hanging in Section II.C.3, namely, to consider how of form functions $f(A, B)$ of two arbitrary observables A and B.

We may well ask what is the difficulty in forming functions of the form $f(A, B)$? Why can we not use the same procedure as we used before, to form arbitrary functions of a single variable? Let us give an intuitive discussion of the underlying problem.

We recall that we defined the state $A(\alpha)$ as the state in which the system is left immediately after an observation of α with an A meter. If we now apply a B meter to the system in this state, the resulting possibilities are expressed in the numbers of the form

$$p(B, A(\alpha), E).$$

These numbers, of course, refer to the state $A(\alpha)$, rather than the state α, so that the reading of this B meter may or may not give information about the state α [or, more accurately, the probability distributions of the form $p(B, \alpha, E)$].

On the other hand, if we first apply a B meter to the system in the state α, the system is left immediately after the interaction in the state $B(\alpha)$. If we now apply an A meter to this state, the results are expressed in the numbers

$$p(A, B(\alpha), E).$$

Once again, these numbers may or may not give information about the A values on the state α.

If we think about this for a moment, we see that these two possible successive observations of A and B give information pertaining to the original state α if and only if the two resulting probability distributions involved are always the same; that is, if and only if

$$p(A, B(\alpha), E) = p(B, A(\alpha), E)$$

for every set E. In this case, it is natural to get the same result that we get by measuring the two observables A and B successively (in either order) with the result we would get from a single meter measuring the state α at a single instant; it is further natural to call the observable corresponding to this single meter the *product* of A and B, and denote it either as AB or BA. This new observable, then, is defined by the property that

$$p(AB, \alpha, E) = p(A, B(\alpha), E) = p(B, A(\alpha), E) = p(BA, \alpha, E). \qquad (5)$$

In this case we say that the observables A and B *commute*, or that if Eq. (5) is not satisfied for all states α, the A and B do not commute, and A and B are not simultaneously measurable.

The distinction between macrophysics and microphysics lies essentially in this: that in macrophysics, the probability distributions p are such that Eq. (5) is satisfied for all pairs of observables A and B, while microphysics there are pairs of noncommuting observables. Indeed, much more than this is customarily assumed in microphysics; it is usually assumed that any observable A will fail to commute with some observable, unless A is a constant observable. The noncommutation of observables in microphysics, which means the impossibility of obtaining simultaneous information about A and B on a state α by performing successive measurements with A and B meters, may indeed be regarded as one way of formalizing the very definition of microscopic systems, which we recall involved the fact that the measurement of microscopic systems introduces effects on the system that are "large" in comparison with the system itself.

Thus, in particular, we see that in microphysics we cannot expect to be able to form functions of a pair of arbitrary observables A and B. In fact, we have just seen (intuitively) that we cannot even in general define the product of two arbitrary observables in a meaningful way. Once again, this means that there is no single observable which, when applied to a state α, gives the same information as we obtain by measuring A and B successively, in both orders.

It is clear that one way to measure the deviation of a pair of observables from commutativity, or simultaneous observability, is to consider their *commutator* $(BA - AB)$. This observable presents no ambiguities of definition, and it is immediately evident that two observables commute if and if their commutator is the zero observable. Another way of stating this is that two observables commute if and only if the (absolute value of the) expectation of $(AB - BA)$ is zero in every state, that is, if

$$|m_\alpha(AB - BA)| = 0$$

for every state α. Therefore, if A and B do not commute, we must have

$$|m_\alpha(AB - BA)| > 0$$

for every state α. This may be regarded as an abstract form of the famous Heisenberg uncertainty relations.

EXERCISE: Prove that (a) two questions Q_1, Q_2 commute if and only if there exists an observable A and two sets E_1, E_2 of real numbers such that $Q_1 = Q_{E_1}^A$ and $Q_2 = Q_{E_2}^A$.

(b) Two observables A, B commute if and only if $Q_E{}^A$ and $Q_F{}^B$ commute for all sets E, F. From this, prove the following:

Corollary. Two observables B_1, B_2, commute if and only if there exists an observable A and functions f_1, f_2 such that

$$B_1 = f_1(A), \qquad B_2 = f_2(A)$$

or more generally, B_1, B_2, . . . commute it and only it there exists an observable A and functions f_1, f_2, , , . such that $B_i = f_i(A)$ for all i.

This corollary was originally stated and proved by von Neumann, and we shall use it in an important way in Section IV.

III. The Usual Form of Quantum Theory

We have seen in the preceding sections how system descriptions, both macrophysical and microphysical, are expressed in terms of statements of the form $p(A, \alpha, E)$ and we were able to derive quite a number of properties of such statements. These statements were not enough to characterize either macrophysics of microphysics. In the last section we saw how to make an essential distinction between them on the basis of simultaneous observability. Even when these conditions are added, there is no reason to suppose that we have uniquely characterized either microphysics or macrophysics in the sense that the sets of states and observables admit only one kind of possible structure.

Thus, somewhere in the development of treatments of microphysics of the kind we have been giving, we reach a stage where we must say what our systems *are*: We must represent them concretely in some way and then henceforth feel ourselves free *to make use of the properties of the objects doing the representing*. We do not do this blindly or arbitrarily (at least not completely arbitrarily) in quantum physics; we have decades of successful experience in quantum theory to guide us. We have now reached the point where we must do this, and tie our formalism down to something which will be recognizable by a quantum physicist.

It should be stated at the outset that setting up a representation of this kind, even when we know what we want our representation to look like, is far from trivial. In the case of quantum theory we can do no more than briefly indicate the main features of the representation within a single chapter of a text devoted primarily to other matters; the mathematical concepts alone would require hundreds of pages for their proper development. Therefore, the present section is more in the nature of a few brief hints to put our formalism and its representation in perspective. We shall not even stop to define the basic concepts involved, and the reader unfamiliar with them may well skip this section entirely. If nothing else, however, this section will indicate to the novice what he must learn in order to fully understand the deeper significance of the quantum theoretic formalism, a significance which is largely unappreciated even by those facile in the techniques of quantum physics. The reader interested in these developments will find a somewhat less sketchy treatment by Mackey [1963].

The representation of our formalism proceeds by means of a brusque identification of the set of questions with the set of projections (idempotent linear operators) on an infinite-dimensional, separable Hilbert space over the field of complex numbers. By means of the spectral theorem, then, we can identify the observables with the self-adjoint linear transformations on that Hilbert space. From this it follows, though in a highly nontrivial fashion, that the pure states of the system must correspond exectly with the unit vectors of our Hilbert space. It further follows (again from the spectral theorem) that if α is a pure state, and if φ is the corresponding unit vector in Hilbert space, then the expectation $m_\alpha(A)$ of any observable A must be represented by the inner product $(A\varphi, \varphi)$; here we use the symbol A to denote both the observable and the self-adjoint transformation which represents it.

It is worth a moment to look in more detail at what the spectrum of an observable A means in this representation. Let A be an observable, and suppose that λ lies in $\sigma_p(A)$. According to our Hilbert space representation, the observable A corresponds to a self-adjoint linerar transformation (which we may again denote by A). By definition of point spectrum, there is a state α such that

$$p(A, \alpha, \{\lambda\}) = 1$$

and this state corresponds to a unit vector φ of our Hilbert space. But it clear from the definitions that

$$\begin{aligned} p(A, \alpha, \{\lambda\}) &= p(\lambda\, A, \alpha, \{1\}) \\ &= m_\alpha(A/\lambda) \\ &= (A/\lambda \cdot \varphi, \varphi). \end{aligned}$$

But since our correspondence preserves expectations,

$$(A/\lambda \cdot \varphi, \varphi) = 1$$

or

$$\lambda^{-1}A(\varphi) = \varphi$$

or

$$A\varphi = \lambda\varphi,$$

which is simply a statement that φ is an eigenvector of A corresponding to eigenvalue λ. Thus, a number in the point spectrum of the observable A must be an eigenvalue of the corresponding linear transformation. This argument can be turned around, from which it follows that the point spectrum of an observable A is identical with the set of eigenvalues of the corresponding linear transformation.

The *ad hoc* but plausible assumption of the representation of our formalism in Hilbert space terms carries with it one further advantage, which must be mentioned now. Thus far, we have defined *algebraic* operations on the set of states, but have not talked at all about topological or metric properties; we have not discussed what could be meant by saying that one state is close to another, or that a sequence of states converges to a limit. Indeed, we could not easily do this in our original fomalism without making many *ad hoc* assumptions. But by using our representation, and the identification of the pure states with unit vectors in Hilbert space, such a topology comes along at no extra charge (in fact, several such topologies actually come along). This is an observation we shall exploit in deriving one of our genetically significant theorems in Section IV.A.

We cannot leave this subject without adding one final word about our representation. We have identified our states with vectors in a separable, infinite-dimensional complex Hilbert space, and let it go at that. In one sense, we are justified in doing so, because mathematically there is only one such space; all such Hilbert spaces are isomorphic. Nevertheless, these spaces can be presented differently. For instance, if M is any Euclidean space, we can generate such a Hilbert space by considering the totality of all complex-valued functions f on M such that the integral

$$\int_M ff^* \, d\tau$$

is defined. In this particular space [called $L^2(M)$] the inner product of two vectors f, g is defined as

$$(f, g) = \int_M fg^* \, d\tau,$$

and the expectation value of an observable A represented in this space is

$$(Af, f) = \int_M f^* Af \, d\tau.$$

These L^2 spaces are the natural habitat of ordinary quantum mechanics; this is why the unit vectors in this space, representing states of microphysical systems, are called *wave functions*. If we are given a microphysical system which we can define classically in terms of positions, momenta and energies of particles, then we may take M to be the classical phase space of the system; this gives the standard representation of position observables x_i as multiplications by x_i, and of the corresponding momenta p_i as the transformation

$$p_i \longrightarrow (i/h)\, \partial/\partial x_i,$$

which represents the standard formalism of quantum mechanics. Once again for the details of these identifications we must refer the reader elsewhere, as we must for the formulation of the equations of motion (Schroedinger's equation); the book by Mackey [1963] is a good place to start for those interested in the detailed discussion.

The Universality Postulate of Microphysics and Its Consequences

We need one more link in our chain before we can return to the fundamental genetic problems which motivated this entire discussion of microphysical systems. This link involves the theoretical limits of extracting information from microphysical systems *in any way*, and boils down to the following:

Postulate of Universality. The only way that a (microphysical) system can convey information, in any form, is in terms of the values assumed on the states of the system by the system observables (or more accurately, by the probability distributions of these values).

This universality postulate has been the subject of much argument which need not concern us here. It will suffice for us to note that it is commonly accepted as true, even by those who do not like it, for the kind of system with which we will be dealing. We will therefore accept it, and note some of its profound consequences.

The first and decisive consequence, for our purposes, is the following. All of our theory has been based on the interaction of the states of a system with a special class of system, the elements of which we called *meters. A priori*, it is conceivable that the states of our system may interact with other systems than our meters in entirely new ways. The universality postulate explicitly denies this, and asserts that the interaction of the states of our system *with any system whatsoever* is governed by the rules we have derived above. In short, the system observables, defined by interactions with meters, are the only resources available to the states of a system for conveying information in arbitrary interactions with other systems. If this is so, the subjective element is banished from microphysics, and the consequences of this are far reaching indeed, as we shall see.

A second consequence of the universality postulate which is worth mentioning is of a technical character. In the preceding sections we have defined a variety of operations by which new states are algebraically constructed out of old, and new observables algebraically constructed out of old. We have tacitly assumed that our constructions were *unique:* that there could not be more than one state, or observable, with the property we have constructed.

The universality postulate comes to our rescue here in the following way: It asserts that the only way that two states α_1, α_2 can *possibly* be different (that is, distinguishable) is for there to be an observable A and a set E such that

$$p(A, \alpha_1, E) \neq p(A, \alpha_2, E)$$

and the only way that two observables A_1, A_2 can *possibly* be different is for there to be a state α and a set E such that

$$p(A_1, \alpha, E) \neq p(A_2, \alpha, E).$$

In other words, two states must be distinguished by an observable in order to be different; two observables must be distinguishable *on a state* in order to be different. If the reader will look back at the algebraic constructions we have defined, he will have no trouble at all in inferring their uniqueness from the universality postulate.

IV. The Genetic Systems

Let us now return to the transmission of primary genetic information, as expressed by Propositions 1–5 of Section I. These propositions tell us that some portion of the structure of a cell contains information relating to the genetic instructions controlling phenotypic properties, relating to the replication of these instructions. Postulate I of Section I associated the transmission of these kinds of information with essential microphysical events. The question is: Does our analysis of microphysical systems, carried out in some detail in the preceding sections, allow us to put meaningful flesh on these bones?

From the universality postulate, it follows that any structure or entity capable of storing or transmitting such information *must be a state of a microphysical system*. Thus, the totality of such information in a cell must be carried by a family of states. *A priori*, there is no need for these states all to belong to the same system (at least, in the naive sense in which this term is customarily used; we shall return to this point several times), but in any event, the least we can say is that primary genetic information must be carried in the states of certain microphysical subsystems of the total cellular system. Let us denote this aggregate of states, then, by a specific symbol $\mathfrak{S}$, which we may as well identify as the states of the genetic system (whatever other significance the states of $\mathfrak{S}$ may have).

The universality postulate says much more than this, however. It says that the only way that the states of $\mathfrak{S}$ can convey information lies in the values (or probability distributions) of certain appropriate observables, and that this information can only be extracted by having the states of $\mathfrak{S}$

interact with other subsystems of the cell. These "other subsystems," of course, play the role of cellular "meters" for the observables involved; the universality postulate assures us that the interaction of the states of $\mathfrak{S}$ with these cellular meters follows the rules we have laid down in the preceding sections, even though these meters have nothing to do with the kind of meter which we would use to observe the states in question.

As we have noted, there are three distinct kinds of information involved; each of these three kinds of information must then be represented by a *family of observables*. The values attained by the observables of any of these families (or their probability distributions) are then the necessary vehicle by which a particular state of $\mathfrak{S}$ conveys the information which must be transmitted to other systems in the cell. We thus have at the outset three separate (though not necessarily disjoint) families of observables, which we may designate as follows:

a. G, the set of observables conveying phenotypic information.
b. R, the set of observables conveying replication information.
c. L, the set of observables conveying "locus" information.

We reemphasize that the existence of these sets of observables follows directly from first principles, and has nothing to do with the physical nature of the states of $\mathfrak{S}$ or of the observables involved.

Can we say anything else about the sets G, R, L from first principles? We shall single out G for special attention, although what we have to say now will apply equally well to the other sets R and L.

Prima facie, there are two possibilities for the structure of the set G, namely:

1. G contains at least one pair of noncommuting observables.
2. G is a set of mutually commuting observables.

Have we any grounds for distinguishing between this initial pair of alternatives? According to the uncertainty relations, Possibility 1 implies that a simultaneous observation of the noncommuting observables involved (call them A and B) is impossible, and that any attempt to observe a state of $\mathfrak{S}$ with A meters and B meters consecutively will cause unpredictable changes in the information transmitted. We have in fact derived a lower bound for this unpredictability in Section II.H. Of course, this lower bound would not matter if only rough measurements were being made. But everything we know about biological interactions bespeaks their high specificity or, stated another way, their great accuracy. Indeed, biological meters are far more reliable than our own in distinguishing between alternative closely related states. All of this bespeaks a very high level of precision indeed in distinguishing states.

Now if the set G contained a pair A, B of noncommuting observables under these circumstances, it would follow that a B observation, say, would significantly and unpredictably alter the A observation on that state and would give a different answer, in general, than on previous observations. Thus, *different* (and in fact, unpredictably different) A information would be fed into the genetic "receptors," and this, from the standpoint of the cell, would result in a mistranslation of the primary genetic information involved. Worse than this, the information originally carried by the states of $\mathfrak{S}$ would be irretrievably lost as a result of successive observations by meters corresponding to noncommuting observables.

Since primary genetic information is not in general lost in this fashion, we can come but to one conclusion: the set G (and also the sets R and L) must consist entirely of mutually commuting observables.

Continuing to restrict attention to the set G, we are now in a position to apply the corollary proved in Section II.H, which states that, given any family of commuting observables, there is a single observable (call it A) of which the observables of the family are all functions. This means, at least theoretically, that all the information carried in a *set* of mutually commuting observables can already be carried by the observable A. On this basis, we can make our second postulate.

Postulate II. The phenotypic information carried by a state α of $\mathfrak{S}$ is completely determined by the value assumed on α (or its probability distribution) by a single observable A. This observable A will be referred to as the *phenotypic observable.*

A similar postulate can be made for the replication and locus information, defining single observables B, C, respectively, which we shall call the *replication observable* and the *locus observable.*

There is, of course, no necessity to assume that A, B, C, always commute, and as we shall see, certain interesting and suggestive results will follow from *not* assuming at the outset that they commute on all states of $\mathfrak{S}$. We shall consider some of the interesting ramifications of the relationships which are possible between A, B, and C in Section V.

A. The Phenotypic Observable A

We have identified a single observable A which carries in principle the phenotypic information in a genetic system of arbitrary structure, which satisfies the basic postulates we have laid down. The states of our system which carry phenotypic information, then, are precisely those states on which this observable A is defined. The properties of A will then have a profound bearing on the kind of primary genetic activity which can be exhibited by our system; it should be quite evident that different choices of this observable will lead to quite different genetic consequences. Thus, we must now investigate the manner in which the structure of A influences the manner in which phenotypic information can be transmitted; the idea, of course, is to compare the results of choosing different structures for A with the known properties of transmission of pheontypic information in real biological systems, and thus, semiempirically, to narrow down the kinds of observables which could be involved in these real systems. This type of semiempirical method is well known in the standard techniques of quantum chemistry.

What do we mean by the "structure" of an observable A? We obviously mean, "What is the nature of the spectrum of A?" Therefore, we must proceed by enumerating all the different possibilities and attempt to elicit the genetic consequences of each.

The simplest question we can ask concerning the observable A is whether it (or its spectrum) is bounded or not. If A is unbounded, we have an immediate implication for the transmission of genetic information, which is summed up in a theorem originally due (in its Hilbert space form) to Hellinger and Töplitz (see Reisz and Nagy, 1955).

Theorem. An observable A is defined on every state α (that is, $m_\alpha(A)$ is defined for every state α) only if A is bounded.

The proof of this simple but important theorem may be found in any book dealing with the theory of linear operators on separable Hilbert spaces. Stated another way, this theorem asserts that there exist states which look in all ways like states of $\mathfrak{S}$, except that they carry no phenotypic information. This looks very much like the formal analog of the existence of "nonsense" messages using the familiar structural ideas of the DNA code; that is, DNA words which convey no genetic information under any circumstances; which are utterly untranslatable into phenotypic characteristics. If the observable A is bounded, however, there can be no such thing as "nonsense"; every state of $\mathfrak{S}$ does convey a primary phenotypic message (though of course whether the message is *significant* or not within a particular metabolic context is quite another matter). Hence, we have an immediate indication of how the spectral properties of A can manifest themselves in phenotypic terms.

Now let us make some more specific assumptions regarding the spectrum of A. We saw in Section II.F that the spectrum $\sigma(A)$ consists of two parts; the point spectrum $\sigma_p(A)$ and the continuous spectrum $\sigma_c(A)$. We may at the outset make three separate assumptions about $\sigma(A)$, namely:

1. $\sigma_c(A) = \varnothing$;
2. $\sigma_p(A) = \varnothing$;
3. both $\sigma_c(A)$ and $\sigma_p(A) \neq \varnothing$.

Making use of the general discussion of Section III, we can pull together some general facts concerning the spectra of observables. The relevant ones are the following:

a. $\sigma(A)$ is always a (topologically) closed subset of the real numbers, and of course, by definition, $\sigma(A)$ is bounded if and only if A is bounded.

b. $\sigma_p(A)$, if not empty, is a denumerable set (possibly even finite) of real numbers. To each number λ in $\sigma_p(A)$ there is a positive integer attached, called its *multiplicity*. This number is defined as the dimension of the subspace of eigenvectors belonging to the eigenvalue λ. If the multiplicity of λ is unity, or in other words if there is only a single eigenvector corresponding to the eigenvalue λ, then λ is called *simple* or *nondegenerate*. If the multiplicity of λ exceeds unity, then λ is *degenerate*. The consequences of having degenerate eigenvalues in the spectrum will be discussed in Section IV.B.

B. Limit Points in $\sigma(A)$ and Resolving Power

Whatever the spectral structure of A, it is known from general operator theory that if A is bounded, then $\sigma(A)$ is a compact (that is, closed and bounded) subset of the real line. Therefore, it follows from the Bolzano–Weierstrass theorem that if $\sigma(A)$ is an *infinite* set (with A bounded), there exists at least one point of accumulation in $\sigma(A)$. On the other hand, if $\sigma(A)$ is a *finite* set (with A bounded), then at least one of the eigenvalues of A must be of infinite multiplicity (otherwise the direct sum of the corresponding subspaces would be finite dimensional). In fact, it readily follows that the only situation in which every point of $\sigma(A)$ can be an isolated point with unit multiplicity is that in which A is unbounded with pure point spectrum. For if either Assumptions 2 or Assumption 3 holds, then $\sigma(A)$ is nondenumerable whether A is bounded or not, and $\sigma(A)$ must then always possess at least one point of accumulation.

We must now discuss the manner in which these spectral properties of A can manifest themselves biologically.

To begin with, let us observe that quantum physics, as such, is generally concerned with making statements of an existential nature in terms of the spectral properties of observables. That is, in quantum physics, we predominately find statements of the following form: Given a number λ_0 in the point spectrum of an observable, there *exists* an observing apparatus which can separate, from an ensemble of states of a system containing the given observable, all and only those states in which the given observable assumes the value λ_0; or given a number μ_0 in the continuous spectrum of an observable, there *exists* an apparatus which can perform this separation to any prescribed degree of accuracy. Statements such as these are, however, not of immediate interest in biology, for biological systems come already equipped with a definite apparatus for observing the states of the system which bears the genetic information. What we need to study, therefore, is not the theoretical optimum for such measurements, but the relation between the properties of particular measuring instruments and the spectrum of the observable which the instrument is designed to measure.

Thus, let there be given a definite physical apparatus, which has been designed to measure the values assumed by some observable in specific states of a given system. The feature of this apparatus which is important for us is the fact that, inherent in the construction of the apparatus, we find a definite limit in the accuracy with which states can be separated when the closeness of their eigenvalues exceeds a certain limit. This limit is usually referred to as the resolving power of the apparatus in question. In purely classical situations, the resolving power of a macroscopic measuring instrument (such as a lens or prism) can be defined in a precise manner, as follows: To each apparatus we can associate a number $\varepsilon > 0$, such that whenever the difference between the values assumed by some observable in two states of an observed systems is greater than ε, then the two states can *always* be distinguished by the apparatus; if this difference is less than ε, then the two states will *never* be distinguished by the apparatus. In quantum theoretical systems, however, we must replace the notion of a precisely defined resolving power by essentially probabilistic ideas: Given any two states of a quantum mechanical system, under observation by a particular observing apparatus, there exists a *definite probability* that the apparatus will distinguish between them. If the eigenvalues corresponding to the two states in question are sufficiently remote, then this probability will be close to unity; if they are sufficiently proximate, the probability of resolution will be close to zero.

C. An Example: Lysogeny

We may now attempt to apply these ideas to the genetic observable A, with respect to the definite observing apparatus with which biological systems determine the values assumed by A on the various states of the genetic system G. We may safely assume that the resolving power (in the quantum theoretic sense) of this apparatus is high, but it is nevertheless finite. In more precise terms, for each point $\lambda_0 \in \sigma(A)$, there exists a neighborhood $N(\lambda_0)$ of λ_0 such that for any number λ in $[\sigma(A)] \cap [N(\lambda_0)]$ the probability that the apparatus will fail to distinguish between λ and λ_0 is appreciable (this probability in fact increases monotonically to unity as the radius of $N(\lambda_0)$ approaches zero). Thus, we conclude that no matter how high the resolving power of the apparatus in questionm it must nevertheless be exceeded in any neighborhood of a point of accumulation of $\sigma(A)$. As we have seen, the only situation in which it is possible for every genetic observation to be completely precise (up to a negligible probability of error) is when the genetic observable A is unbounded with pure point spectrum, and satisfying the further condition that no two eigenvalues of A lie so close together that there is an appreciable probability that the observing apparatus will fail to distinguish between the corresponding states.

We have thus seen that if $\sigma(A)$ possesses points of accumulation, then no matter how high the resolving power of the genetic observing system, it may nevertheless happen that some cell contains two states of G the eigenvalues of which lie so close together that the probability of these states being separated by the apparatus is significantly less than unity. Let us now attempt to describe the possible behavior of biological systems in which this situation obtains. For simplicity, we envision a large number of identical such systems, each of which contains two states ψ, ψ' of G, such that the corrsponding eigenvalues λ, λ' lie so close together that the propability p of their being distinguished by the cell's observing system is much less than one. There is no *a priori* reason to suppose that the *phenotypes* (that is, the metabolic implementation of the genetic information carried by ψ, ψ') are related; let us therefore assume that the phenotype corresponding to ψ' is *lethal* (that is, results in a destruction of the metabolism of a system bearing it). We remark explicitly that since each member of the ensemble of biological systems under consideration is assumed initially viable (that is, expresses the phenotype corresponding to ψ), the probability p defined above becomes in this case the probability that a particular member of the ensemble will be capable of distinguishing ψ' from ψ.

The observed behavior of such an ensemble will be very simple. Most of the sytems will remain viable (that is, exhibit the ψ phenotype); occasionally a system will die. In fact, if there are N systems in the ensemble at a time t, and N is a sufficiently large number, then pN of the systems can be expected to exhibit the lethal phenotype at time t.

An example of this type of behavior seems to be provided by various lysogenic strains of bacteria. In these bacteria, there is apparently present in the genome a character called *prophage*, which becomes activated on a purely statistical basis, and alters the phenotypes of bacteria in such a manner as to cause them to produce particles of active bacteriophage, and hence to cause their own lysis and death. If we suppose that the "prophage" locus corresponds to a state ψ' satisfying the conditions described above, then we obtain a situation which seems to account for the experimental facts of lysogeny quite well. In particular, the two fundamental characteristics of lysogenic bacteria are immediate consequences of this situation. These properties are:

a. The ability of every cell of a lysogenic population to produce virulent bacteriophage.

b. The complete immunity of bacteria of a lysogenic population to superinfection by virulent phage homologous to those arising by activation of the prophage locus.

That Property a follows from the above discussion is evident. Property b can be accounted for in a number of different ways, depending on the role that phage DNA is assumed to play as a genetic agent. As one possibility, we may assume that the infective phage DNA actually contains a state ψ' of the genetic system G. We further assume that the bacteria of a lysogenic strain contains, as part of their genome, a state ψ of G which lies very close to ψ' and which is not possessed by nonlysogenic bacteria. Therefore, a nonlysogenic strain will immediately (with a high probability) be able to recognize the state ψ' and carry out its metabolic instructions (that is, lyse), while the bacteria of a lysogenic strain will be able to distinguish the state ψ' from the state ψ, which is already present in their genome, with a small probability. It will be observed that this picture is in good qualitative agreement with the actual behavior of different bacterial strains when exposed to bacteriophage. Moreover, the probability that a lysogenic bacterium will distinguish between the states ψ and ψ' is not affected by the number of copies of ψ' which are present, but only on the distance between the corresponding eigenvalues. Therefore, the superinfection of lysogenic strains, which on the basis of the above picture means precisely the incorporation of further copies of ψ' into the strain, will not increase the frequency of lysis in such strains beyond the "spontaneous" rate.

D. Degeneracy, Perturbation, and Allelism

Thus far, we have only considered the manner in which points of accumulation of $\sigma(A)$, coupled with the finite resolving power of the observing apparatus, can lead to biologically interesting situations. We now turn to a discussion of the implications of the existence of multiple eigenvalues in $\sigma(A)$. In this case, it will be remembered, there may exist many distinct (that is, orthogonal) states corresponding to the same eigenvalue of A; the number of such states being the multiplicity of the eigenvalue. In quantum physics, an eigenvalue with multiplicity greater than one is called *degenerate*.

The situation in the case of a degenerate eigenvalue is quite different from that which obtains in the neighborhood of a point of accumulation of $\sigma(A)$. In the former case, the several states corresponding to a degenerate eigenvalue can *never* be separated by an observation (always with reference to the single observable under consideration, of course), however precise this observation may be; in the latter case, there always exists a finite probability that closely neighboring states can be separated, the probability increasing with increasing resolving power of the observing apparatus.

However, although degenerate states (the states corresponding to degenerate eigenvalues) cannot be separated by observations under ordinary circumstances, it is possible for them to manifest themselves by interaction with other systems, This type of situation, in which an observed system is not to be considered in isolation, but in interaction with its environment, is usually referred to as a *perturbation* of the system. We must therefore study the manner in which these perturbations may give rise to biologically significant effects. Before we undertake this study, we shall briefly describe those aspects of general perturbation theory which we will require.

Speaking roughly, a system $\mathfrak{T}'$ which is being perturbed can be regarded, from the formal standpoint, as equivalent to a system $\mathfrak{T}$ in isolation, where the observables which define $\mathfrak{T}'$ differ from those of $\mathfrak{T}$ by a "small" amount; this is the basic idea of perturbation theory. In greater detail, let $\mathfrak{T}$ denote a definite unperturbed system, and let $A_1, A_2, \ldots$ be the observables which generate $\mathfrak{T}$ (in a suitable algebraic sense). We shall denote the perturbed system by $\mathfrak{T}'$, and we shall assume that $\mathfrak{T}'$ is generated by the observables $A_1', A_2', \ldots$. The relation between A_i and A_i', for each i is assumed to be given by

$$A_i' = A_i + \varepsilon A_i^{(1)} + \varepsilon^2 A_i^{(2)} + \cdots, \tag{6}$$

where ε is a (small) parameter, and the $A_i^{(j)}$ are suitable self-adjoint transformations, dependeing on the perturbation in question. In quantum physics, the first two terms of Eq. (6) generally suffice.

We now inquire what is the relation between the states of $\mathfrak{T}$ and $\mathfrak{T}'$, respectively. In the present situation, we are interested only in the effects of the perturbation on a single observable (the genetic observable) which we shall assume to be given by Eq. (6); consequently, we restrict our attention to eigenstates of that observable. A full analytic answer to the question of how these states transform under perturbation is long and complicated, and depends ultimately on the properties of the transformations $A^{(j)}$. Therefore, we shall content ourselves with a qualitative description of the pertinent facts. (For a fuller treatment, we refer the reader to the literature; Riesz and Nagy [1955] present a useful discussion of perturbation theory. However, the treatment of this theory to be found in most textbooks dealing with quantum mechanics is very specialized.)

The states of $\mathfrak{T}'$ will in general be different from those of $\mathfrak{T}$; however, if the perturbation is sufficiently small, they will not be too much different. Therefore, the eigenvalues corresponding to the perturbed and unperturbed states will also not differ too much from each other in this situation. What is of particular interest for us is the behavior of the degenerate states under perturbation. On the basis of what has already been said, it can be seen intuitively that if ψ_1, ψ_2 are two states corresponding to a single degenerate eigenvalue λ of A, there is no reason to suppose that the corresponding states

ψ_1', ψ_2' of A' should still correspond to a single eigenvalue. The precise calculations verify this intuitive picture. There are many physical examples of this type of effect, such as the splitting of various lines in atomic spectra under the application of a magnetic field (Zeeman effect).

However, the perturbation of an observable may produce other effects besides the splitting of degenerate eigenvalues. These other effects depend upon the spectral structure of the observable in question, and the nature of the perturbation. Thus, under appropriate conditions, eigenstates which were initially distinct in the unperturbed system may become degenerate in the perturbed system; points of the continuous spectrum can become points of the discrete spectrum and conversely (see especially Friedrichs [1965] for a discussion of the possible effects of perturbing an observable with a nonempty continuous spectrum). Physical manifestations of most of these effects are known, although simple examples are not abundant.

What are the possible genetic implications of these considerations? To begin with, let us recognize explicitly that the capacity of any single biological system (that is, cell) to utilize genetic information is very strictly limited. That is, there are many states of the genetic observable which cannot be deciphered by the system in question, in the sense that such states do not enable the cell to assume a corresponding phenotype. This limitation has been proved repeatedly by many different kinds of experiments; whether it is caused by inherent deficiencies in the observing apparatus of cells or is due to other effects it is not possible to say. We merely wish to emphasize here that the capacity of a biological system to utilize genetic information is sharply and characteristically circumscribed. This limitation will be important for the following discussion.

Let us begin by considering a state ψ of the genetic system G, where ψ corresponds for the moment to a *simple* eigenvalue λ of A (that is, we can write $A\psi = \lambda\psi$). If we now perturb the system G, then we in effect replace the genetic observable A by a new observable A', where A' is given by an expression of the form in Eq. (6). The state ψ will then be replaced by a state ψ' which is an eigenstate of A', corresponding to a value λ' (that is, we have $A'\psi' = \lambda'\psi'$). Now λ' is in general different from λ; an exact expression for λ' can be written down if the perturbation is known [see Riesz and Nagy, 1955].

The operational significance of the perturbation is thus as follows: Subsequent to the perturbation, the observing apparatus of the cell will report the value λ' for the state in question. If the perturbation is very small so that λ' lies so close to λ that the resolving power of the observing apparatus is not exceeded, then (with high probability) the system will behave as if no perturbation has taken place. On the other hand, if the perturbation is sufficiently large so that the observing apparatus can differentiate between λ

and λ' with an appreciable probability, then a phenotypic indication of this state of affairs is to be expected.

The actual phenotypic behavior of the system is contingent on whether or not λ' is sufficiently close to another eigenvalue μ of the unperturbed genetic operator A. In any case, however, we can assert that the phenotype corresponding to the eigenvalue λ has disappeared from the system. If λ' is so close to μ that the observing apparatus will report the perturbed state ψ' as being the unperturbed state φ corresponding to the eigenvalue μ, then the system will assume the phenotypic trait corresponding to μ if it is possible for it to do so; if this is not possible (that is, if the state φ corresponds to genetic information which the cell in question cannot utilize), or if λ' is not sufficiently close to another eigenvalue of A, then the only phenotypic manifestation of the perturbation will be the disappearance of the trait corresponding to λ from the phenotype of the system.

Let us notice that we have assumed, throughout the above discussion, that the perturbation under consideration is acting throughout the entire time in which biological observation can take place. An example of such a perturbation might be the following: Suppose we consider, *provisionally*, that the genetic information is carried by molecules of DNA. If we substitute various analogs for the purine and pyrimidine bases occurring in DNA, we may obtain molecules which are still capable of carrying genetic information, but may not carry the same information as the unsubstituted molecules did. The structural alteration arising from such substitutions may be considered as inducing a perturbation of the observable A, at least with respect to the particular state under consideration; such a perturbation will remain in effect as long as the molecule retains its integrity; that is, for practical purposes, such a perturbation acts infinitely long.

On the other hand, if we have a perturbation which acts only momentarily, so to speak, we get somewhat different results. Let us denote the perturbed state by ψ', and let us suppose that ψ' arose through perturbation of the state ψ. The general theory of such momentary perturbations tells us that, when the perturbation is removed, ψ' will revert to a state φ of the original system. There is no reason why we should have $\varphi = \psi$; in fact it follows from the general theory that there exists a finite probability (*transition probability*) that the initial state ψ will be transformed to any other state of the unperturbed system as a result of the perturbation (subject, of course, to the various conservation rules of quantum mechanics; transitions which violate these rules are called *forbidden*, and do not ordinarily occur in nature.

Perturbations of this (momentary) type may perhaps be induced by radiation, although the gross structural alterations and crosslinking of polymers induced by high-frequency radiation are almost certainly of a permanent nature. It is readily seen, nevertheless, that the effect of a transition induced by a "temporary" perturbation will correspond intuitively to what the geneticist calls a *point mutation*. It therefore follows that the perturbation of a state of $\mathfrak{S}$ corresponding to a degenerate eigenvalue may give rise to transitions whose effects bear a close resemblance to the generation of *multiple alleles* (or pseudoalleles) of the corresponding gene. If the generation of allelic series can indeed occur in this manner, then we obtain as an immediate corollary the important result that not every gene (that is, state of $\mathfrak{S}$) can give rise to a series of alleles. In order for allelism to be possible, it is necessary that the corresponding eigenvalue of the genetic observable be degenerate; the multiplicity of this eigenvalue is then the maximal number of alleles possible in the series. We draw the further obvious conclusion, from the fact that allelic series exist, that the genetic observable A *cannot* then possess only simple eigenvalues.

Thus far in this section, we have discussed only the possibility of the splitting of degenerate eigenvalues in the point spectrum of A. We noted, however, that perturbations can have other effects upon the spectrum; a perturbation can cause a transition of a state of A from the point spectrum to the continuous spectrum and vice versa. In genetic terms, this type of transition might manifest itself in a change of the apparent "stability" of certain genes, since as we have seen, a transition from continuous to point spectrum will in general decrease errors arising from the resolving power of the genetic observing apparatus, while a transition in the opposite direction will tend to increase the probability of such errors. Many of the *position effects* of classical genetics are of this type; for example, numerous examples are known in which genes which have been translocated to a new chromosomal environment exhibit an alteration in their mutation rate.

Moreover, under certain circumstances, a perturbation can cause initially distinct states, corresponding to distinct eigenvalues, to correspond to a single degenerate eigenvalue of the perturbed observable. This is an effect inverse to the splitting of a degenerate state. This kind of effect in the genetic observable would be more difficult to identify phenotypically, since its manifestation would vary enormously, depending on the specifics of the phenotypes involved. However, it is possible that some of the phenomena collectively called *gene conversion* could be interpreted in this light.

V. The Operators *B* and *C*, and Their Interrelationships with the Genetic Observable *A*

In Section IV we introduced the three systems required for the bearing of the primary phenotypic, replicative, and spatial (locus) information required of a genetic system, and designated these systems by G, R, and L, respectively. Each of them must, according to first principles, correspond to a family of observables. We saw in that section that the persistence of genetic information of each of these three kinds led to the hypothesis that these families each consist of mutually commuting observables. From this it followed that, for each of these families, there would be a single observable of which all the observables in the family were functions, and hence that each kind of genetic information could in principle be carried by a single observable. Thus each of the families G, R, L could be replaced by single observables, which we denoted by A, B, C, respectively. The past few sections were devoted to an analysis of the manner in which the spectral properties of A could manifest themselves biologically; in the present section we shall briefly consider the same question applied to the observables B and C separately, as well as the question of the commutativity or compatibility of the set of observables $\{A, B, C\}$.

We have already discussed in detail the meaning of any biological observable possessing a point spectrum or continuous spectrum, and it is an easy exercise for the reader to adapt these discussions to the particular observables B, C. Therefore we shall discuss in detail the possible effects of perturbations on these observables, together with their biological and genetic effects.

Let us then assume, as genetic experience seems to indicate, that each state of $\mathfrak{S}$ (that is, each carrier of primary genetic information, or each gene) always occupies a more-or-less definite locus, relative to the other states which collectively comprisee the genome. If ψ is our state and θ designates its locus and if the observable C has point spectrum, then we must have

$$C\psi = \theta\psi.$$

Let us now suppose that a momentary perturbation is applied, causing a transition $\psi \longrightarrow \varphi$. If φ corresponds to the same eigenvalue θ of C as did ψ, but corresponds to a new eigenvalue of the genetic observable A, then the effect will be the manifestation of new phenotypic information at the same locus as before, corresponding to the generation of a new allele. On the other hand, if we have for the new state φ

$$C\varphi = \theta'\varphi$$

with $\theta \neq \theta'$, then we have the manifestation of altered phenotypic information together with a shift of locus. Effects of this nature bear a close resemblance to the generation of *pseudoalleles* of the original gene. It will be recalled that the only distinction between a true allele and a pseudoallele is that in the former case no crossing over occurs, while in the latter case crossing over can occur (generally with low probability).

In the case of a permanent perturbation, a degenerate eigenvalue of C will likewise manifest itself as an apparent migration of a particular gene, or of its alleles (depending on the effect of the perturbation on A) into a sequence of linked loci. The number of such loci is dominated by the multiplicity of the eigenvalue θ under consideration. The fact that these new loci are invariably strongly linked means that many theoretically possible transitions between the states of the system L are in fact forbidden; this in turn may be an important clue into the actual physical realizations of the systems involved.

With respect to the observable B, similar considerations apply. However, one particular fact is worthy of note. Namely, if we subject the replicative system R to a permanent perturbation, but one which does not change the A eigenvalue, then this means that *the perturbed state may replicate in an aberrated fashion*, or not at all, since it is no longer a state of the replicative system R, even though the genetic information for phenotypic qualities has been preserved. This provides one way for the discrimination of effects caused by temporary or permanent perturbations; a temporary perturbation causes a reversion back to a *bona fide* state of R, which will then by hypothesis replicate perfectly; a permanent perturbation need not so replicate.

From the above, it will be seen that many genetically important aspects of the response of the observables A, B, C to perturbations depends on the extent to which a perturbation of one of the systems G, R, L is a perturbation of the others. This is a question of the extent to which the corresponding observables A, B, C can be independently perturbed. We propose to call two observables *strongly linked* if it is imposible to perturb one of them without perturbing the other; *unlinked* if they can always be separately perturbed. Between these two extremes there is an entire continuum of possibilities. This question of *linkage* or coupling between the observables of a microphysical system, which is of obvious importance in considerations of system structure, seems never to have been raised in the physical literature, but it is of crucial importance in an understanding of the deeper structure of primary genetic mechanisms and their response to perturbations. This is but one indication of how biological questions can raise important new concepts within the structure of physics itself.

One final question remains to be considered: namely, the compatibility or commutativity between the observables A, B, C themselves. It might be supposed at the outset that our arguments regarding the necessity of preserving genetic information of each of the three types under consideration requires also that A, B, C themselves commute, and hence that these three observables may themselves be replaced by a single observable of which they are all functions. This is indeed one strong possibility. However, in this case it does not follow automatically, because we need to take into account the relative resolving powers of the three different observing systems involved in determining the phenotypic, replicative, and positional states. If these resolving powers are sufficiently different, then noncommutativity would not be as disastrous between A, B, and C as it would be within one of the families G, R, L. However, the question of the commutativity of the three observables under consideration is closely related to the degree of linkage between them; if they do commute they must be strongly linked, and this, as we have seen, would have important and detectable biological consequences.

VI. Interpretation

We have seen that a wide variety of genetic effects can be accounted for simply on the basis of Postulates I and II, together with the most elementary properties of microphysical systems. All this, then, is utterly independent of any specific *physical* characterization of the states which carry genetic information, and would remain true for any genetic system satisfying Postulates I and II. Our analysis could have been carried a great deal further without requiring any specification whatever of the states involved.

Nevertheless, it is plausible and desirable to make contact with the enormous amount of information on primary genetic mechanisms obtained from molecular biology, and centering around the involvement of nucleic acids in genetic processes. The obvious way to try and do this is to identify specific observables of nucleic acid molecules, considered as microphysical or quantum theoretic systems, with the observables to which our analysis has led us. But if we try to do this, we run up against some major difficulties, which it is instructive to examine more closely; they are typical of the kind of difficulty we run into whenever we try to reconcile alternative descriptions of the same system from different points of view. Some of these difficulties can be regarded as technical; but when technical difficulties become sufficiently severe, we can wonder whether they do not in fact represent conceptual difficulties as well. We shall take up a number of these difficulties, one at a time.

We have developed the microphysical formalism entirely in terms of meters and measuring instruments. In our discussion we have contemplated an enormously rich profusion of observables belonging to any system. Quantum theory itself, however, was developed only on relatively simple systems, and for such systems it is true that only a very small number of observables is necessary to characterize them completely. For such systems, we know in principle how to compute the properties of arbitrary observables when the defining observables, corresponding to the state variables of classical mechanics, are known.

But this raises two questions:

a. Can the more complex systems involved in primary genetic processes actually be characterized (in any effective or useful sense) solely in terms of the classical state variables involving positions, momenta, energies (and, in microphysical systems, spins) at all?

b. Even assuming the answer to question a is yes, can we in fact identify the observables we want with specific functions of these state variables?

Question b is more technical, and we will consider it first. Anyone familiar with quantum physics knows that for systems more complicated than the hydrogen atom (that is, systems not possessing spherical symmetry) the equations of motion cannot be solved exactly, and recourse must be had to a variety of approximations: approximate wave functions, approximate energy operators, and so forth. The conventional approximations which are made, say, in quantum chemistry, have of course, proved most useful, and in any event they are the only recourse. But it is evident that no approximation can be valid across the full spectrum of physical properties of a system; if φ is the correct wave function, and if φ' is an approximation to it, then no matter how close φ is to φ', there will be an observable A for which

$$(A\,(\varphi - \varphi'), \varphi - \varphi')$$

is as large as we please (exactly the same difficulty arises in macrophysics also, and is one reason why we cannot use approximations to solve classical problems such as the three-body problem). Thus, the use of standard techniques of approximation in dealing with unknown observables must always be suspect. And approximate information is all that is available.

With respect to Question a, let us make another observation. The kinds of systems with which physicists and chemists conventionally deal are easy to associate with the kinds of entities we call "substances." They are ordinarily single particles or aggregates of such particles, which may interact with each other or with radiation, in readily visualized ways. But there is nothing in the microphysical formalism itself, as we developed it, which restricts us to this kind of system. Indeed, the formalism itself suggests the existence of perfectly good microphysical systems which cannot be associated with "substances" in a naive way. Biology itself is replete with "functional units," which are associated with specific substances in complicated ways, but which have crucial properties of their own: active sites, binding sites, receptors, and so forth. Why should these functional terms not be themselves regarded as microphysical systems in their own right, which are nonsubstantial in the sense that there is no specific "substance" to which they correspond (even though their observables are related, in complicated ways, to the observables of "real substances")?

The microphysics of such "nonsubstantial" systems would represent a radical innovation for microphysics, one which was not required for dealing with the relatively simple systems of physics and chemistry, but which seems natural and even necessary in biology. And it is by no means clear how the conventional quantum theoretic formalism, dominated by notions of energy and its expression in Hamiltonian terms, is to be applied to such systems; what is the "energy" of an active site of an enzyme? This becomes clear when we try to relate the various genetic observables we have described to standard quantum theoretic terminology.

For instance, it is clear that (barring mutation) genetic information is "conserved," in the physical sense. If we are dealing with an energetically closed system, quantum mechanics tells us that the observable corresponding to any conserved quantity must commute with the energy observable (Hamiltonian) of the system. But if an observable commutes with the Hamiltonian, it must either be already a function of the Hamiltonian, or else both it and the Hamiltonian are functions of some other observable. The latter possibility is discounted by quantum theory; the Hamiltonian already contains all the information about the system. But on intuitive grounds, there is nothing to prevent states with the same energy E from carrying different genetic information, or states with different energies (and even different Hamiltonians) from carrying the same information. Thus, we cannot even say that the genetic observables commute with the Hamiltonian; we have to ask: Hamiltonian of what?

Thus, as with all novel approaches, we find that we have kicked open a hornet's nest of new questions, which in this case touch on the basic aspects of the microphysical formalism, and the manner in which that formalism is related to the world. All we can say at the moment, however, is that the picture emerging from purely microphysical considerations is consistent with that obtained from molecular biology. A good deal more will have to be done before we can replace this "consistency" by an inferential chain from microphysics to DNA.

References

Friedrichs, K. O. [1965]. "Perturbations of Spectra in Hilbert Space." American Mathematical Society, Providence.

Mackey, G. W. [1963]. "Mathematical Foundations of Quantum Mechanics." Benjamin, New York.

Riesz, F., and Nagy, B. [1955]. "Functional Analysis." Ungar, New York.

Chapter 4

EXCITABILITY PHENOMENA IN MEMBRANES

D. Agin

Department of Physiology
University of Chicago
Chicago, Illinois

I. Introduction

In this chapter we will consider a number of topics concerning membrane excitability which should be of interest to the research student in mathematical biology. The problem of membrane excitation has traditionally attracted mathematicians, physicists, and physical chemists, in addition to biologists. Unfortunately, those who are interested in the construction of elegant physical theories or mathematical models, in the sense of classical theoretical physics, soon find themselves coping with bone-rattling frustrations. Biological membranes are barriers of uncertain structure, surrounded by solutions of uncertain structure, and penetrated by particles with uncertain behavior. The theoretician must be something of a physicist, chemist, and biologist, and usually a part-time magician, if he hopes to understand membrane excitability. Perhaps his only reward is that which comes from tangling with a fascinating puzzle.

It is not possible here to present more than an introductory survey. Rather than arrange once more information already available in recent reviews and monographs, I have chosen to focus on some theoretical problems which will be important in the near future. The hope, frankly, is to stimulate research workers, particularly the younger ones, to work on these problems. The categorized bibliography should be helpful to correct and supplement the ideas presented here and the serious student is urged to make use of it.

II. Squid Axon and the Hodgkin–Huxley Equations: Concerning Models and Theories

The excitable biological membrane about which there is most experimental information is undoubtedly the membrane of the squid axon [3, 15–17, 27, 32, 34]. For the past 20 years most research workers have attempted to interpret electrophysiological experiments in terms of the Hodgkin–Huxley theory proposed in 1952 [18–23]. It will be assumed that the reader has some familiarity with the experimental observations which formed the basis for this theory, and a critical review of relevant material by Cole [3], Hodgkin [17], and Hodgkin and Huxley [21] is suggested.

The mathematical formulation proposed by Hodgkin and Huxley in 1952 was stated in terms of a system of differential equations. Differential equations appearing in physical theory ordinarily result from microscopic considerations of variation; that is, they are based essentially on "first principles" of physics. In the Hodgkin–Huxley system of differential equations, a fourth-order nonlinear system, only one of the differential equations is based on first principles; the other three equations are entirely empirical. This introduction of empirical differential equations into what is considered a physical theory has promoted a great deal of confusion, particularly among nonphysicists, and it is important to consider the situation in detail.

The first mathematical observation, of course, is that if the Hodgkin–Huxley system of four first-order equations is converted into its unwieldy, single fourth-order equivalent, nearly all terms containing derivatives are completely arbitrary; that is, the set of terms which can produce the same integral is unbounded. Whether one chooses to call the Hodgkin–Huxley formulation theoretical or empirical or quasitheoretical or quasiempirical is not important; what is important is that we realize at the outset here that we have a satisfactory mathematical *model* for the squid axon, but we do not have a satisfactory mathematical *theory*. The formulation of such a theory remains an important challenge for mathematical biologists and biophysicists.

III. Descriptive Equations for the Axon Membrane

Early experiments showed that nerve fibers behaved in many ways like electrical transmission lines with imperfect insulation [3]. It was therefore reasonable to attempt to apply the equations of transmission-line theory [3, 32, 35] to the behavior of axons. The reader should be familiar with the experiments revealing the overshoot of the action potential and its dependence on extracellular sodium concentration [22]. This information coupled with the apparent applicability of transmission-line theory soon led to the detailed description made by Hodgkin and his collaborators [18–23]. However, two important technical advances were required and both of these were made independently by others [3, 28]. The transmission-line equation which had been used was a partial differential equation: the action potential was a propagated wave. The first technical achievement, the "space clamp," caused the action potential to become a standing wave (by reducing potential variation along the axon to insignificant values). The second achievement, the "voltage clamp," made possible a direct measurement of ionic currents and their dependence on ion concentrations by eliminating the transmembrane capacitative current.

The current across the membrane at any point under ordinary conditions is assumed to be given by

$$I_{\mathrm{m}} = C\,\partial V/\partial t + \sum I_i, \tag{1}$$

where I_m is the membrane current, C is the membrane capacitance, V is the membrane potential, $\sum I_i$ is the current through the membrane resistance (presumably carried by ions), and t is time.

If the axon is considered as a transmission line (35), according to cable theory,

$$I_{\mathrm{m}} = A\,\partial^2 V/\partial z^2, \tag{2}$$

where z is the longitudinal axis of the axon and A is a constant. The complete partial differential equation for a propagated disturbance is then

$$A\,\partial^2 V/\partial z^2 = C\,\partial V/\partial t + \sum I_i. \tag{3}$$

If one assumes a propagated wave at constant velocity,

$$\partial^2 V/\partial z^2 = B\,\partial^2 V/\partial t^2, \tag{4}$$

where B is a constant. Using this in Eq. (3),

$$A'\,d^2V/dt^2 = C\,dV/dt + \sum I_i. \tag{5}$$

Under space clamp conditions, in the absence of an external current source,

$$\partial V/\partial z = 0, \tag{6a}$$

$$I_{\mathrm{m}} = 0 = C\,dV/dt + \sum I_i. \tag{6b}$$

With appropriate feedback devices [35], I_m can be clamped to some arbitrary value. If the value is high enough, a repetitive series of action potentials is produced in response to the constant applied current. Under ordinary conditions with only a space clamp, a current pulse of short duration is used as a stimulus and produces only a single action potential. In both cases—space clamp alone and space clamp plus current clamp—the responses are called "membrane" action potentials.

Finally, by means of a space clamp and voltage clamp [35], one can eliminate the time derivative in Eq. (6b) and proceed to "separate" the ionic currents by varying external ion concentrations. With this information, Hodgkin and Huxley sought to provide an expression for I_i that would predict experimental observations when used with Eqs. (3) and (6b).

For the case of a short duration (for example, 50 μsec) current pulse as stimulus producing a subsequent membrane action potential, the magnitude of the current pulse determines the starting potential displacement (that is, the initial value of V).

Hodgkin and Huxley assumed that the total ionic current was simply a summation of individual species currents and that each species current could be expressed simply as the product of a conductance and a driving force. The final system of 1952 equations was as follows:

$$dV/dt = -C^{-1}[g_K(V - V_K) + g_{Na}(V - V_{Na}) + g_L(V - V_L)], \tag{7}$$

$$g_K \equiv \bar{g}_K n^4, \tag{8a}$$

$$g_{Na} \equiv \bar{g}_{Na} m^3 h, \tag{8b}$$

$$g_L \equiv 0.3, \tag{8c}$$

$$dn/dt = \alpha_n(1 - n) - \beta_n n, \tag{9}$$

$$dm/dt = \alpha_m(1 - m) - \beta_m m, \tag{10}$$

$$dh/dt = \alpha_h(1 - h) - \beta_h h, \tag{11}$$

$$\alpha_n \equiv 0.01(V + 10)/\{\exp[(V + 10)/10] - 1\}, \tag{12}$$

$$\beta_n \equiv 0.125 \exp(V/80), \tag{13}$$

$$\alpha_m \equiv 0.1(V + 25)/\{\exp[(V + 25)/10] - 1\}, \tag{14}$$

$$\beta_m \equiv 4 \exp(V/18), \tag{15}$$

$$\alpha_h \equiv 0.07 \exp(V/20), \tag{16}$$

$$\beta_h \equiv 1/\{\exp[(V + 30)/10] + 1\}, \tag{17}$$

$$V_K = 12, \qquad g_K = 36, \tag{18a}$$

$$V_{Na} = -115, \qquad g_{Na} = 120, \tag{18b}$$

$$V_L = -10.613, \qquad C = 1, \tag{18c}$$

where

$V = (E - E_r)$ represents membrane potential displacement (mV);
t represents (msec);
C represents membrane capacitance (μF cm^{-2});
g_K represents potassium conductance (mmho cm^{-2});
g_{Na} represents sodium conductance (mmho cm^{-2});
g_L represents leakage conductance (mmho cm^{-2});
$V_K = (E_K - E_r)$ relative potassium equilibrium potential (mV);
$V_{Na} = (E_{Na} - E_r)$ relative sodium equilibrium potential (mV);
$V_L = (E_L - E_r)$ represents relative leakage equilibrium potential (mV);
E represents membrane potential (mV);
E_r represents resting membrane potential with the inside take as positive (mV);
E_K represents potassium equilibrium potential (mV);
E_{Na} represents sodium equilibrium potential (mV);
E_L represents leakage equilibrium potential (mV);
n is a dimensionless variable associated with potassium conductance;
m is a dimensionless variable associated with sodium conductance;
h is a dimensionless variable associated with sodium conductance;
$\bar{g}_K$ represents potassium conductance parameter (mmho cm^{-2});
$\bar{g}_{Na}$ represents sodium conductance parameter (mmho cm^{-2}).

These are the equations for a nonpropagated or "membrane" action potential. There is no question that they accurately describe nearly everything seen in the laboratory. The model can be stated as follows:

$$dV/dt = -C^{-1} \sum I_i = (-C)^{-1} \sum g_i(V - V_i), \tag{19}$$

$$dg_i/dt = f_i(g_i, V). \tag{20}$$

Equation (19) is theoretical. Equation (20), however, is entirely empirical: both the differential form and the functions f_i are arbitrary.

The Hodgkin–Huxley model is thus essentially a mathematical *model* rather than mathematical *theory*. On occasion one is confronted with the naive argument that since the Hodgkin–Huxley model works so well it must have some validity. The following set of equations, with numerical values based on the identical experimental observations made by Hodgkin and Huxley, produces behavior not distinguishable in any important way from their model:

$$dV/dt = (-C)^{-1}(0.8)[\alpha_2\phi_2 - \alpha_1\phi_1 + 0.3V], \tag{21}$$

$$d\phi_1/dt = -0.8\beta_1(V + \phi_1), \tag{22}$$

$$d\phi_2/dt = -0.8\beta_2(V + \phi_2), \tag{23}$$

$$\alpha_1 = \alpha_2 + \frac{\{1 + 26/[1 + 5\exp(0.06V)]\}}{[1 + \exp(0.5V + 5)]}, \tag{24}$$

$$\alpha_2 = (V + 120)/[1 + \exp(0.5V + 5)], \tag{25}$$

$$\beta_1 = 0.01(V - 20)[\exp(0.1V - 2) - 1] + 0.1, \tag{26}$$

$$\beta_2 = 0.03(V - 20)[\exp(0.05V - 1) - 1]. \tag{27}$$

This is a *third*-order system of differential equations. It is an *entirely* empirical mathematical model; none of the equations is theoretical. The reader may find it useful in a counterargument for those who do not understand empirical equations. The best summary of research on the Hodgkin–Huxley model is the monograph by Cole [3]. The serious student should program both sets of equations and examine their properties with a digital or analog computer. In the latter case, one must be careful that inherent computer noise, which will appear as an incremental total current, does not "excite" the system without any applied stimulus.

The best available analysis of the mathematical properties of the Hodgkin–Huxley model is that made by FitzHugh [7–11]. and the interested student should consult his work.

IV. Research Objectives

We can agree, then that the Hodgkin–Huxley model is largely empirical. The task for the theoretician is not to produce another such model. An empirical model, no matter how accurately it simulates the behavior of a nerve fiber, does not tell us very much about the physical processes responsible for that behavior. The same is true for "electronic" models, which are only empirical mathematical models stated with hardware. What is needed is a quantitative theory based on elementary physicochemical assumptions and which is detailed enough to produce calculated membrane behavior identical with that observed experimentally. At the present time such a theory does not exist. Its construction will be an important advance, for like all theories its lifetime will be short and in the process of destroying it we will learn things which are new and exciting.

A useful theory must account for the large body of detailed experimental information in terms of assumptions concerning the movements of ions across potential barriers. It is not reasonable to expect a fully developed theory to emerge at once. At the present time theoretical research should focus on important subsidiary problems. Two of these stand out in importance far above the others:

1. How does the axon membrane achieve its extraordinary relation between current flow and potential difference?

2. What processes are responsible for the remarkable time dependence of this relation?

Concerning the first question, perhaps the most important property of the axon is its transient negative conductance. It can be shown analytically that if a system which exhibits a negative conductance in some region of the I–V plane is associated with appropriate reactances, a number of physiologically significant phenomena such as excitation threshold and repetitive activity become possible. It is therefore important to examine the physical theory of electrodiffusion with negative conductance phenomena in mind. An important experimental observation is that under certain conditions axons can show a *steady-state* negative conductance [29]. Since this is mathematically and physically a simpler situation, we will first discuss steady-state electrodiffusion phenomena. Our objective is to determine, if possible, under what conditions a negative conductance may appear.

V. The Movement of Charged Particles across Potential Barriers

There are two general approaches to the detailed description of the movement of charged particles across potential barriers. In the first approach the barrier or membrane is considered as a discrete phase of finite thickness in which electrodiffusion equations are assumed to apply. These equations can be derived rigorously from statistical mechanics [84] or less rigorously by methods similar to that of Planck [95]. In order to solve the resulting differential equations, important assumptions concerning boundary conditions must be made.

The second approach essentially treats the membrane as a discontinuity and descriptive equations are formulated in terms of transition probabilities across an energy barrier. The chief virtue of this approach is that it usually produces explicit quantitative statements which can be tested experimentally.

Other popular approaches are those involving irreversible thermodynamics and "rate theory." They can usually be classed with either the continuous or discontinuous approaches mentioned above, and they will not be treated here in any specific detail.

Which particular approach should be used must depend on the thickness of the barrier and the mechanism of particle migration. If the probability of an ion being stopped by a collision while crossing the membrane is much greater than the probability of its being reflected at the boundary, a diffusion theory should be used. Another way to state this is that if the mean free path of the ions is small compared with the distance over which the internal potential changes by kT, diffusion theory is appropriate. On the other hand, if the mean free path of ions is large compared with the distance over which the internal potential changes by kT, so that ions cross the "membrane" on the average without making collisions, then a discontinuous theory should be used. It must be remembered that a moving charged particle is interested only in potential barriers, that is, the actual physical dimensions of the membrane are not relevant.

With a few interesting and instructive exceptions [49, 50], most theoretical work in membrane biophysics has involved a continuous approach.

VI. The Equations of Electrodiffusion

The physical model which is used is that of a plane barrier separating two ionic solutions of large volume. It is assumed that electric field and ion concentrations in the solutions have zero gradients. The following equations are used for the one-dimensional electrodiffusion regime of the membrane:

$$\partial c_i/\partial t = -\partial J_i/\partial x, \tag{28}$$

$$J_i = -D_i\, \partial c_i/\partial x - u_i z_i c_i\, \partial \psi/\partial x, \tag{29}$$

$$\partial^2 \psi/\partial x^2 = -F/\varepsilon \sum z_i c_i\,, \tag{30}$$

$$I_i = z_i F J_i, \tag{31}$$

$$D_i = RTu_i/F, \tag{32}$$

where

c_i represents concentration of the ith species (moles cm^{-3});
t represents time (sec);
J_i represents flux of the ith species (moles cm^{-2} sec^{-1});
x represents distance (cm);
D_i is a diffusion coefficient of the ith species (cm^2 sec^{-1});
u_i is a mobility coefficient (cm^2 sec^{-1} V^{-1});
z_i represents valence (dimensionless);
ψ represents electric potential (V);
F is the Faraday constant (C $moles^{-1}$);
ε is the dielectric constant (C cm^{-1} V^{-1});
R is the gas constant (J $moles^{-1}$ deg^{-1});
T is the Kelvin temperature (deg).

Consider a membrane of finite thickness δ, bounded on each side by electrolyte solutions of large volume. The solvent flux is represented by

$$J_0 = -D_0\, \partial c_0/\partial x. \tag{33}$$

In order to simplify, we will consider only the solvent and monovalent ions, that is, $z_i^2 = 1$ $(i \neq 0)$, In the stationary state, concentrations and potentials are time invariant, so that

$$dc_i/dx = -J_i/D_i - z_i c_i (F/RT)\, d\psi/dx, \tag{34}$$

$$d^2\psi/dx^2 = -F \sum z_i c_i/\varepsilon. \tag{35}$$

Let

$$\sum c_i = N, \qquad F \sum z_i c_i = \rho, \qquad d\psi/dx = -E, \tag{36}$$

and we have

$$dN/dx = -\sum J_i/D_i + E\rho/RT, \tag{37}$$

$$d\rho/dx = -\sum I_i/D_i + F^2 EN/RT, \tag{38}$$

$$dE/dx = \rho/\varepsilon. \tag{39}$$

Using Eq. (39) in (37) yields

$$RT\, dN/dx = -RT \sum J_i/D_i + \tfrac{1}{2}\varepsilon\, d(E^2)/dx. \tag{40}$$

This important and fundamental balance equation relates the Maxwell pressure gradient to the osmotic pressure gradient. Continuing,

$$RT(N_\delta - N_0) = -RT\delta \sum J_i/D_i + \tfrac{1}{2}\varepsilon(E_\delta{}^2 - E_0{}^2). \tag{41}$$

We must assume that the membrane is electrically neutral in the macroscopic sense, that is,

$$\int_0^\delta \rho\, dx = 0. \tag{42}$$

This means, from Eq. (39), that

$$E_\delta = E_0, \tag{43}$$

$$\sum J_i/D_i = -(N_\delta - N_0)/\delta, \tag{44}$$

$$RT\, dN/dx = RT(N_\delta - N_0)/\delta + \varepsilon E\, dE/dx. \tag{45}$$

There are several significant possibilities concerning Eq. (45).

1. It is clear from Eq. (45) that if there are no osmotic pressure gradients anywhere in the system,

$$N_\delta = N_0, \qquad dN/dx = 0 \tag{46}$$

and the electric field must be constant. We might call this a condition of total mechnical equilibrium.

2. If the total mass densities on the two sides of the membrane are identical, but internal osmotic pressure gradients exist,

$$RT\, dN/dx = \varepsilon E\, dE/dx. \tag{47}$$

This can be called a condition of *quasi-mechanical equilibrium*. The student should satisfy himself that both the first and second cases are consistent with the Gibbs–Duhem equation of themodynamics. Because many authors have ignored Eq. (45), a large number of contrived and fallacious analyses abound in the literature.

One important feature of Eq. (45) is that it shows unequivocally that the Maxwell pressure gradient acts as a perturbation on the mass–density gradient, a perturbation which is proportional to the dielectric constant.

The objective is to determine a relation between ionic current density and the potential difference across the barrier. In the case of total mechanical equilibrium, the electric field is constant. For any ionic species,

$$I_i = -RTu_iz_i\, dc_i/dx - Fu_iz_i^2c_i\, d\psi/dx; \tag{48}$$

$$d\psi/dx = -E \qquad \text{and} \qquad V = -\int_0^\delta E\, dx \qquad \text{so that} \qquad d\psi/dx = V/\delta,$$

$$\psi_\delta \equiv V, \qquad \psi_0 \equiv 0, \qquad (c_i)_\delta \equiv c_i'', \qquad (c_i)_0 = c_i',$$

and so forth. Then Eq. (48) can be written as

$$I_i = -RTu_iz_i \exp(-z_iF\psi/RT)\,(d/dx)[c_i \exp(z_iF\psi/RT)] \tag{49}$$

$$I_i = -RTu_iz_i \frac{[c_i'' \exp(z_iFV/RT) - c_i']}{\int_0^\delta \exp(z_iF\psi/RT)\, dx}, \tag{50}$$

$$I_i = -\frac{Fu_iV}{\delta} \frac{[c_i'' \exp(z_iFV/RT) - c_i']}{[\exp(z_iFV/RT) - 1]}. \tag{51}$$

This is the familiar constant field equation relating current to potential difference. The I–V relation is monotonic. In this type of system it is useful to consider the "barrier" as simply the region of nonuniform electrolyte concentration. In all other cases the barrier should be considered as the region of nonzero charge density.

The other two cases, quasi-mechanical equilibrium and the general case, are not as simple as the first and they have hardly been explored. One yields an elliptic integral and the other involves a Painlevé equation without a closed solution. A complete analysis would be a useful project for an enterprising student. The procedure is to determine $E(x)$ or $\psi(x)$ and use this to solve Eq. (48) or Eq. (50).

For the case of quasi-mechanical equilibrium, the equations are

$$\sum J_i/D_i = 0, \tag{52}$$

$$RT\, dN/dx = \varepsilon E\, dE/dx, \tag{47}$$

so that

$$RTN = RTN_0 + \tfrac{1}{2}\varepsilon(E^2 - E_0^2). \tag{53}$$

From Eqs. (47) and (38), multiplying by N and ρ, respectively, and subtracting,

$$F^2N\,dN/dx - \rho\,d\rho/dx = \sum (I_i/D_i)\rho, \tag{54}$$

$$\tfrac{1}{2}F^2\int_0^\delta d(N^2) - \tfrac{1}{2}\int_0^\delta d(\rho^2) = \sum (I_i/D_i)\int_0^\delta \rho\,dx = 0. \tag{55}$$

Therefore,

$$\rho_\delta^{\,2} = \rho_0^{\,2}. \tag{56}$$

Equation (56) is an important requirement for boundary conditions on the total charge density. From Eqs. (38), (39), and (53),

$$\varepsilon\frac{d^2E}{dx^2} = -\sum\frac{I_i}{D_i}\frac{F^2E}{RT}\frac{N_0 + \varepsilon(E - E_0^{\,2})}{2RT} \tag{57}$$

so that

$$\tfrac{1}{2}\varepsilon\left(\frac{dE}{dx}\right)^2 = \sum\frac{I_i}{D_iE} + \frac{F^2}{RT}\left[\left(\frac{\tfrac{1}{2}N_0 - \varepsilon E_0^{\,2}}{4RT}\right)E^2 + \frac{\varepsilon E^4}{8RT}\right]. \tag{58}$$

This yields an elliptic integral.

In the general case, from Eqs. (37)–(39),

$$N = -\sum (J_i/D_i)x + (\varepsilon/2RT)E^2 + a,$$

where, for $x = 0$,

$$N = N_0 \qquad \text{and} \qquad E = E_0. \tag{59}$$

Then,

$$N = -\sum\frac{J_i}{D_i}x + \frac{\varepsilon}{2RT}E^2 + N_0 - \frac{\varepsilon}{2RT}E_0^{\,2}, \tag{60}$$

$$\varepsilon\frac{d^2E}{dx^2} = -\sum\frac{I_i}{D_i} + \frac{F^2E}{RT}\left[-\sum\frac{J_i}{D_i}x + \frac{\varepsilon E^2}{2RT} + N_0 - \frac{\varepsilon E_0^{\,2}}{2RT}\right]. \tag{61}$$

This is a Painlevé equation (the conversion to canonical form is left to the student) and is of considerable mathematical interest, since the term in x produces a moving singularity.

In each case it should be useful to use a series approximation for $E(x)$ to determine $I_i(V)$. The student should keep in mind that the necessary condition imposed by Eq. (42) is a most critical restriction which is often overlooked.

A complete numerical solution for Eqs. (28)–(30) has been recently provided for both the stationary state and time-dependent phenomena [57, 58]. Although the physical problem involved semiconductor electrodiffusion, the situation is analogous and the interested student should consult this work. As might be expected, a negative conductance was nowhere observed.† Since semiconductor models usually involve "fixed charges" and since such systems are also of biological interest, we can at this point discuss the electrodiffusion equations in terms of a fixed-charge model. In order to simplify the analysis, we will consider a barrier which contains a uniform distribution of fixed negative charges.

In the stationary state and for monovalent ions, we have

$$dN/dx = -\sum J_i/D_i + Ep/RT, \tag{62}$$

$$dp/dx = -\sum I_i/D_i + F^2EN/RT, \tag{63}$$

$$dE/dx = (p + w)/\varepsilon, \tag{64}$$

$$\int_0^\delta (p + w)dx = 0, \tag{65}$$

where w (C cm^{-3}) is the fixed-charge concentration. Combining Eqs. (64) and (62), we find that

$$\sum J_i/D_i = wV/RT\delta - (N_\delta - N_0)/\delta, \tag{66}$$

$$dN/dx = -wV/RT\delta + (N_\delta - N_0)/\delta + \varepsilon E/RT\, dE/dx - wE/RT. \tag{67}$$

For the case of total mechnical equilibrium,

$$dN/dx = 0, \qquad N_\delta = N_0$$

$$E\, dE/dx = w/\varepsilon(E - V/\delta). \tag{68}$$

An explicit function $E(x)$ is more easily obtained from Eq. (63),

$$dp/dx = \varepsilon d^2E/dx^2 = -\sum I_i/D_i + (F^2N_0/RT)E, \tag{69}$$

which gives

$$E = \frac{RT}{F^2N_0}\sum\frac{I_i}{D_i} + \left[E_0 - \frac{RT}{F^2N_0}\sum\frac{I_i}{D_i}\right] \times\left[e^{-x/\lambda} + \frac{(e^{x/\lambda} - e^{-x/\lambda})}{(e^{\delta/\lambda} - e^{-\delta/\lambda})}(1 - e^{-\delta/\lambda})\right], \tag{70a}$$

$$\lambda \equiv (RT\varepsilon/F^2N_0)^{1/2}, \tag{70b}$$

where λ is a characteristic length. Equation (70) can then be used to solve Eq. (35) in order to obtain $I_i(V)$. The more general cases, of course, are not this elementary, and as far as I am aware they have not been treated analytically.

VII. Physical Systems with Negative Conductance

It should be apparent that an exact statement of the physical model to be used for mathematical analysis is not simply accomplished. If we postulate a barrier which is simply a phase between two boundaries, without any

†The reason for this expectation is discussed later on.

unique properties, we should not expect any result more unique than that obtained for a simple electrolyte solution. Such solutions never show the nonlinearities produced by biological membranes. Stationary concentration gradients are sufficient to produce rectification, but not a negative conductance. A phase exhibiting merely a space variation of mobility and diffusion coefficients is also not satisfactory: a system of varying fixed resistances in series cannot show a negative conductance. The problem is not that the basic electrodiffusion equations are inapplicable, but that in the past they have been applied to physical models without unique properties. There are various real physical systems which show negative conductance behavior, and we will review them briefly here.

A. Tunnel Effects

Tunneling phenomena are usually considered with respect to specific types of semiconductor regimes (for example, Esaki tunnel diode) but the quantum theoretical principle which is involved is not restricted to particular kinds of systems. The essential idea is quite simple: Whether a particle is considered as a mathematical entity defined by the Schroedinger wave equation or mentally imaged as a wave-like disturbance of some sort, in both cases one concludes that a potential barrier of finite height becomes transparent as its thickness becomes small compared with the particle wavelength. Particle currents *through* the energy barrier (that is, tunneling) as well as the ordinary ones over it are therefore possible.

If the height and thickness of the energy barrier and the particle wavelength are known, it is possible to estimate the tunneling probability and from that the tunnel current as a function of applied emf. This is then a component of the total particle I–V curve, since the total particle current is the sum of the tunnel current and the ordinary current. In certain physical systems the tunnel current is not only large, but has a well-defined maximum, so that a negative conductance appears.

Although the large masses of ions should preclude any significant probability for *ion* tunneling in biological membranes [37], there is really not enough information about the energy barrier to completely discount the possibility. Electron tunneling, however, may easily exist in these membranes and may be of some importance for biochemical reactions associated with cell surfaces.

B. Oxide Films

An apparently unexplained electronic dc negative conductance has been observed in metal–oxide–metal sandwiches prepared from evaporated metal films 150–1000 Å thick [125]. The amorphous oxide film is considered to be too thick for a tunnel effect. Several properties are of some interest:

1. The instantaneous I–V curve does not show a negative conductance.
2. The conductance appears to depend on the potential difference and not the electric field, since variation of film thickness produces no important change in behavior.
3. Changes in temperature over a wide range alter the absolute magnitude of current, but not the shape of the I–V curve.

C. Passivated Iron

The negative conductance (and consequent electrical oscillations) shown by passivated iron has been of interest to biologists for many years [134] and also involves an oxide layer, in this case between a metal and an electrolyte solution. The system has been treated in great detail [138]. It is a rather remarkable fact that although the conduction process in the metal–oxide–metal sandwich is believed to be electronic; conduction in the metal–oxide–electrolyte system is apparently now considered to be ionic. Ord and Bartlett [138] have recently suggested the following for passive iron:

1. An oxide layer exists on iron when it is passive.
2. An electric field drives ions through this layer.
3. The resistance to current flow is nonohmic.
4. In the steady state the thickness of the oxide layer varies linearly with the passivating potential from about 10 to 70 Å.
5. When the system is in a nonsteady state, there is a significant space charge within the oxide layer.

D. Teorell Oscillator

Another important and well-known system which shows a negative conductance is the Teorell hydraulic oscillator [119, 120, 144, 145]. As far as I am aware, this has been presented in its most concise form by Yamamoto [146] for a single glass capillary. The Teorell oscillator demonstrates that a rather simple coupling of electrodiffusion and mechanical hydrostatic pressure can produce a well-defined negative conductance. This system should serve as an archetype for study of nonlinear electrokinetic cross-phenomena and merits much more general attention than it has received, particularly since linear cross phenomena are probably of restricted relevance for biological processes.

E. Black Lipid Membranes and Organic Films

Under certain conditions, black lipid membranes separating two aqueous electrolyte solutions may exhibit a negative conductance [137]. In all cases known to me, some organic material must be added to the system in order to produce the effect. However, according to the recent report of Stafeev *et al.* [143], films of only cholesterol or cetyl alcohol between mercury electrodes show a pronounced negative conductance. The authors suggest that these films are bimolecular and that a voltage-dependent "melting" process may be involved.

VIII. A Steady-State Model Involving Fixed Charges

If it is apparent that a useful theoretical model for the negative conductance of biological membranes will probably have a unique component; it should also be apparent that there are many possibilities for such models. Their evaluation must depend on the degree to which they predict new experimental observations. This criterion is often ignored or deemphasized, but it remains (in biology as in physics) a mandatory requirement of any useful theory. The elementary theoretical model presented in this section may serve as a prototype for the student.

Assume a barrier containing negative charges, both fixed and mobile, in a regime where the electric field is (in the steady state) everywhere constant. From Eq. (48), for a monovalent cation species, we have

$$I_i = \frac{Fu_i(V_i^0 - V)}{\left(\int_{\prime}^{\prime\prime} dx/c_i\right)}, \tag{71}$$

where V_i^0 is the potential difference at which the species current vanishes and the primes refer to the barrier boundaries. We now ask the following question: What relations between the densities of positive and negative ions produces a result which explains a variety of observations? The following postulate emerges:

$$c_i/c_i^* = b/w, \tag{72}$$

$$c_i = w - c_i^*, \tag{73}$$

where

c_i^* represents the concentration of bound cations (moles cm^{-3}) in any region;
c_i represents the concentration of mobile cations (moles cm^{-3});
b represents the concentration of mobile anions;
w represents the concentration of fixed anions.

The mobile anions are assumed to be confined to the barrier. Since they carry little or no current in the steady state, they presumably are distributed according to

$$b = b^0 \exp(\bar{z}FVx/RT\delta), \tag{74}$$

where b^0 (mole cm^3) is a boundary value, $\bar{z}$ (dimensionless) a valence. Combining Eqs. (72)–(74) yields

$$(c_i)^{-1} = (b^0)^{-1} \exp(-zFVx/RT\delta) + [w(x)]^{-1} \tag{75}$$

and from Eq. (71) we obtain

$$I = \frac{Fu(V^0 - V)}{-(RT\delta/\bar{z}FVb^0)[\exp(-\bar{z}FV/RT) - 1] + \int_{\prime}^{\prime\prime} dx/w(x)}. \tag{76}$$

The integral in the denominator depends on the fixed anion distribution and is not a function of V. When V is large and positive, a limiting resistance appears and is given by

$$R_{\text{lim}} = \int_{\prime}^{\prime\prime} dx/Fu\, w(x). \tag{77}$$

For a step change from V^0, the instantaneous resistance should be

$$R_\infty = R_{\text{lim}} + (RT\delta/\bar{z}F^2V^0ub^0)[1 - \exp(-\bar{z}FV^0/RT)] \tag{78}$$

and when $V^0 = 0$,

$$R_\infty = R_{\text{lim}} + \delta/Fub^0. \tag{79}$$

Equation (76) is then

$$I = \frac{(V^0 - V)}{(-RT/\bar{z}FV)[R_\infty{}^0 - R_{\text{lim}}][\exp(-\bar{z}FV/RT) - 1] + R_{\text{lim}}}. \tag{80}$$

Typical experimental values of $R_\infty{}^0$ and R_{lim} for the squid axon are 30 Ω cm^2 and 25 Ω cm^2, respectively. The computation of Eq. (80) has been given elsewhere [147]: a marked negative conductance region appears, and for $V^0 = 0$ the results are in good quantitative agreement with the experimental observations of Moore [29].

The dc or chord resistance R_0 is given by

$$R_0 = -(RT/\bar{z}FV)[R_\infty{}^0 - R_{\text{lim}}][\exp(-\bar{z}FV/RT) - 1] + R_{\text{lim}}. \tag{81}$$

To examine Eq. (81), the most convenient data are those given by Hodgkin and Huxley [21] for their axon #17. For this axon,

$$V^0 = -55 \text{ mV}, \qquad I^0 = -1.2 \text{ mA cm}^{-2},$$

$$R_\infty{}^0 = 45\ \Omega \text{ cm}^2, \qquad R_{\text{lim}} = 41\ \Omega \text{ cm}^2.$$

These values have been used to compare Eq. (81) with axon #17 [147]. The agreement between theoretical predictions and experimental observations is apparently good.

In Eqs. (74) and (77) the mobile anions were assumed to obey a Boltzmann distribution without restrictive boundary conditions. Suppose, however, the *average* number of mobile anions remains fixed at some value B,

$$\int_{\prime}^{\prime\prime} b\,dx = B\delta; \tag{82}$$

this imposes a condition on b^0, so that using Eq. (74) we obtain

$$b^0 = \frac{\bar{z}FVB}{RT[\exp(\bar{z}FV/RT) - 1]}, \tag{83}$$

$$b = \frac{\bar{z}FVB\exp(\bar{z}FVx/RT\delta)}{RT[\exp(\bar{z}FV/RT) - 1]}, \tag{84}$$

and the current the becomes

$$I = \frac{(V^0 - V)}{\left(\frac{-\delta}{FuB}\right)\left(\frac{RT}{\bar{z}FV}\right)^2[\exp(\bar{z}FV/RT) - 1][\exp(-\bar{z}FV/RT) - 1] + \int_{\prime}^{\prime\prime}\frac{dx}{Fuw(x)}}. \tag{85}$$

In order to compute this equation, it is useful to assume that w is constant and that $B = w$. We then have

$$A \equiv I\delta/Fuw$$

$$= \frac{(V^0 - V)}{-(RT/\bar{z}FV)^2[\exp(\bar{z}FV/RT) - 1][\exp(-\bar{z}FV/RT) - 1] + 1}. \tag{86}$$

A symmetrical negative conductance appears, not unlike that which has been observed experimentally in black lipid membranes [137].

The most important theoretical prediction by the elementary model which has been outlined concerns the instantaneous resistance, R_∞. According to Eq. (78), the instantaneous resistance is related in a specific manner to the limiting resistance R_{lim} and to V^0. This relation can be tested quantitatively in the laboratory.

IX. The Nonstationary State

The fundamental equations appropriate for the nonstationary state are the continuity equation, the Poisson equation, and the current equation:

$$(\partial/\partial x)\sum I_i(x, t) = -(\partial/\partial t)\sum Fz_i c_i(x, t), \tag{87}$$

$$(\partial/\partial x)E(x, t) = (F/\varepsilon)\sum zc_i(x, t), \tag{88}$$

$$I_i = -RTu_i(\partial/\partial x)c_i(x, t) + Fu_i z_i^2 c_i(x, t)E(x, t). \tag{89}$$

If Eq. (88) is differentiated with respect to time and used with Eq. (87),

$$(\partial/\partial x)\sum I_i = -\varepsilon(\partial/\partial t)(\partial E/\partial x) = -\varepsilon(\partial/\partial x)(\partial E/\partial t), \tag{90}$$

and an integration yields the Maxwell current equation

$$\underset{\text{total}}{I(t)} = \sum I_i(x, t) + \varepsilon(\partial/\partial t)E(x, t). \quad (91)$$

The total current density through the barrier is thus the sum of a displacement current, $\varepsilon\,\partial E/\partial t$, and a convection current, $\sum I_i(x, t)$. Although an applied potential difference across the barrier may be established with a time constant (determined by the external series resistance) of microseconds, the interior electric field reaches its final profile with a time constant of $\varepsilon/\bar{\sigma}$, where $\bar{\sigma}$ is a mean specific conductivity. In the squid axon membrane this latter time constant should be of the order of milliseconds. Equation (91) provides an immediate distinction between the charging process associated with the establishment of the interior electric field and the redistribution process associated with the convection current, a distinction discussed by Planck [95] and more recently by Cole [150]. An important point revealed by Eq. (90) which should be emphasized is that when the charging process is completed, that is, when $\partial E/\partial t = 0$, the following must be true:

$$\partial \sum I_i/\partial x = 0 \qquad (\partial E/\partial t = 0). \quad (92)$$

Therefore, if a redistribution process occurs following the charging process, the total convection current will be time dependent only. The individual ionic currents, however, may depend on both time and position, namely,

$$\partial \sum I_i/\partial x = 0 \qquad (\partial I_i/\partial x \neq 0). \quad (93)$$

Equation (93) may be said to describe a quasi-steady state, while $\partial I_i/\partial x = 0$ would imply a true steady state (assuming an already terminated charging process).

There has been very little analytical work on time-dependent electrodiffusion phenomena. The complete numerical analysis by De Mari [57, 58] is the only treatment known to me which does not involve severe restrictive assumptions. The unusual time-dependent behavior of nerve and muscle membranes poses an important challenge to the theoretician, and here again the question of uniqueness should be considered. It makes little sense to expect the equations for an ordinary salt solution to produce transient negative conductance behavior. Such behavior is never observed empirically. Efforts should be directed only to the formulation of physical models with unique properties and final theoretical statements which can be tested in the laboratory. Several models have been proposed [151–154, 156–159], but much work remains to be done.

Bibliography

The following categorized bibliography is both incomplete and highly selective. It contains items which I have found useful and provocative and which I hope will assist the student. Some reviews and monographs which I thought would be listed elsewhere in these volumes have been omitted. Reviews, monographs, and symposia are marked with an asterisk (*), and particularly important items are marked with a dagger (†).

1. The Squid Axon and the Hodgkin–Huxley Equations

*1. Adelman, W. J. (ed.) [1965]. Physical and mathematical approaches to the electrical behavior of excitable membranes, *J. Cell. Comp. Physiol. Suppl. 2* **66**.

2. Agin, D., and Rojas, E. [1963]. A third-order system for the squid axon membrane, *Proc. 16th Conf. Eng. Med. Biol.* p. 4.

†3. Cole, K. S. [1968]. "Membranes, Ions and Impulses." Univ. of California Press, Berkeley, California.

4. Cole, K. S., Antosiewicz, H. A., and Rabinowitz, P. [1955]. Automatic computation of nerve excitation, *J. Soc. Indust. Appl. Math.* **3**, 153.

5. Cooley, J. W., and Dodge, F. A., Jr. [1966]. Digital computer solutions for excitation and propagation of the nerve impulse, *Biophys. J.* **6**, 583.

6. Derksen, H. E. [1965]. Axon membrane voltage fluctuations, *Acta Physiol. Pharmacol. Neerl.* **13**, 373.

7. FitzHugh, R. [1955]. Mathematical models of threshold phenomena in the nerve, membrane, *Bull. Math. Biophys.* **17**, 257.

8. FitzHugh, R. [1960]. Thresholds and plateaus in the Hodgkin–Huxley nerve equations, *J. Gen. Physiol.* **43**, 867.

9. FitzHugh, R. [1961]. Impulses and physiological states in theoretical models of nerve membrane, *Biophys. J.* **1**, 445.

10. FitzHugh, R. [1966]. Theoretical effect of temperature on threshold in the Hodgkin–Huxley model, *J. Gen. Physiol.* **49**, 989.

†11. FitzHugh, R. [1968]. Mathematical models of excitation and propagation in nerve, *in* "Bioelectronics" (H. Schwan, ed.). McGraw-Hill, New York.

12. George, E. P., and Johnson, E. A. [1961]. Solutions of the Hodgkin–Huxley equations for squid axon treated with TEA and in potassium-rich media, *Aust. J. Exp. Biol. Med. Sci.* **39**, 275.

*13. Harmon, L. D., and Lewis, E. R. [1966]. Neural Modeling, *Physiol. Rev.* **46**, 513.

14. Hearon, J. Z. [1964]. Application of results from linear kinetics to the Hodgkin–Huxley equations, *Biophys. J.* **4**, 69.

*15. Hodgkin, A. L. [1951]. The ionic basis of electrical activity in nerve and muscle, *Biol. Rev.* **26**, 339.

*16. Hodgkin, A. L. [1958]. Ionic movements and electrical activity in giant nerve fibers, *Proc. Roy. Soc.* (*London*) **B 148**, 1.

*17. Hodgkin, A. L. [1964]. "The Conduction of the Nervous Impulse." Thomas, Springfield, Illinois.

18. Hodgkin, A. L., and Huxley, A. F. [1952]. Currents carried by sodium and potassium ions through the membrane of the giant axon of *Loligo*, *J. Physiol.* (*London*) **116**, 449.

19. Hodgkin, A. L., and Huxley, A. F. [1952]. The components of membrane conductance in the giant axon of *Loligo*, *J. Physiol.* (*London*) **116**, 473.

20. Hodgkin, A. L., and Huxley, A. F. [1952]. The dual effect of membrane potential on sodium conductance in the giant axon of *Loligo*, *J. Physiol.* (*London*) **116**, 497.

†21. Hodgkin, A. L., and Huxley, A. F. [1952]. A quantitative description of membrane current and its application to conduction and excitation in nerve, *J. Physiol.* (*London*) **117**, 500.

22. Hodgkin, A. L., and Katz, B. [1949]. The effect of sodium ions on the electrical activity of the giant axon of the squid, *J. Physiol.* (*London*) **108**, 37.
23. Hodgkin, A. L., Huxley, A. F., and Katz, B. [1952]. Measurement of current-voltage relations in the membrane of the giant axon of *Loligo*, *J. Physiol.* (*London*) **116**, 424.
24. Hoyt, R. C. [1963]. The squid axon. Mathematical models, *Biophys. J.* **3**, 399.
†25. Huxley, A. F. [1959]. Ionic movements during nerve activity, *Ann. N.Y. Acad. Sci.* **81**, 221.
26. Lewis, E. R. [1965]. Neuroelectric potentials derived from an extended version of the Hodgkin–Huxley model, *J. Theor. Biol.* **10**, 125.
*27. Loewenstein, W. R. (ed.) [1966]. Biological Membranes: Recent progress. *Ann. N.Y. Acad. Sci.* **137**, 403–1048.
28. Marmont, G. [1949]. Studies on the axon membrane. I. A new method, *J. Cell. Comp. Physiol.* **34**, 351.
†29. Moore, J. W. [1959]. Excitation of the squid axon in isosmotic potassium chloride, *Nature* **183**, 265.
*30. Mullins, L. J. (ed.) [1965]. *Newer Properties of Perfused Squid Axons*, *J. Gen. Physiol. Suppl. 5* **48**.
†31. Noble, D. [1966]. Application of the Hodgkin–Huxley equations to excitable tissues, *Physiol. Rev.* **46**, 1.
†32. Plonsey, R. [1969]. "Biolectric Phenomena." McGraw-Hill, New York.
33. Segal, J. R. [1958]. An anodal threshold phenomenon in the squid giant axon, *Nature* **182**, 1370.
†34. Tasaki, I. [1968]. "Nerve Excitation." Thomas, Springfield, Illinois.
†35. Taylor, R. E. [1963]. Cable theory, *in* "Physical Techniques in Biological Research" (W. L. Nastuk, ed.). Academic Press, New York.
36. Verveen, A. A., and Derksen, H. E. [1968]. Fluctuation phenomena in nerve membrane, *Proc. IEEE* **56**, 906.

2. *Electrodiffusion*

37. Agin, D. [1967]. Electroneutrality and electrodiffusion in the squid axon, *Proc. Nat. Acad. Sci.* **57**, 1232.
38. Bakshi, P. M., and Gross, E. P. [1968]. Kinetic theory of nonlinear electrical conductivity, *Ann. Phys.* **49**, 513.
39. Bass, L. [1964]. Electrical structures of interfaces in steady electrolysis, *Trans. Faraday Soc.* **60**, 1656.
40. Bass, L. [1964]. Potential of liquid junctions, *Trans. Faraday Soc.* **60**, 1914.
41. Buck, R. P. [1968]. Transient electrical behavior of glass membranes. I. Theory of DC pulse processes, *J. Electroanal. Chem.* **18**, 363.
42. Buck, R. P. [1968]. Transient electrical behavior of glass membranes. II. The impedance, *J. Electroanal. Chem.* **18**, 381.
43. Buck, R. P. [1968]. Transient electrical behavior of glass membranes. III. Experimental, *J. Electroanal. Chem.* **18**, 387.
*44. *Ber. Bunsengesel. Phys. Chem.* [1967]. Stofftransport durch Membranen in Chemic und Biologie, **71**, 749–932.
45. Billig, E., and Landsberg, P. T. [1950]. Characteristics of compound barrier layer rectifiers, *Proc. Phys. Soc.* (*London*) **63A**, 101.
*46. Boltaks, B. I. [1963]. "Diffusion in Semiconductors." Academic Press, New York.
*47. Bird, R. B., Stewart, W. E., and Lightfoot, E. N. [1960]. "Transport Phenomena." Wiley, New York.

48. Christopher, H. A., and Shipman, C. W. [1968]. Poisson's equation as a condition of equilibrium in electrochemical systems, *J. Electrochem. Soc.* **115**, 501.
49. Ciani, S. [1965]. A rate theory analysis of steady diffusion in a fixed charge membrane, *Biophysik* **2**, 368.
50. Ciani, S. and Gliozzi, A. [1967]. An irreversible themodynamics treatment of electrodiffusion in uniform membranes, *Biophysik* **3**, 281.
51. Cohen, J., and Cooley, J. W. [1965]. The numerial solution of the time-dependent Nernst–Planck equations, *Biophys. J.* **5**, 145.
52. Conti, F., and Eisenman, G. [1965]. Steady-state properties of an ion exchange membrane with fixed sites, *Biophys. J.* **5**, 511.
53. Conti, F., and Eisenman, G. [1966]. Steady-state properties of an ion-exchange membrane with mobile sites, *Biophys. J.* **6**, 227.
54. Coster, H. G.. L., George, E. P., and Simons, R. [1969]. The electrical characteristics of fixed charge membranes: solutions of the field equations, *Biophys. J.* **9**, 666.
55. Coulson, J. M. [1966]. The effects of electric fields on transport phenomena, *Trans. Inst. Chem. Eng.* **44**, T388.
56. Croitoru, Z. [1965]. Space charges in dielectrics, *Progr. Dielect.* **6**, 104.
57. De Mari, A. [1968]. An accurate numerical steady-state one-dimensional solution of the p–n junction, *Solid State Electron.* **11**, 33.
58. De Mari, A. [1968]. An accurate numerical one-dimensional solution of the p–n junction under arbitrary transient conditions, *Solid State Electron.* **11**, 1021.
59. Dignam, M. J. [1968]. Ion transport in solids under conditions which include large electric fields, *J. Phys. Chem. Solids* **29**, 249.
60. Dignam, M. J., and Gibbs, D. B. [1969]. The electric field dependence of the net activation energy for migration of ionic species across an activation barrier, *J. Phys. Chem. Solids* **30**, 375.
*61. Eisenman, G., Sandblom, J. P., and Walker, J. L. [1967]. Membrane structure and ion permeation, *Science* **155**, 965.
62. Finkelstein, A., and Mauro, A. [1963]. Equivalent circuits as related to ionic systems, *Biophys. J.* **3**, 233.
63. Fromhold, A. T., and Cook, E. L. [1967]. Diffusion currents in large electric fields for discrete lattices, *J. Appl. Phys.* **38,** 1546.
64. Garton, C. G. [1946]. The distribution of relaxation times in dielectrics, *Trans. Faraday Soc.* **42A**, 56.
65. Gavis, J. [1967]. Nonlinear equation for transport of electric charge in low dielectric constant fluids, *Chem. Eng. Sci.* **22**, 359.
66. George, E. P., and Simons, R. [1966]. Solutions of the field equations in the fixed-charge model of cell membranes, *Aust. J. Biol. Sci.* **19**, 459.
*67. Girifalco, L. A. [1964]. "Atomic Migration in Crystals." Blaisdell, New York.
*68. Gurney, R. W. [1936]. "Ions in Solution." Cambridge Univ. Press, London and New York.
†69. Harned, H. S., and Owen, B. B. [1959]. "The Physical Chemistry of Electrolytic Solutions." Reinhold, New York.
70. Harris, J. D. [1956]. On the diffusion of ions in membranes, *Bull. Math. Biophys.* **18**, 255.
†71. Helfferich, F. New York. [1962]. "Ion Exchange." McGraw-Hill, New York.
*72. Henisch, H. K. [1957]. "Rectifying Semiconducting Contacts." Oxford Univ. Press, London and New York.
†73. Kirkwood, J. G. [1954]. Transport of ions through biological membranes from the standpoint of irreversible themodynamics. *in* "Ion Transport Across Membranes" (H. R. Clanke, ed.). Academic Press, New York.

74. Laity, R. W. [1963]. Diffusion of ions in an electric field, *J. Phys. Chem.* **67**, 671.
*75. Lakshminarayanaiah, N. [1965]. Transport phenomena in artificial membranes, *Chem. Rev.* **65**, 491.
†76. Lakshminarayanaiah, N. [1969]. "Transport Phenomena in Membranes." Academic Press, New York.
77. Landsberg, P. T. [1951]. The theory of direct-current characteristics of rectifiers, *Proc. Roy. Soc.* (*London*) **206A**, 463.
78. Landsberg, P. T. [1952]. On the diffusion theory of rectification, *Proc. Roy. Soc.* (*London*) **213A**, 226.
*79. Levich, V. G. [1962]. "Physicochemical Hydrodynamics." Prentice-Hall, Englewood Cliffs, New Jersey.
80. MacDonald, J. R. [1955]. Note on theories of time-varying space-charge polarization, *J. Chem. Phys.* **23**, 2308.
81. MacGillivray, A. D. [1968]. Nernst–Planck equations and the electroneutrality and Donnan equilibrium assumptions, *J. Chem. Phys.* **48**, 2903.
*82. Manning, J. R. [1968]. "Diffusion Kinetics for Atoms in Crystals." Van Nostrand–Reinhold, New York.
*83. Manning, M. F., and Bell, M. E. [1940]. Electrical conduction and related phenomena in solid dielectrics, *Rev. Mod. Phys.* **12**, 215.
†84. Mazo, R. M. [1967]. "Statistical Mechanical Theories of Transport Processes." Pergamon Press, Oxford.
85. Maserjian, J. [1967]. Theory of conduction through thin insulating films with ionic space charge, *J. Phys. Chem. Solids* **28**, 1957.
*86. Mills, R. [1965]. Diffusion in electrolytes, *J. Electroanal. Chem.* **9**, 57.
87. Mott, N. F. [1939]. The theory of crystal rectifiers, *Proc. Roy. Soc.* (*London*) **171A**, 27.
*88. Newman, J. [1967]. Transport processes in electrolytic solutions, *Advan. Electrochem. Electrochem. Eng.* **5**, 87.
89. Oberlander, S., and Sikorski, S. On the mathematical description of electrical processes in semiconductors. I. Phase space analysis in subspaces of dimension lower than three, *Phys. Status Solidi* **25**, 345.
90. Oberlander, S. [1969]. On the mathematical description of electrical processes in semiconductors. II. Phase space anlaysis in subspaces of dimension three, *Phys. Status Solidi* **33**, 277.
91. Oshida, I. [1960]. Theory of electrochemical diodes, *J. Phys. Soc. Japan* **15**, 2288.
92. Papadakis, A. C. [1967]. Theory of transient space-charge perturbed currents in insulators, *J. Phys. Chem. Solids.* **28**, 641.
93. Parsegian, A. [1969]. Energy of an ion crossing a low dielectric membrane: solutions to four relevant electrostatic problems, *Nature* **221**, 844.
*94. Pickard, W. F. [1965]. Electric force effects in dielectric liquids, *Progr. Dielectrics* **6**, 1.
95. Planck, M. [1890]. Uber die Erregung von Elektricitat und Warme in Elektrolyten, *Ann. Phys. Chem. Neue Folge* **39**, 161.
96. Pohl, H. A. [1958]. Some effects of non-uniform fields on dielectrics, *J. Appl. Phys.* **29**, 1182.
†97. Prim, R. C., III [1951]. Some results concerning the partial differential equations describing the flow of holes and electron in semiconductors, *Bell Syst. Tech. J.* **30**, 1175.
98. Roberts, G. G. [1967]. Equilibrium space-charge distributions in semiconductors, *Brit. J. Appl. Phys.* **18**, 749.
99. Scales, J. L., and Ward, A. L. [1968] Effects of space charge on mobility, diffusion and recombination of minority carriers, *J. Appl. Phys.* **39**, 1692.

*100. Schlogl, R. [1964]. "Stofftransport durch Membranen." Steinkopf, Darmstadt.
101. Schlogl, R. [1969]. Nonlinear transport behavior in very thin membranes, *Quart. Rev. Biophys.* **2**, 305.
†102. Shockley, W. [1949]. The theory of p–n junctions in semiconductors and p–n junction transistors, *Bell Syst. Tech. J.* **28**, 435.
*103. Teorell, T. [1953]. Transport processes and electrical phenomena in ionic membranes, *Progr. Biophys. Biophys. Chem.* **3**, 305.
*104. Tredgold, R. H., "Space Charge Conduction in Solids." Elsevier, New York.
105. Vaidhyanathan, V. S. [1965]. Statistical mechanical theory of electrolyte diffusion inside a charged membrane, *J. Theoret. Biol.* **8**, 344.
106. Vaidhyanathan, V. S. [1965]. Statistical theory of transport of electrolytes across a charged membrane, *J. Theoret. Biol.* **9**, 478.
107. Vaidhyanathan, V. S. [1966]. Some comments on steady state transport of ions across a charged biological membrane, *J. Theoret. Biol.* **10**, 159.
*108. Van Roosbroeck, W. [1950]. Theory of the flow of electrons and holes in germanium and other semiconductors, *Bell Syst. Tech. J.* **29**, 560.
*109. Vetter, K. J. [1967]. "Electrochemical Kinetics." Academic Press, New York.
110. Wright, G. T. [1961]. Mechanisms of space-charge limited currents in solids, *Solid State Electron.* **2**, 165.

3. Negative Conductance

*111. Agin, D. [1972]. Negative conductance and electrodiffusion in excitable membrane systems. *In* "Advances in Membranes" (G. Eisenman, ed.), Vol. I. Dekker, New York.
112. Becker, J. A., Green, C. B., and Pearson, G. L. [1947]. Properties and uses of thermistors—thermally sensitive resistors, *Bell Syst. Tech. J.* **26**, 170.
113. Boer, K. W., and Quinn, P. L. [1966]. Inhomogeneous field distribution in homogeneous semiconductors having an N-shaped negative differential conductivity, *Phys. Status Solidi* **17**, 307.
114. Burgess, R. E. [1953]. The influence of mobility variation in high fields on the diffusion theory of rectifier barriers, *Proc. Phys. Soc. (London)* **66B**, 430.
*115. Caplan, S. R., and Mikulecky, D. C. [1966]. Transport processes in membranes, *in* "Ion Exchange" (J. A. Marinsky, ed.). Dekker, New York.
†116. Counningham, W. J. [1958]. "Introduction to Nonlinear Analysis." McGraw-Hill, New York.
117. Das, P., and Marom, E. [1966]. Possibility of Gunn effect in inhomogeneous semiconductors, *Phys. Lett.* **20**, 444.
118. Degn, H. [1968]. Theory of electrochemical oscillations, *Trans. Faraday Soc.* **64**, 1348.
119. Drouin, H. [1969]. Experimente mit dem Teorellschen Membran-oszillator, *Ber. Bunsenges. Phys. Chem.* **73**, 223.
120. Drouin, H. [1969]. Uber nichtlineare Elektroosmose, *Ber. Bunsenges. Phys. Chem.* **73**, 590.
121. Enderlein, R. [1968]. On a certain type of instability of electrical domains, *Phys. Status Solidi* **28**, 519.
122. Esaki, L. [1958]. New phenomenon in narrow germanium p–n junctions, *Phys. Rev.* **109**, 603.
123. Franck, U. F. [1956]. Models for biological excitation processes, *Progr. Biophys. Biophys. Chem.* **6**, 171.
124. Heinle, W. [1968]. Principles of a phenomenological theory of Gunn-effect domain dynamics, *Solid State Electron.* **11**, 583.

125. Hickmott, T. W. [1962]. Low-frequency negative resistance in thin anodic oxide films, *J. Appl. Phys.* **33**, 2669.
126. Hines, M. E. [1960]. High-frequency negative resistance circuit principles for Esaki diode applications, *Bell Syst. Tech. J.* **39**, 477.
127. Iida, M., Nojima, S., and Kurosu, T. [1968]. A simplified theory of negative resistance diodes, *Japan. J. Appl. Phys.* **7**, 1078.
128. Knight, B. W., and Peterson, G. A. [1966]. Nonlinear analysis of the Gunn effect, *Phys. Rev.* **147**, 617.
129. Knight, B. W., and Peterson, G. A. [1967]. Theory of the Gunn effect, *Phys. Rev.* **155**, 393.
130. Kobatake, Y., and Fujita, H. [1964]. Flows through charged membranes. I. Flip-flop current vs. voltage relation. II. Oscillation phenomena, *J. Chem. Phys.* **40**,2 212.
131. Krishnamurthy, B. S., and Sinha, K. P. [1968]. Theory of field dependent mobility in semiconductors, *Ind. J. Pure Appl. Phys.* **6**, 401.
132. Kroll, K. E. [1967]. Slowly propagating high field domains in semiconductors with negative differential conductivity, *Phys. Status Solidi* **24**, 707.
133. Lampert, M. A. [1969]. Stable space-charge layers associated with bulk, negative differential conductivity, *J. Appl. Phys.* **40**, 335.
134. Lillie, R. S. [1936]. The passive iron wire model of protoplasmic and nervous transmission and its physiological analogies, *Biol. Rev.* **11**, 181.
135. McCumber, D. E., and Chynoweth, A. G. [1966]. Theory of negative conductance amplification and of Gunn instabilities in "two-valley" semiconductors, *IEEE Trans. Electron Devices* **ED 13**, 4.
136. Mizuno, H. [1966]. Some aspects of tunneling through junctions, *Japan. J. Appl. Phys.* **5**, 1008.
137. Mueller, P., and Rudin, D. O. [1963]. Induced excitability in reconstituted cell membrane structure, *J. Theoret. Biol.* **4**, 268.
138. Ord, J. L., and Bartlett, J. H. [1965]. Electrical behavior of passive iron, *J. Electrochem. Soc.* **112**, 160.
139. Ridley, B. K. [1963]. Specific negative resistance in solids, *Proc. Phys. Soc.* **82**, 954.
140. Simmons, J. G. Generalized formula for the electric tunnel effect between similar electrodes separated by a thin insulating film, *J. Appl. Phys.* **34**, 1793.
141. Simmons, J. G., and Verderber, R. R. [1967]. New conduction and reversible memory phenomena in thin insulating films, *Proc. Roy. Soc.* (*London*) **A301**, 77.
142. Spyropoulos, C. S. [1965]. The role of temperature, potassium, and divalent ions in the current–voltage characteristics of nerve membranes, *J. Gen. Physiol.* **48** (5, pt. 2), 49.
143. Stafeev, V. I., *et al.* [1968]. Negative resistance of very thin organic films between metal electrodes, *Sov. Phys. Semicond.* **2**, 642.
144. Teorell, T. [1959]. Electrokinetic membrane processes in relation to properties of excitable tissues I. Experiments, *J. Gen. Physiol.* **42**, 831.
145. Teorell, T. [1959]. Electrokinetic membrane processes in relation to properties of excitable tissues II. Theoretical, *J. Gen. Physiol.* **42**, 847.
146. Yamamoto, K. [1965]. Negative resistance and electrokinetic cross-phenomenon in ionic solution, *J. Phys. Soc. Soc. Japan* **20**, 1727.

4. Theoretical Models

147. Agin, D. [1972]. Negative conductance and electrodiffusion in excitable membrane systems, *in* "Advances in Membranes" (G. Eisenman, ed.), Vol. I. Dekker, New York.

148. Agin, D. [1969]. An approach to the physical basis of negative conductance in the squid axon, *Biophys. J.* **9**, 209.
149. Agin, D., and Schauf, C. [1968]. Concerning negative conductance in the squid axon, *Proc. Nat. Acad. Sci.* **59**, 1201.
†150. Cole, K. S. [1965]. Electrodiffusion models for the membrane of squid giant axon, *Physiol. Rev.* **45**, 340.
151. Goldman, D. [1964]. A molecular structural basis for the excitation properties of axons, *Biophys. J.* **4**, 167.
152. Liquori, A. M. [1968]. Physicochemical model of nerve membrane, *Farmaco* (*Pavia*) *Ed. Sci.* **23**, 999.
153. Mackey, M. C. [1971]. Kinetic theory model for ion movement through biolgicaol membranes. I. Field dependent conductances in the presence of solution symmetry, *Biophys. J.* **11**, 75.
154. Mackey, M. C. [1971]. Kinetic theory model for ion movement through biological membranes. II. Interionic selectivity, *Biophys. J.* **11**, 91.
†155. Moore, J. W. [1968]. Specifications for nerve membrane models, *Proc. IEEE* **56**, 895.
156. Offner, F. F. [1969]. Ionic forces and membrane phenomena, *Bull. Math. Biophys.* **31**, 359.
157. Ofiner, F. F. [1970]. Kinetics of excitable membranes, *J. Gen. Physiol.* **56**, 272.
158. Vaidhyanathan, U. S., and Phillips, H. M. [1966]. Molecular theory of nerve potentials, *J. Theoret. Biol.* **10**, 460.
159. Wei, L. Y. [1969]. Molecular mechanisms of nerve excitation and conduction, *Bull. Math. Biophys.* **31**, 39.

AUTHOR INDEX

Numbers in parentheses are reference numbers and indicate that an author's work is referred to, although his name is not cited in the text. Numbers in italics show the page on which the complete reference is listed.

C

D

E

F

G

H

SUBJECT INDEX